Financial Management in Construction Contracting

The book's companion website is at
www.wiley.com/go/rossfinancialmanagement
and offers invaluable resources for both students and lecturers:

- PowerPoint slides for lectures on each chapter
- Excel worksheets to practice what you learn
- Sample valuations and cashflows

Financial Management in Construction Contracting

Andrew Ross
School of the Built Environment
Liverpool John Moores University

Peter Williams
Consultant and Lecturer

WILEY-BLACKWELL

A John Wiley & Sons, Ltd., Publication

Blackwell Publishing was acquired by John Wiley & Sons in February 2007.

Blackwell's publishing program has been merged with Wiley's global
Scientific, Technical and Medical business to form Wiley-Blackwell.

Registered Office
John Wiley & Sons, Ltd, The Atrium, Southern Gate, Chichester, West
Sussex, PO19 8SQ, UK

Editorial Offices
9600 Garsington Road, Oxford, OX4 2DQ, UK
The Atrium, Southern Gate, Chichester, West Sussex, PO19 8SQ, UK
2121 State Avenue, Ames, Iowa 50014-8300, USA

For details of our global editorial offices, for customer services and for information about
how to apply for permission to reuse the copyright material in this book please see our
website at www.wiley.com/wiley-blackwell.

Library of Congress Cataloging-in-Publication Data

Ross, Andrew, 1960–
Financial management in construction contracting / Andrew Ross.
     p.   cm.
   Includes bibliographical references and index.
   ISBN 978-1-4051-2506-2 (pbk.)
1. Construction industry-Accounting.   2. Construction industry-Finance.   I. Williams, Peter.
II. Title.
   HF5686.B7R67 2012
   624.068′1-dc23

                                                          2012023207

A catalogue record for this book is available from the British Library.

Wiley also publishes its books in a variety of electronic formats. Some content that appears in
print may not be available in electronic books.

Cover image courtesy of Shutterstock.com
Cover design by Sandra Heath

Set in 9/11.5pt Interstate Light by SPi Publisher Services, Pondicherry, India

1   2013

Contents

This book's companion website is at
www.wiley.com/go/rossfinancialmanagement
and offers invaluable resources for both students and lecturers:

- PowerPoint slides for lectures on each chapter
- Excel worksheets to practice what you learn
- Sample valuations and cashflows

About the Authors

Andrew Ross is Head of Postgraduate Programmes in the School of the Built Environment, Liverpool John Moores University. He teaches construction project financial management to undergraduate and post graduate students and has successfully supervised many PhD students as well as acting as external examiner to numerous UK and overseas Universities for undergraduate, postgraduate and research degree courses.

Peter Williams is a Consultant and Lecturer with extensive practical experience in building, civil engineering and surveying. Formerly a chartered builder, chartered quantity surveyor and principal lecturer, he is now a writer, researcher, lecturer and consultant with particular interests in contracts and finance, delay analysis and health and safety management.

Preface

For the love of money is the root of all evil

1 Timothy 6:10
The Holy Bible: King James Version (1769)

It could be argued that the love of money (or greed) has become an endemic evil in modern society and that this cannot be better exemplified than by reference to the worst excesses of the 2008 banking crisis. Whilst there may well be ideological, religious, political and other evils in society, few would argue that those who seek to profit at the expense of others are not evil. However, it is not money itself but the love of money which is the problem and it is greed that has impacted on the standards of humanity seen in modern society.

So 'what's new?' you might ask – the world is full of scams, cons, fraud, identity theft, unfair and underhand practices and downright dishonesty, and these manifestations of the worst in human nature can be seen at all levels of society from governments to utility companies to commercial enterprises and even within families.

The construction industry is not insulated from the worst excesses of the 'money motive' either, whether in the name of profit, tax avoidance, business survival or sheer greed. Whilst it would be a sweeping generalisation to say that construction is 'evil', there is no doubt that sharp practice, deception and dishonesty are features of the industry.

The question arises as to 'when does sharp practice become dishonesty?' and 'when does dishonesty become fraudulent?' Sharp practice is not a crime – fraud is – but there is certainly a 'grey area' regarding dishonesty. Dishonesty may have some moral justification (e.g. stealing from the rich to give to the poor) and there may well be degrees of dishonesty that fall short of a criminal act, but behaviour that is knowingly dishonest may not pass the test of what is expected of a 'reasonable and honest person' in a court of law.

Deception? Well, that's another question! This is the act of deceiving someone with a view to mislead, distort or falsify and may involve equivocation, concealment of the truth, exaggeration of facts or figures, understatement of the true situation or plain telling lies and may or may not lead to a crime.

Sharp practice and deception short of the criminal might well be called 'questionable practice' and there is no doubt that the construction industry is no stranger to either. Examples of questionable practice are not difficult to find. Some clients (and their professional advisers) are familiar with the 'art' of deception and this can be seen in the way that their projects are tendered and documented. Main contractors are frequently accused of 'subbie-bashing' often, but not always, with justification. On the other hand, subcontractors are not averse to making 'a fast buck' when the opportunity arises and there have been several well documented cases of corrupt practices in the materials supply side of the industry.

The worst examples of questionable practice in the industry might nevertheless attract the soubriquet of 'crime', not in a legal sense, but in terms of a crime against the industry. This might happen where the greed of a main contractor squeezes a subcontractor to the point of insolvency or where a small businessman loses his home when a bank calls in an overdraft at the precise time when financial support is most needed. As a consequence, the industry loses skills that will be gone forever and the rare gift of entrepreneurship is lost, never to be replaced.

It would be disappointing, however, if the reader went away with the idea that the construction industry was a 'den of iniquity' - it is merely a microcosm of society. Exciting, challenging, rewarding, risky, insular and reactionary are all adjectives that could be used to describe an industry that is capable of breathtaking triumphs of architectural and engineering genius delivered by resourceful and talented people. Latham (1993)[i], however, highlighted the 'fly in the ointment' - the mistrust that exists in the construction industry, especially when it comes to payment. It is this mistrust that conditions the relationships between all participants in the construction supply chain and influences peoples' behaviour.

Against this background, this book is concerned with financial aspects of contracting from the perspective of the authors' combined experience in the industry of some 70 years. The aim is to explain 'how things operate' in the 'real world' of contracting to those who wish to understand and if this means opening the 'black box' of financial practice both good and bad then so be it. The book is not intended to be an exposé of the 'evils' of the industry but is meant to illustrate good practice in financial control whilst at the same time being honest about some of the questionable practices that can and do happen.

The law of the land protects society from dishonesty and fraud and professional standards seek to ensure that members of the professions behave ethically. Cynics, however, would perhaps argue that 'there are no ethics when it comes to money' - and there may be some truth in this - but at some point the construction professional must take a stance on both personal and professional ethics.

This stance is affected by the professional codes of conduct or core values that are shared by members of a profession, it is affected by the community of practice that exists within the different strands of the professions and also by employing organisations. It was no surprise that Sir Michael Latham entitled his first report 'Trust and Money', as they both go to the heart of difficulties that exist in construction practice.

Euphemisms abound when describing some of the practices that can be found in the industry - commercial opportunity, muscling, loading, discounting, using information asymmetries to protect positions, overmeasuring, opportunism, protecting positions - are all phrases used within the book and are practices evidenced within financial management practice. There is no doubt that some of these practices are unethical and even bordering on fraudulent.

The authors thought long and hard about the ethical stance of the textbook and decided that they would describe good and bad practice, as it is only by learning about bad practice that students and practitioners can put in place safeguards for clients and supplying organisations.

---

[i] Latham, M. (1993) Trust and Money, Interim report of the joint government/industry review of procurement and contractual management in the UK construction industry. HMSO, London.

A lot of practice can be described as normative, that is it has always been done this way. These practices are often adopted by new employees without question, sometimes through an understandable desire to fit in, or it might be because of ignorance of best practice or through fear of doing something outside the norm.

Construction projects are always unique and the processes of design and construction are always subject to elements of uncertainty and change. The procurement routes and contract conditions used in the industry are structured in such a way to provide guidance to the parties about how to manage the financial and programme consequence of such change. The relationships between organisations designing and constructing these projects are also determined by procurement processes and contract conditions. The consequence of this is that the skill of the practitioner in securing the best return for the organisation's efforts is held in great value. Interpretation, communication and negotiation skills are at the heart of construction practice.

However, as in every walk of life, there are practitioners who are overly opportunistic in their interpretation of the 'rules' and they use their skills to maximise the financial returns and minimise risks for the organisations they work for. The authors had an interesting debate about what to include within this book, as some of the practices described in the text could be considered as 'sharp' or to use a euphemism 'commercial'. We have both observed practices which are unethical and have both had implicit and explicit pressures to act in a way that crosses a personal ethical line. This line is drawn from one's own moral principles and each reader will have their personal and subjective view about what is ethical.

Included within the book are descriptions of practices which may be considered unethical by some and these have been included, not to promote them as good practice, but to educate the reader as to their existence. The development of safeguards can only be undertaken by acknowledging commercial practices. The contractor's quantity surveyor plays a vitally important role in ensuring fairness to subcontractors and also to make sure that our industry's reputation is enhanced. Each reader will make their own contribution to these ends and it is up to them to draw their own ethical line.

Finally, the reader should be careful to disassociate the 'questionable practices' mentioned in this book from illegal practices, such as price fixing and cover pricing, which have been successfully prosecuted by the UK Office of Fair Trading.

Andrew Ross and Peter Williams
Liverpool and Chester

1

Finance in the construction industry

1.1 Introduction

To anyone walking past a construction site the scene can perhaps be best described as 'organised chaos'. The site will be fenced off, or there may be a hoarding around the site, and there will invariably be a variety of plant, equipment and scaffolding in evidence as well as stacks of bricks, heaps of sand and gravel; there will be partially completed work and work under construction and there will be cabins and site offices too.

Financial Management in Construction Contracting, First Edition. Andrew Ross and Peter Williams.
© 2013 Andrew Ross and Peter Williams. Published 2013 by John Wiley & Sons, Ltd.

Clearly, all of this activity has a monetary value but the means of arriving at this value may not be immediately obvious to the untrained eye. The mechanisms of valuation and financial reporting of completed and partially completed building projects under construction are explained in this book, as are the means of assessing the value of the work in progress, the valuation of materials on site and the determination as to whether a contract is making a profit or loss. The book is also concerned with why the work has to be valued and how such valuations are conveyed or 'reported' to interested parties outside the contracting organisation.

The scene painted above would be typical of many sites irrespective of whether the contractor is large or small or whether the contract is for building, civil engineering, maintenance or any other type of construction work. However, one thing that contractors large and small have in common, whether they are limited companies or unincorporated, is the need, at some point in time, to determine the value of such partially completed contracts. This is necessary so as to enable a set of annual accounts to be prepared for submission to HM Revenue and Customs (HMRC) and, for most companies of any substance, to file their accounts annually at Companies House.

Construction is a multifaceted industry and construction projects are invariably not straightforward. The processes of tendering, contract award, work on site, completion and handover are often complex and fraught with difficulties and sometimes disputes. There are many influences that bear on the presentation of true picture of the financial position of construction projects, not least the culture of the industry itself.

1.2   The purpose of this book

The purpose of this book is to explain how the financial position on construction contracts is reported, how work in progress is valued, how this information is reported to management and how this is reflected in the annual accounts of the business. The book also explains why things are done as they are and brings into question certain practices that might be considered less than desirable.

To achieve this, it is necessary to understand some basic accounting terminology and practice, how the construction industry and its system of contracts works, how tenders for construction projects are put together and how financial information flows in a construction business.

The book is written for undergraduate and postgraduate students and for practitioners working in the construction industry; it is written in a language that this audience will hopefully recognise and understand. It is not written for accountants or bankers, although some of the insights revealed in the book may help them to better understand how the industry operates and why. We have tried to avoid accountancy 'jargon' and where this has been unavoidable we have tried to explain, in layman's terms, what it all means.

Above all, the authors believe that the book is an honest representation of 'how things are' in the reporting of the financial position of construction contracts and make no apologies for being brutally frank about some of the 'questionable practices' that the industry suffers from. This is not to say that we endorse such practices – far from it – but good practice cannot flourish without awareness of the bad.

1.3 Construction contracting

The subject matter of this book concerns the financial management of construction projects. To be more specific, the focus is on the 'contracting' side of the construction industry – that is to say where projects are undertaken by contractors who are engaged by clients (employers) to carry out a building or civil engineering project for a stated price or for a price to be determined on completion. The principles and issues discussed apply equally to main contractors and specialist subcontractors but the financial management of speculative housing developments, carried out by contractor-developers, is handled somewhat differently and is not, therefore, covered by this text.

All contractors – whether small, medium or large – need to know and understand the financial situation of their projects in order to recognise when things are going wrong and be able to take remedial action before it is too late. However, many contractors and subcontractors in the construction industry, especially the smaller ones, are simply not 'in the loop' when it comes to the financial aspects of their business. They see a healthy order book, they see cash coming in, they see a healthy bank balance and they assume that all is well. This may be far from the case, however, and disaster may be waiting just around the corner. The reason is that what they 'see' is not the 'true' position and, hopefully, the reasons for this will become clearer as the chapters unfold.

One of the great problems in understanding what goes on financially in contracting is that construction contracts of any significant size are complex. The way that contracts are priced, the design changes and unexpected events that take place during construction, the natural human tendency to argue over money and the endemic financial instability of many of the firms that operate in the construction industry all contribute to the complex nature of the financial aspects of construction projects. Add to this the singular culture of the industry, the problems caused by the separation of design from construction, the complex contractual and procurement arrangements employed and the 'grey water' becomes very 'murky'!

A large part of the work of a contractor's quantity surveyor is to provide financial data in order to show the financial position of projects under his/her control. This is usual practice in most medium and large sized contractors but much less so in smaller firms and specialist 'trade' contractors. The whole idea of contracting is to win contracts and make money and the quantity surveyor acts in a quasi-accountancy role to provide information for line managers to run projects efficiently and within budget and to capitalise on opportunities to 'make money' when the occasion arises.

1.4 Work in progress

Ask any accountant what the main problem is in contracting and the answer will be 'the valuation of work in progress'. Work in progress is the *bête noir* of construction accounting and Barrett (1981) pointed out that *no area of accounting has produced wider differences in practice than the computation of the amount at which stocks and work in progress are stated in financial accounts.*

At any given point, a contracting company will have a number of projects running that are incomplete; this means that there will inevitably be a significant amount of work in progress. On one particular day in the year the annual accounts will be 'struck' and the work in progress will have to be reported. To know the true financial position of the business at such a point, the work in progress has to be valued. This has to be done in a consistent fashion across all contracts and must be done in line with defined and accepted standards of accounting practice in order to ensure that the annual accounts state a true and fair view of the company.

Taken in its narrow meaning, 'work in progress' is the term used to describe work carried out on site that has not yet been invoiced. In other words, it is work done and materials delivered to site after a valuation has been carried out and before the next one is done. Consequently, work in progress represents an amount of money that has not been agreed or certified for payment and is, therefore, subject to question, disagreement or dispute. Accountants see work in progress as a problem because it is frequently the case that the amount received is less than that expected; this can have a serious impact on cash flow and the availability of working capital.

With respect, it is likely that many accountants and bankers are unaware that there may well also be a problem with work done that *has been* certified for payment. This may arise due to a lack of understanding about the way that construction tenders are priced and the influence this has on the valuation of work carried out on site. Consequently, albeit that the work may have been valued by the employer's quantity surveyor and certified for payment, it is quite possible that the valuation will not be a 'true value' because of the way that the contractor has priced his tender in order to reduce negative cash flows and maximise the commercial opportunities provided by the contract. These and other related issues are explained in later chapters.

Consequently, 'work in progress' could be viewed in a broad sense to mean all the work done on a contract to date, whether certified or not and whether paid for or not, because despite payments made on account during a contract, the valuations made are not 'true values', the payments on account are not binding (only the final account is) and the eventual settlement on the contract may be no more than a 'horse deal'.

Notwithstanding this, 'work in progress' has a particular meaning in the annual accounts more in line with the narrow meaning referred to earlier. 'Work in progress' is a truncated version of more long-winded terms that appear in a set of annual accounts including 'stocks and work in progress', 'stocks and long term contracts', 'amount recoverable on contracts' and so on. It all means the same. The 'stocks' aspect is not so important in construction as in other industries. Traditionally, contractors always carried stocks of materials in their 'builder's yard' – for emergencies, small jobs and as a store for over-ordered materials from contracts. Nowadays, holding stocks of materials represents vital working capital tied up and most contractors employ 'just-in-time' ordering methods for their sites.

The 'work in progress' aspect is the important bit!

1.5 Reporting

Whilst there is no denying the importance of the issue of 'work in progress', this book is concerned with much more than that. In the final analysis this book is about reporting. At one end of the scale the quantity surveyor is reporting the financial position on a construction project and at the other end the accountant is reporting the financial position of the company as a whole. In between is a flow of information that is influenced by many factors and it is the quality of this information that determines whether or not the financial position, either on the project or in the accounts, is true and correct.

The importance of reporting the true position on individual projects is vital from a business survival point of view but it is also important in terms of filing tax returns, filing annual returns to Companies House and informing shareholders about the business and how it is doing. Consequently, a clear picture is needed for management control and for giving all sorts of outsiders a true view of the affairs of the business. As will be discovered later in the book, this is far from easy to do and a distorted impression of what is going on financially may well be the outcome of any lack of understanding, questionable practices and frail reporting systems.

More than thirty years ago, Barrett (1981) observed that *inconsistent financial reporting and failure to identify the true financial position of contracts is unfortunately all too frequent*. Much has changed since then in that there are now higher standards of corporate governance and greater transparency in financial reporting. The fact remains, however, that the reporting of the financial position on contracts is at best problematic and at worst misleading and this stems from the nature of construction contracting, ignorance of best practice and the human tendency to 'gild the lily' in order to make things look better than they really are.

1.6 Structure of the book

The book is structured in three main parts:

Part 1 – External environment, which provides the context in which contracting firms operate including:

- How the contracting side of the industry works.
- The problems the industry faces and their impact on contracting.
- The risks and uncertainties that face firms working in contracting.
- How contractors are financed and what the problems are.
- The system of contracts and payments that operates in construction contracting.
- The corporate governance and accounting standards and practices that apply.

Part 2 – Internal environment, which explains:

- How contracting firms are governed financially.
- How contractors are organised so as to operate effectively.
- How contractors go about obtaining work.
- How contractors budget for and control their finances.

Part 3 – Project environment describes:

- The contractual and procurement mechanisms whereby contractors are paid for the work they do.
- How work in progress is valued and certified for payment.
- How money and resources are budgeted for at project level.
- The financial control systems needed to effectively manage project risks.
- How physical and financial progress is reported.
- How the profitability of contracts is reported and how losses on projects are recognised.

Above all, the book is structured in such a way as to provide an understanding of corporate reporting standards and practices so that a true and fair view of a company is presented in the context of the contracts that it carries out.

1.7 The construction industry

The construction industry is similar to other manufacturing industries in that a product is produced and sold to a client. An organisation has to procure resources from the market place, combine them with other resources, add value and then dispose of the final product to make a return on its investment. To understand financial management in construction it is important to understand the context within which construction organisations work.

The UK construction industry has been the subject of much interest over the years and numerous investigations and reports have been published describing the problems of the industry, the tensions that exist between those involved in the construction process and the outdated and unfair practices that characterise the way that the industry conducts its business.

1.7.1 Industry reports

The Latham and Egan reports are perhaps the best known in that long succession of investigations, many of which identified similar problems and made similar recommendations. There have been a total of 13 reports since 1944 that have investigated and produced recommendations about the industry. Langford and Murray (2003) provide an indepth critique of each of the reports. These reports are:

- The Simon committee report (1944)
- The Phillips report on Building (1948–1950)
- The Emmerson report (1962)
- The Banwell report (1964)
- The Tavistock studies (1965 and 1966)
- Large Industrial Site Report (1970)
- The Wood Report (1975)
- Faster Building for Industry: NEDO(1983)
- Faster Building for Commerce: NEDO (1988)
- Constructing the Team: The Latham Report (1994)
- Technology Foresight Report: Progress through Partnership (1995)
- Rethinking Construction: The Egan report (1998)
- Never waste a good crisis: Wolstenholme report (2009).

Sir Michael Latham and Sir John Egan set a series of challenges to the industry and gave us an 'official' view of what are now seen as some of its strengths and many of the weaknesses. The reports have been widely discussed, welcomed, criticised and – in some respects – ignored. At this point, a brief review of the critical issues they raised may provide a useful background to understanding the industry and its complexities discussed in later chapters of the book. A brief comparison of some of their recommendations with those found in earlier reports reveals how deep-seated the problems are.

1.7.2 Industry reform: origins and responses

In 1991, Sir Michael Latham was commissioned, in a joint venture by Government and the industry, to conduct a review of the 'Procurement and Contractual Arrangements in the UK Construction Industry'. An interim report, *Trust and Money*, was published for consultation in December 1993, and the final report, *Constructing the Team*, in July 1994 (Latham, 1993, 1994). These reports identified a wide range of weaknesses in current procedures. Most of these had already been separately recognised in the industry, and indeed discussed in some of the earlier official reports identified above, but Latham linked them together and set out an agenda for reform.

The Latham Report made over thirty specific recommendations, which can be summarised in a few categories:

- Government should take the lead in improving clients' knowledge and practice, particularly of how to brief designers and select procurement methods.
- The whole design process should be reviewed and the link between design and construction improved.
- Building contracts should be simpler, clearer, more standardised and less prone to lead to disputes.
- There should be simpler faster means to resolve disputes where they do occur.
- There should be a Construction Contracts Bill, outlawing some unfair practices, the introduction of adjudication as the normal method of dispute resolution and the establishment of trust funds for payment.
- Government should maintain lists of approved consultants and contractors for public sector work.
- The traditional methods of tendering should be revised and improved.
- Training and research programmes should be rationalised and improved.
- The industry should aim for a 30% reduction in costs by the year 2000.

Although many of the Latham recommendations were accepted and followed up, including the enactment of a Construction Act, only a few years later the Deputy Prime Minister, John Prescott, set up another committee – this time a 'task force' – under Sir John Egan of the British Airports Authority to advise:

'from the client's perspective on the opportunities to improve the efficiency and quality of delivery of UK construction, to reinforce the impetus for change and make the industry more responsive to customer needs'. (Strategic Forum for Construction, 2002)

The Egan report, *Rethinking Construction* published in 1998 (Egan, 1998), was shorter and sharper but more radical than Latham. The language was different, the criticism harsher and it implied a total change in the industry's culture. Many of the recommendations of the two reports were in effect very similar, but whereas Latham seemed to look for reform within the old traditions, Egan was proposing a revolution, or so it seemed.

The Egan Report identified what it called *'five key drivers for change'*:

- committed leadership;
- a focus on the consumer;
- integrated processes and teams;

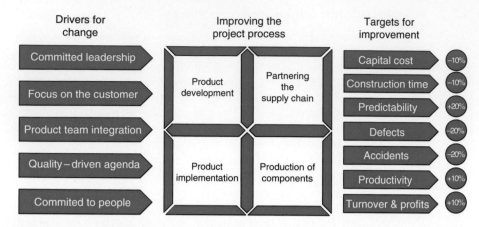

Figure 1.1 The Egan 'drivers' and 'targets' (Strategic Forum for Construction, 2002, p. 13).

- a quality-driven agenda;
- commitment to people.

Egan set specific targets:

- an annual reduction of 10% a year in construction costs and time;
- a reduction in defects by 20% a year;
- a radical change in industry methods in order to create an integrated project;
- dramatically improved working conditions;
- improved management and supervisory skills.

The Egan 'drivers' and 'targets' are illustrated in Figure 1.1.

The immediate impact of Egan was considerable. The report was widely discussed; new bodies were set up to push the ideas forward, such as the Movement for Innovation (known as M4I), which was linked to the already existing 'Best Practice Programme'. The Government moved towards forcing all public sector and publicly supported bodies, such as its own Departments, the health service and the housing associations to become 'Egan compliant'. The Auditor General's office produced its own report, *Modernising Construction* (NAO, 2001), showing how the Egan principles were to be applied throughout the public sector.

The Egan Task Force felt that for the industry to reach its full potential, it needed to change its culture and structure to support the improvement. It recommended that the industry should provide decent and safe working conditions and improve management and supervisory skills at all levels. Furthermore, it felt that better results could be achieved through long-term relationships based on clear performance measures and sustained improvements in quality and efficiency by continuing to learn and improve as a team, rather than competitively tendering and having to create a new team for every project.

There was no suggestion that construction companies should lead the change and, instead, the emphasis was placed on construction clients to show leadership and put forward 'demonstration projects' to show the recommendations of the report in practice. The Government in particular was invited to lead public sector bodies to become best practice clients. The report resulted in the development of

How have we changed Egan's targets?

Figure 1.2   Industry performance post-Egan (Reproduced by permission of Constructing Excellence Ltd).

a set of Construction Industry Key Performance Indicators, which are now published annually by Constructing Excellence.

The impact of the Egan report was the focus of a study by a Constructing Excellence review team lead by Andrew Wolstenholme of Balfour Beatty Construction. This report considered the progress made by the industry on the various performance measures suggested by Egan, collected data from over 500 demonstration projects, hosted focus groups and analysed almost 1000 questionnaire returns. Figure 1.2 illustrates the results of the review team's findings which show that, whilst the demonstration projects were 100% successful against target, the industry:

● had failed to meet virtually every target;
● had reverted to traditional practice;
● had only partially and superficially adopted the Egan principles.

By analysis of the reasons for change, Wolstenholme identified three main categories of barriers to change and improvement:

● economic model blockers
● business model blockers and
● capability blockers.

Table 1.1 indicates the economic and business blockers highlighted in the Wolthenholme Report.

1.7.3   Housing Grants, Construction and Regeneration Act 1996

The reasons why the Government intervened by introducing the Construction Act can be rooted back to the original review by Latham in 1994. Latham recognised the importance of the industry to the national economy and also appreciated that the industry was essentially made up of thousands of small enterprises that were economically vulnerable due to small profit margins and heavy reliance on careful cash flow management. Unfortunately, insolvencies in construction are high when compared to the economy as a whole. If the recommendations for developing integrated supply chains, focused on high quality and efficient delivery in a non-adversarial way, were to be achieved, a level of trust which was not present was required.

Table 1.1   Economic and business blockers.

Blocker	Definitions
Lack of Cohesive Industry Vision	A lack of joined-up thinking in Government and our industry about how the built environment contributes to the UK's long-term prosperity and the aim of achieving a sustainable, low-carbon economy.
Few Business Drivers to Improve	For much of the supply chain, there are too few business or economic drivers to deliver meaningful change. They are prepared to accept stable, though unexciting returns, rather than attempt changes that are seen as being 'too difficult'.
Construction 'Does not Matter'	The low impact of construction costs and outcomes on the client's business case means that in some sectors construction 'does not matter'.
No Incentives for Change	Most client business models are focused on short-term gain and do not reward suppliers who can deliver long-term sustainable solutions.
Construction is Seen as a Commodity Purchase	Too many clients focus on the upfront costs of construction, rather than the value created over the lifetime of an asset. Few suppliers, other than those involved in PFIs, have any continued interest in the operation of the building and therefore no incentive to raise quality standards.
Industry Culture is Driven by Economic Forces	Even where clients plan for the long term, few have avoided cuts during the current downturn. Many clients and suppliers appear to have abandoned partnering behaviour (if they ever adopted it in the first place) and returned to transactional relationships.

(Constructing Excellence, 2009:15).

In Egan's report in 1998 he stated that 'The extensive use of subcontracting has brought contractual relations to the fore and prevented the continuity of teams that is essential to efficient working.'

The required collaboration and trust between all parties in the construction process was hampered by a culture of blame, claim and counter claim, disputes and poor payment practices. The National Audit Office, Constructing Excellence and Government client bodies all endorsed the integrated team approach but it was not until publication of the National Audit Office Report, *Improving public services through better construction* (NAO, 2005), that the issue of unfair payment practices was raised. The NAO Report stated that:

'Unfair payment practices, such as unduly prolonged or inappropriate cash retention, undermine the principle of integrated team working and the ability and *motivation of specialist suppliers to invest in innovation and capacity. Departments should have the appropriate visibility of the entire supply chain. Understanding how specialist subcontractors and particularly small and medium sized enterprises are engaged, evaluated and managed can contribute considerably to the achievement of value for money. For example, Departments should have in place effective and fair payment mechanisms … to provide more certainty to suppliers' payments dependent upon delivery to time, cost and quality.'*

Latham brought together interested parties in 2004/2005 to review the Construction Act with a view to specifically improving its adjudication and payment provisions. A consultation paper on *Improving payment practices in the construction industry* was published by the Department of Trade and Industry (DTI, 2005).

The main aim of the Act with reference to payment was to:

- provide a right to interim, periodic or stage payments, making clear when payments become due, their amount and a final date for payment;
- prevent the payer from withholding money from the 'sum due' after the final date for payment unless he has given a withholding notice;
- provide a statutory right for the payee to suspend performance where a 'sum due' is not paid, or properly withheld, by the final date for payment;
- prohibit 'pay-when-paid' clauses which delay payment until it is received by the payer.

The Housing Grants, Construction and Regeneration Act 1996 contained wide ranging provisions regarding:

'An Act to make provision for grants and other assistance for housing purposes and about action in relation to unfit housing; to amend the law relating to construction contracts and architects; to provide grants and other assistance for regeneration and development and in connection with clearance areas; to amend the provisions relating to home energy efficiency schemes; to make provision in connection with the dissolution of urban development corporations, housing action trusts and the Commission for the New Towns; and for connected purposes.'

Part 1 of the Act was concerned with grants for private and social housing and contained wide ranging provisions dealing with the management of social housing provision. The provisions of the Act that were of more interest in relation to the subject matter of this book were contained in Part 2. This part of the Act in effect related to amendments needed in construction contracts so as to:

- give each party to a construction contract the right to refer a dispute to adjudication and require parties to include terms in their contract relating to adjudication that comply with section 108 (2) to (4);
- provide that contractors are entitled to stage payments – section 109;
- provide that contracts should have an 'adequate mechanism' for determining what should be paid and when – section 110 (1); and
- require that the payer should issue a notice in advance of each payment of the sum he proposes to pay – section 110 (2).

The impact of the Housing Grants, Construction and Regeneration Act 1996 was that:

● it formalised the approach to payment;
● it identified the need for a construction contract; and that
● such a contract should have a payment mechanism which was in accordance with sections 109 and 110.

1.8 Industry output

The output of the construction industry varies enormously, from repairs to a roof to redevelopment of city centres. The large majority of projects are of relatively low value and short duration and, proportionly, repair and maintenance to existing buildings forms a major part of the industry's workload. The demand for the industry's services varies significantly over time due to Government interest rate policy, public sector spending programmes and economic confidence.

The construction industry is similar to other industries in that it is difficult to define its boundaries and classifications. The Office of National Statistics (ONS) compiles data on the output of the industry and adopts an approach to classification based on a Standard Classification of Economic Activity which was revised in 2007 and published in 2009. A Standard Industrial Classification (SIC) was first introduced into the United Kingdom in 1948 for use in classifying business establishments and other statistical units by the type of economic activity in which they are engaged.

The SIC classification provided a framework for the collection, tabulation, presentation and analysis of data; Table 1.2 indicates the different SIC categories. There are three major sections within category F, these are:

Table 1.2   SIC Categories.

SIC (2007)			
A	Agriculture, forestry and fishing	L,M,N	Real estate activities, professional and scientific activities
B	Mining and quarrying	O	Public admin and defence
C	Manufacturing	P	Education
D,E	Electricity, Gas supply	Q	Human health and social work
F	Construction	R,S	Arts entertainment and recreation
G	Wholesale and retail trade; repair of motor vehicle	T	Activities as households as employers
I	Accommodation and food service activities	U	Activities of extra territorial organisations
H,J	Transport and storage Information and communication		
K	Financial and Insurance activities		

- 41: Construction of buildings;
- 42: Civil Engineering; and
- 43: Specialised construction activities.

Within these categories there are 27 subdivisions, which are shown in Table 1.3.

The economic output of the construction industry is measured annually by the ONS which currently uses the SIC (2003) categories. The data are collected from a variety of sources, including from questionnaires sent to almost 100 000 construction organisations who report upon their construction output and from information held by government departments. The data are collected, grouped under standardised headings and then reported to allow annual comparisons to be made.

Table 1.3   SIC Category F – Construction.

Section F: Construction	
41	**Construction of Buildings**
41.1	Development of building projects
41.2	Construction of residential and non-residential buildings
41.2.01	Construction of commercial buildings
41.2.01	Construction of domestic dwellings
42	**Civil engineering**
42.1	Construction of roads and motorways
	Construction of railways and underground railways
	Construction of bridges and tunnels
42.2	Construction of utility projects
	Construction of utility projects for fluids
	Construction of utility projects for electricity and communication
42.9.1	Construction of water projects
42.9.9	Construction of other civil engineering projects
43	**Specialised construction activities**
43.11	Demolition
	Site preparation
	Testing, drilling and boring
43.2	Electrical installation
	Plumbing, heat and air conditioning
43.3	Building completion and finishing
	Plastering
	Joinery
	Painting and glazing
	Painting
	Glazing
	Other
43.9	Other specialised completion activities
	Roofing
	Scaffold erection

Table 1.4   Construction demand.

Current prices		Great Britain									£ million	
		New housing				Other new work						
							Other new work exc. infrastructure					
		Public new housing	Private new housing	All new housing	Infrastructure	Public new work exc. infrastructure	Private industrial	Private commercial	All other new work	All new work	Total Public	Total Private
Total	1997	1245	7608	8852	4971	4538	5595	13320	28427	37278	10754	54950
Total	1998	1159	7229	8388	5503	5423	5216	16764	32905	41293	12085	62114
Total	1999	1203	7125	8328	5173	5084	4505	16102	30864	39191	11460	58596
Total	2000	1126	7323	8450	6179	5949	4577	17104	33808	42259	13254	62812
Total	2001	1344	7865	9208	6399	6423	4500	18019	35340	44547	14166	65724
Total	2002	1406	9803	11210	7008	9304	4019	18672	39001	50211	17718	71495
Total	2003	1690	11611	13301	6203	9770	4294	17452	37720	51021	17663	71077
Total	2004	2160	15040	17200	4722	10793	4631	21395	41543	58742	17675	82609
Total	2005	2475	16258	18730	6974	10624	6140	23553	47291	66021	20073	93242
Total	2006	3356	16572	19929	5306	9541	6376	30627	51851	71779	18203	105426
Total	2007	3733	16037	19769	6965	11393	5836	32115	56309	76078	22091	110297
Total	2008	3081	9200	12283	7897	14672	4346	23353	50267	62550	25650	87166
Total	2009	3107	6393	9500	11032	14709	2654	12886	41280	50780	28848	63213
Total	2010	3559	10303	13863	10019	13751	2204	14215	40191	54053	27329	66913

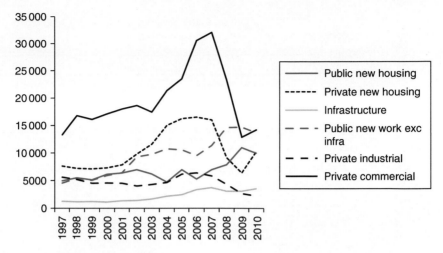

Figure 1.3 Construction demand.

Table 1.4 shows an example of the data reported by the ONS in the 2010 Annual Construction Statistics. These data give an indication of the demand for construction and the reader can easily identify the two major categories of public and private sectors. Data are collected for new housing both public and private, infrastructure work and public commercial and industrial work. The data in Table 1.4 are illustrated graphically in Figure 1.3.

1.9 Industry clients

Construction is largely a bespoke industry. Clients typically specify their requirements to an architect or other designer, or to a contractor with design expertise, and the project is then built 'on demand'. There are exceptions to this where a building is constructed in anticipation of a future demand, such as speculative housing or speculative office and retail developments. Demand for construction products varies year on year due to disposable incomes, income distribution, public sector finances and population and relative prices.

The division between public sector and private sector clients is a fairly straightforward distinction to make but with the advent of utilities and rail privatisation, the private finance initiative, arms-length management organisations and registered social landlords the traditional distinctions are sometimes difficult to identify.

The public sector includes Government departments, local authorities, health trusts, universities and defence. The Government is a major client of the construction industry and can act as an important regulator of demand. Figure 1.4 gives an indication of the change in public spending over a ten-year period. The significance of the Government as a client is in no doubt, but what determines how much is spent?

This is a simple question with a many-faceted answer. Fundamentally there are two dominating factors:

● first there is the perceived demand for a new building or service and the Government's view on its affordability;
● second there is the Government's perception of the impact of its spending programme on the economy as a whole.

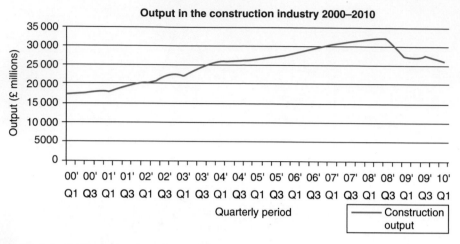

Figure 1.4 Public sector spending 2000–2010.

Private sector clients are more difficult to categorise but four broad groupings may be suggested:

- clients for small buildings;
- major clients developing for their own occupation;
- property developers;
- private house buyers.

Each of these categories has distinctly different needs, experience of the industry and access to finances.

1.9.1 Clients for small buildings

A great deal of construction output is for clients in need of small buildings; these may be for individuals or organisations that require new premises or extensions and alterations to existing buildings.

The nature of such clients varies enormously but the feature that they all generally share is that they are inexperienced and unused to the construction industry and its way of working. The work that this type of client commissions tends to form the core business of construction organisations that operate locally. Clients within this group which require larger projects would usually approach an architect who takes them through the planning process and sometimes procures the builder to undertake the works. Many clients, however, approach a contractor directly to undertake the design and construction of their projects.

1.9.2 Major clients developing for their own occupation

These types of client are large, powerful and influential and have driven the agenda for improvement which has made the industry change its ways of operating significantly over the last twenty years. Examples of clients that fall into this group are the British Airports Authority, retailers such as Marks and Spencer, Tesco, Waitrose and Sainsbury's, hotel groups such as InterContinental Hotels Group and Whitbread and food retailers such as MacDonald's and Yum! (Pizza Hut and KFC).

These clients are experienced and often have their own in-house project management teams. They are repeat clients and often used the promise of future work to derive discounted prices from construction consultants and organisations. They also have a good knowledge of how long a construction project will take and how much it will cost.

1.9.3 Property developers

A second major category of client is the property developer, whose influence can be seen in large towns and cities throughout the country. They are generally clients who build with tenants in mind, have a clear idea of the demand for the completed building and undertake careful feasibility studies before deciding to invest.

Examples include Grosvenor, which has recently completed a £1bn redevelopment of Liverpool city centre, The Peel Group and Bruntwood.

1.9.4 Private house buyers

The house buying market is a volatile one and a great deal of analysis is undertaken as it is often used as a barometer to measure the health of the economy generally. This is because private house buyers make up, numerically, the largest group of the construction industry's customers. The demand for houses is very much influenced by affordability (often measured as a ratio between house price and income), the availability of finance, the perception of the value of the market, overall confidence in job security and the availability of houses to purchase. Housing demand is generally in ten-year cycles; at the time of writing house building is showing a slow recovery after a five-year period of decline.

Whilst consideration of the financial aspects of the housing side of the industry is outside the scope of this book, it must be remembered that 'contractors' operate in this sector either as 'developer-contractors' or 'trade contractors' engaged in housing developments. As such, the reporting practices discussed in this book are relevant only to the extent that the contractors in question are engaged to carry out work under a formal building or engineering contract following some sort of formal tendering process.

1.10 Structure of the industry

A study of the structure of any industry gives an indication of the extent to which the concentration of supply or the proportion of an industry's output lies in the hands of a few firms. Large manufacturing industries such as aerospace, motor cars or ship building can be considered to be highly concentrated and this is similarly the case for manufactured construction products such as bricks, cement or glass.

Table 1.5 shows a clear picture of the banding of construction firms by numbers employed and indicates the annual value of work done in each banding category.

1.10.1 Size and distribution of firms

As can be seen from Table 1.5, almost 70% of firms that make up the industry employ up to three people and there are only 62 organisations that employ over 1200, a tiny 0.03%. However, looking at the value of work carried out, this tiny number of large firms is responsible for 22% of the total of new work.

Table 1.5   Number of firms, employment and work done by private contractors, 3rd Quarter 2009.

Size of firm (by number employed)	Number of firms	% of total	Total employment (000)	Value of work (£ million)	% of total
Band 1					
1	75 382	38.85	88.4	609	2.69
2–3	60 625	31.25	114.2	1076	4.74
4–7	33 069	17.04	141.4	1253	5.53
8–13	12 390	6.39	103.2	1507	6.65
Band 2					
14–24	6502	3.35	102.1	1687	7.44
25–34	2104	1.08	56.9	1171	5.17
35–59	2064	1.06	82.3	2191	9.66
60–79	596	0.31	34.9	962	4.24
Band 3					
80–114	484	0.25	40.9	1092	4.82
115–299	531	0.27	83.8	2666	11.76
300–599	154	0.08	60.4	2268	10.01
600–1199	62	0.03	38.0	1152	5.08
Band 4					
1200+	62	0.03	167.9	5038	22.22
TOTALS					
All firms	194 025		1114.4	22 673	

(ONS, 2011, Table 3.3). Contains Parliamentary information licensed under the Open Parliament Licence v1.0 (http://www.parliament.uk/site-information/copyright/open-parliament-licence/).

This trend for concentration – that is the high proportion of the industry's output produced by the few largest businesses – has increased over time. Since 1972, the number of large firms has reduced but their share of total output has increased.

Measurement of output is problematic, however, as there are no categories for firms which employ over 1200 people. The use of specialist subcontractors and labour-only subcontractors makes interpretation of the data even more difficult.

A further subdivision of the industry is available; it is shown in the Table 1.6. This provides a good representation of the nature of the industry as a whole because it categorises:

● numbers employed in the three main categories (or 'trades') of the industry – housing, non-housing and civil engineering;
● numbers employed in the other 'trades' normally associated with the construction industry, such as demolition, roofing, plumbing and plastering and so on.

Further scrutiny of Table 1.6 reveals that the statistics for the specialist contractors identify that very few of the firms in the specialist trades have large numbers of employees. The only exceptions are the electrical contractors and heating and ventilating engineering firms. There are, of course, lots of other data

Table 1.6   Trade and number of employees, 3rd Quarter 2009.

Trade	Employees (000)
Main trades	
• Non-residential building	58.7
• House building	69.3
• Civil engineering	50.1
Total main trades	**178.2**
Other trades	
• Constructional engineers	0.2
• Demolition	7.9
• Test drilling and boring	0.6
• Roofing	15.0
• Construction of highways	21.5
• Construction of water projects	1.3
• Scaffolding	21.8
• Installation of electrical wiring and fitting	90.7
• Insulating activities	7.7
• Plumbing	58.6
• Plastering	4.2
• Joinery installation	21.8
• Floor and wall covering	8.4
• Painting	18.1
• Glazing	8.8
• Plant hire (with operators)	9.3
• Other construction work and building installation and completion	107.9
Total other trades	**403.6**
Total all trades	**582.0**

(ONS, 2011, Chapter 12). Contains Parliamentary information licensed under the Open Parliament Licence v1.0 (http://www.parliament.uk/site-information/copyright/open-parliament-licence/).

on the manufacturing sectors that supply the construction industry that are not included here.

The structure of the industry has changed significantly over the last thirty years. The changes have been due to increased specialisation, costs of entry, transactional costs of employment and the fluctuations in workload.

1.10.2   Risk culture

Construction can be a risky and uncertain process and there are many factors that contribute to this condition including:

- the economic conditions prevailing at the time of a proposed development;
- the various engineers, architects, surveyors and other consultants appointed by the client who all have their own professional allegiances, cultures and attitudes towards contractors;
- the design process and its separation from production;
- the procurement of contractors, subcontractors and suppliers;

- the physical process of construction; and
- the eventual completion and handover of the project.

It is not difficult to understand why a culture of risk transfer has developed in the construction industry. Clients who have previously embarked on construction projects have found that it is an industry which is highly litigious and has poor predictability in terms of costs and time. Projects are often completed late and are frequently over budget. Problems of poor communication abound. There are many high profile instances of construction projects that were late and over budget – Wembley Stadium and the Holyrood Parliament building being two noteworthy examples.

Consequently, clients try to ensure that their risks are transferred to the main contractor, which in turn seeks to transfer risk to subcontractors and suppliers.

It is unsurprising that money is often at the heart of the problems that occur between organisations. In 1971 Lord Denning stated that 'cash flow is the lifeblood of the building trade' and whilst being a simple and obvious statement, the litigious practices that have developed in the construction industry generally revolve around getting paid.

Organisations try to maximise what they are paid, try to get paid as early as possible and then pay what they owe as late as possible. The consequence of this is that construction is beset with disputes and a new 'industry' in construction adjudication has grown up since the advent of the Construction Act in 1996. Construction has more insolvencies than any other sector and more money is spent on litigation relating to payment as a percentage of GDP than any other industry. The culture of claim, counterclaim and blame is a far cry from Latham's dream of 'win–win'.

1.10.3 Specialist contractors

The problems of the construction industry are not limited to clients and contractors as the process of planning, design, construction, refurbishment and maintenance of the built environment is heavily reliant on specialist subcontractors and supply organisations that provide a wide range of services and expertise to clients, designers, consultants and contractors.

These specialisms include heating, ventilating and air conditioning, electrical engineering, fire protection and prevention, demolition, asbestos removal and remediation services, as well as the more easily recognisable 'trades' such as plumbing, plastering, painting and tiling and so on.

James Wates, Chairman of Wates Construction, identified that over 500 representative bodies exist in the construction industry and, of these, some 32 specialist trade organisations are members of the National Specialist Contractors Council (NSCC) which claims to be *the authoritative voice of Specialist Contractors in the UK*. Its members include the National Federation of Roofing Contractors, the Resin Flooring Association, the Federation of Piling Specialists, the Glass and Glazing Federation, the Door and Hardware Federation and the Mastic Asphalt Council. The NSCC brings together the common aims of specialist trade organisations and is recognised as the principal point of contact with specialist contractors for organisations such as the Construction Skills Certification Scheme (CSCS), ConstructionSkills and the Cross-Industry Construction Apprenticeship Task Force (CCATF).

The remit of the Specialist Engineering Alliance (SEA) is *to facilitate greater integration of all the parties within the specialist engineering supply chain* whether consultants, contractors or manufacturers and its membership includes the Association for Consultancy and Engineering (ACE), the Building Services Research and Information Association (BSRIA), the Chartered Institute of Building Services Engineers (CIBSE), the Federation of Environmental Trade Associations (FETA) and the Specialist Engineering Contractors' Group (SEC Group).

The Heating and Ventilating Contractors Association represents its member companies and, in turn, is a member of the Specialised Engineering Contractors' (SEC) Group. The National Federation of Demolition Contractors is a member of the UK Contractors Group (UKCG), which is the main representative association for contractors in the UK and successor to the Major Contractors Group and the National Contractors Federation.

All of these representative bodies believe that they have a unique role to play in helping their members to deliver a quality service, best value and integrated solutions to their clients and, as such, feel that their members have a unique set of issues that need to be represented whether they be of a technological, economic, environmental or contractual nature.

Each of these specialist contractors use different resources – materials, labour and plant – depending upon the nature of work undertaken, the precedence of their operations, the period on site and so on and it is not surprising that different cultures exist between individual specialist organisations and between these organisations and those who employ them, whether main contractor or client.

1.10.4 Payment processes

The nature of the payment process in construction is another reason for cultural difficulties in the industry. Generally speaking, clients employ consultants who value the works and then arrange for payment to the parties to the contract. The contract is usually between a client and a main contractor, albeit that there may be direct payments from client to subcontractors under certain procurement methods.

The valuation process is agreed before the contract commences and appropriate valuation dates arranged. The time lag between valuation and payment is often considered as an adequate safety mechanism against over valuation of the work by consultants but the additional contingencies of retention monies and performance bonds and so on can be considered as further security. However, the costs likely to be incurred by an employer if a contractor becomes insolvent before contract completion are likely to be in excess of this contingency. In such a case, the employer has to look elsewhere for recompense for the loss incurred and this is likely to be from the consultants.

Consequently, consultants are wary of overpayment but contractors are aware of this and their agenda is to seek to maximise their claims for payment. This tension between consultants and contractors can blight relationships between the parties.

Similar tensions arise between subcontractors and main contractors as a consequence of the contract that normally binds them together. These contracts are often bespoke forms written by the main contractor to protect his interests and designed to pass on risk in an inequitable way to the subcontractor. Whilst 'pay-when-paid' and 'pay-what-paid' clauses are now outlawed under the Construction Act 1996, subcontractors are never sure to gain their just entitlement under the contract and, furthermore, are often commercially 'muscled' to provide additional discounts when they have already tendered at highly competitive prices. Such an atmosphere between contracting parties leads to distrust and tense relationships, and it was with a little hidden irony that Latham entitled his first report 'Trust and Money'!

References

Barrett, F.R. (1981), *Cost value reconciliation*, 1st edn. The Chartered Institute of Building, Ascot UK.

Constructing Excellence (2009) Never Waste a Good Crisis: A Review of Progress since *Rethinking Construction* and Thoughts for Our Future. Constructing Excellence, London (http://www.constructingexcellence.org.uk/pdf/Wolstenholme_Report_Oct_2009.pdf; accessed 20 April 2012).

DTI (2005) Improving payment practices in the construction industry, proposals to amend Part II of the Housing Grants Construction and Regeneration Act 1996 and Scheme for Construction Contracts (England and Wales) Regulations 1998. UK Department of Trade and Industry (DTI), London.

Egan, J. (1998) Rethinking Construction: Report from Construction Task Force. UK Department of Transport and the Regions, London.

Latham, M. (1993) Trust and Money: Interim report of the joint government industry review of procurement and contractual arrangements in the United Kingdom Construction Industry. HMSO, London.

Latham, M. (1994) Constructing the team: Joint review of the procurement and contractual arrangements in the UK Construction Industry. HMSO, London.

Murray, M. and Langford, D. (eds) (2003) *Construction Reports, 1944-1998*. Blackwell Science, Oxford.

NAO (National Audit Office) (2001) Modernising Construction: report by the comptroller and Auditor General. The Stationery Office, London.

NAO (National Audit Office) (2005) Improving Public Services through better construction: report by the comptroller and Auditor General. The Stationery Office, London.

ONS (2007) UK Standard Industrial Classification 2007 (UK SIC 2007). Office for National Statistics (ONS) (http://www.ons.gov.uk/ons/guide-method/classifications/current-standard-classifications/standard-industrial-classification/index.html; accessed 20 April 2012).

ONS (2011) Construction Statistics, No. 12, 2011 Edition. Office for National Statistics (ONS) (http://www.ons.gov.uk/ons/rel/construction/construction-statistics/no--12--2011-edition/index.html; accessed 20 April 2012).

Strategic Forum for Construction (2002) Accelerating Change: A report by the Strategic Forum for Construction (Chairman: Sir John Egan). Rethinking Construction, London (http://www.strategicforum.org.uk/pdf/report_sept02.pdf; accessed 20 April 2012).

Wolthenholme, A (2009) Never Waste a Good Crisis: A review of progress since Rethinking Construction and Thoughts for our Future, Constructing Excellence. (http://www.constructingexcellence.org.uk/pdf/Wolstenholme_Report_Oct_2009.pdf).

2

Stakeholders and the regulatory environment

Financial Management in Construction Contracting, First Edition. Andrew Ross and Peter Williams.
© 2013 Andrew Ross and Peter Williams. Published 2013 by John Wiley & Sons, Ltd.

2.1 Accounting

Businesses of all sorts are required to account for their activities each year so that the appropriate tax liability can be calculated and paid over to the public revenue. This requires a set of accounts to be prepared and the accountancy profession has long recognised the need for this information to be presented in a consistent way. Consequently, over the years, guidelines have been developed in order to ensure that:

- **prudence** has been exercised in the preparation of the accounts;
- the accounts represent **a true and fair view** of the affairs of the business.

Company accounts should, therefore, be prepared on the basis of applicable accounting and auditing standards and the auditor's statement will normally confirm this.

Whilst there is no compulsion to comply with the provisions of the Statement of Standard Accounting Practice No. 9 (SSAP9) (ICAEW, 1989), if its principles are not applied then the accounts may well not reflect the true and fair view required under statute and the accounts may well have to be qualified by the auditors.

2.1.1 Accounting reference period

In the same way that individual tax payers have a tax year ending on the 5th April each year, so all companies have their own 'tax year end' when the financial affairs of the business are concluded and the tax liability is calculated. The tax 'year end' is not, however, a fixed date but is one that varies depending upon the date of incorporation of the company. Consequently, the year-end date may be the 30th April, the 30th June, the 31st December, or any other end of month date.

The financial year of companies is referred to as the **accounting reference period**.

2.1.2 Accounting reference date

The **accounting reference date** (ARD) is the financial year end. For instance, if a company is formed (incorporated) on the 15 May 2006, then the accounting reference date will be the 31st May and the first accounts would cover the period 15 May 2006 to 31 May 2007.

The accounts must be filed at Companies House, whether or not they have been agreed by HM Revenue and Customs, within a specified period of the ARD, that is:

- nine months for a private company;
- six months for a public company.

2.1.3 Statutory compliance

All companies, whether sole trader, partnership or private or publicly incorporated, are required to submit accounts to HM Revenue and Customs, normally annually, for taxation purposes. Furthermore, incorporated companies, registered with the Registrar of Companies, are required to provide certain basic accounting information; this is held on file at Companies House and is available publicly on payment of a company search fee.

The relevant law relating to the preparation, contents and filing of annual accounts is the Companies Act 1985 as amended in 1989 and subsequently.

In publicly quoted companies the company's accounts appear in the annual report, which also contains, amongst other things, a statement from the chairman and an auditor's report. The annual report represents the external face of the business and is read by shareholders, professional analysts, stockbrokers, potential investors and so on. This information is publicly available and, as such, must be reliable, as it reports the financial position of a company to those people not concerned with the day to day running of the business.

Public companies are required to publish accounts for their shareholders' scrutiny and this information is generally available in companies' annual reports. Stock exchange regulations require quoted companies to produce figures showing profits half-yearly; this is contained in a 'mini' version of the annual report.

The financial data which appear in the annual report are based on information generated within the company and are meant to convey a picture of what the company is doing at a particular point in time. It is, however, no more than an 'overview' of what is happening in the company and, as will be seen later in the book, a huge amount of detailed information has to be gathered, sifted, organised and analysed so that the accounts can be prepared.

The annual accounts report on the activities and performance of the company during the financial year and show the turnover figure, the profit made and the current value of its assets and liabilities. The accounts will normally include:

- a profit and loss account;
- a balance sheet signed by a director;
- an auditors' report signed by the auditor (if appropriate);
- a directors' report signed by a director or the company secretary;
- notes to the accounts; and
- group accounts (if appropriate).

Medium sized, small and very small companies are allowed to file 'abbreviated accounts' containing less detail. Such companies are defined in the Companies Act as having at least two of the characteristics shown in Table 2.1.

Table 2.1 SME classification.

Characteristic	Small company	Medium sized company
Annual turnover	£5.6 million or less	£22.8 million or less
Balance sheet total	£2.8 million or less	£11.4 million or less
Average no. of employees	50 or less	250 or less

2.1.4 Annual accounts

A set of meaningful accounts is crucial in order to give the outside world **a true and fair view of the state of affairs of the business** at a particular point in time. This is the purpose of the annual accounts which, in small companies, consists of a 'balance sheet' and 'profit and loss account' and, in larger companies, includes other documents such as a 'movement of funds statement' and so on.

The balance sheet and profit and loss account and, in larger businesses, the movement of funds statement (or source and application of funds statement) and the chairman's report, are the key documents in the annual report. These give an 'over-view' of the state of affairs of the business at a particular point in time.

The **balance sheet** is a statement of the assets and liabilities of the concern at the point in time when the accounts are struck and provides a 'snapshot' of the business on one specific day of the year. The **profit and loss account**, on the other hand, is a statement of turnover, cost of turnover and profit for the entire trading year. The **movement of funds statement** shows how much money the firm had at the beginning of the year, how much money has come in and gone out during the year and how much money the firm has at the end of the year.

2.1.5 Audit procedures

Many companies carry out a variety of audits as a control measure to determine the validity of their systems and procedures and to test their reliability and consistency. Health and safety standards, environmental performance and information systems are typical examples of areas of business activity that might be audited. Financial reporting is another.

An audit is an independent assessment of the fairness and reliability of the financial statements presented by the management of the company and whether they comply with International Financial Reporting Standards (IFRS) or Generally Accepted Accounting Principles (GAAP) in the United Kingdom. Audits are conducted by auditors or accountants and may be internal or external.

Internal auditors are employees of the company whose job is to assess and evaluate internal control systems. To maintain independence, they report directly to top management or the board of directors or, in large firms, to a special audit committee. Where there is an audit committee, they will present a report which will be included in the annual accounts of the company.

External auditors are independent accountants who specialise in carrying out audits to assess and evaluate the financial statements made by companies in order that shareholders, tax authorities, banks and other interested parties may be assured of their validity and reliability.

The external auditors prepare an **auditor's report** in order to provide assurance that the accounts are free from material mis-statement and give a true and fair view of the state of the company's affairs at a particular date and its profit/loss at the year end. The auditor's report is included in the annual accounts.

2.2 The Companies Acts

The Companies Acts are United Kingdom statutes that regulate companies falling within its jurisdiction, such as limited companies and companies listed on the London Stock Exchange. The latest is the Companies Act 2006, which gained Royal assent on the 8 November 2006.

2.3 Accounting standards

To ensure that the published accounts of companies meet particular standards of clarity and transparency, guidance as to good accounting practice has traditionally been developed by the accountancy profession. In the early 1970s, accounting standards were developed by the Institute of Chartered Accountants of England and Wales (ICAEW) but later in that decade other accountancy bodies also became involved and the Accounting Standards Committee (ASC) of the Consultative Committee of Accountancy Bodies (CCAB) was formed to recommend good practice standards. These standards were known as Statements of Standard Accounting Practice (SSAPs).

In 1990, the system was reformed under the auspices of the Financial Reporting Council (FRC) which, in 2004, became the single independent regulator of the accounting and auditing profession with responsibility for issuing and enforcing accounting standards through five subsidiary boards, including the Accounting Standards Board (ASB).

Accounting reporting standards issued by the ASB are called Financial Reporting Standards (FRSs), which are first issued for consultation as Financial Reporting Exposure Drafts (FREDs). Whilst FRSs have replaced many of the old generation of accounting standards (the SSAPs), some SSAPs still exist because they were adopted in the early days of the ASB in order to bring them within the statutory definition of accounting standards prescribed by the Companies Act 1985.

Consequently SSAP9 – Stocks and long-term contracts (revised) – which is very important in construction for valuing work in progress, is still in use. However, it is proposed that SSAP9 will be replaced by FRED 28, although this is still in the consultation process.

2.4 UK accounting standards

In the United Kingdom, annual accounts are normally prepared in accordance with Generally Accepted Accounting Practice (UK GAAP). This is a combination of legislation, rules and accounting standards generally accepted by the accounting profession as representing good accounting practice.

Statement of Standard Accounting Practice No. 9 (SSAP9) is one of the standards of good practice to be followed; it relates to the valuation of Stocks and Work in Progress. This is an important standard in the construction business but, although SSAP9 is not necessarily adopted in the management reporting within all contracting organisations, its precepts represent generally accepted good practice.

2.4.1 SSAP9

Statement of Standard Accounting Practice No. 9 (SSAP9) is particularly relevant to the construction industry because it deals essentially with work in progress. However, as indicated in Figure 2.1, the reporting of construction work is complex because, at any one point in time, a contractor may have a portfolio of contracts which are either:

- complete;
- complete but final account not agreed;
- incomplete.

Additional factors further complicate the situation:

- contracts starting and finishing in different tax years;
- long-term contracts which cover several tax years;

These contracts have to be accounted for in the financial reporting of the business from the point of view of:

- turnover
- attributable profit
- asset value.

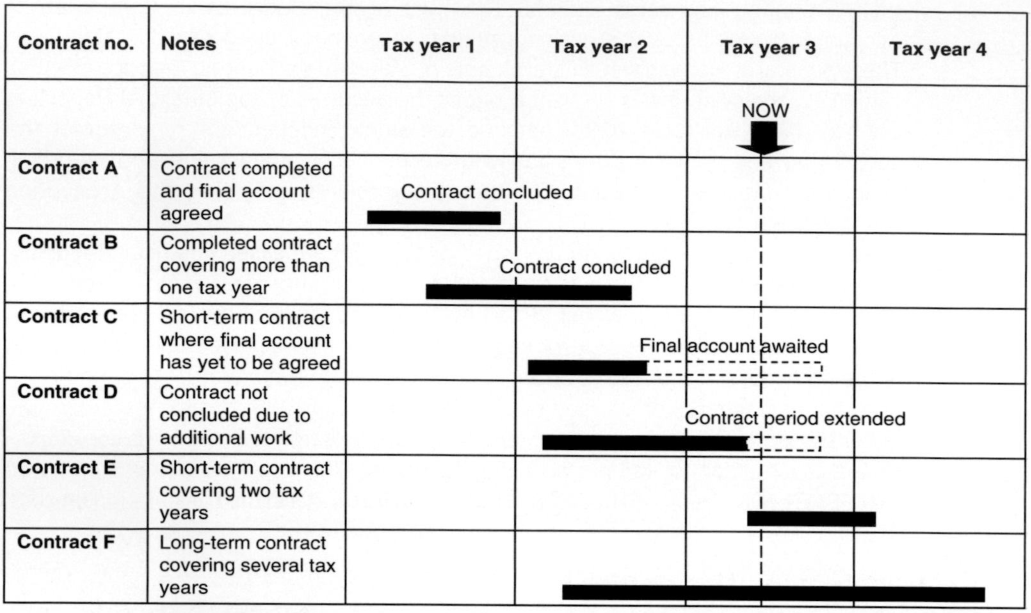

Contract no.	Notes	Tax year 1	Tax year 2	Tax year 3	Tax year 4
				NOW	
Contract A	Contract completed and final account agreed	Contract concluded ▬▬▬			
Contract B	Completed contract covering more than one tax year		Contract concluded ▬▬▬		
Contract C	Short-term contract where final account has yet to be agreed		▬▬▬ Final account awaited - - - -		
Contract D	Contract not concluded due to additional work		▬▬▬ Contract period extended - - -		
Contract E	Short-term contract covering two tax years			▬▬▬	
Contract F	Long-term contract covering several tax years		▬▬▬▬▬▬▬▬▬▬▬▬		

Figure 2.1 Contract portfolio.

Under the provisions of SSAP9, 'the accounting policies that have been applied to stocks and long-term contracts, in particular the method of ascertaining turnover and attributable profit, should be stated and applied consistently within the business and from year to year'.

Whilst there is no compulsion to comply with the provisions of SSAP9, if its principles are not applied then the accounts may well not reflect the true and fair view required under statute and the accounts may well have to be qualified by the auditors.

2.5 International accounting standards

Since 1 January 2005, companies may opt to prepare accounts in accordance with the International Financial Reporting Standards (IFRSs) issued by the International Accounting Standards Board (IASB). These standards differ from those traditionally used in the United Kingdom but they are widely used in the European Union and in other parts of the world, such as Australia and Hong Kong.

The UK GAAP and IASB use different terminology.

2.6 Financial reporting

As stated earlier, very few people know *exactly* how a business is doing financially at any one point in time and this is no different in large or small companies.

In a small company the principal will *think* he/she 'knows' what is going on because he can see all the bills coming in, knows what the tax, PAYE and VAT payments are likely to be, knows what is happening on all his contracts, has close contact with all his/her employees and subcontractors and can see the money coming in and the money going out. This, however, is not a true perspective because, for one thing, the various costs and revenues will not be reconciled to a common date and, therefore, the owner is not able to compare true cost with true value.

In a larger company, the managing director and finance director should know what is happening on a day to day basis but other directors will only find out at board meetings or on a 'need to know' basis. Senior managers will only know parts of the story and other staff, such as contracts managers, site agents/managers and quantity surveyors, will only know what is happening on their projects.

In some contracting firms, the site manager has little or no involvement in financial issues and is left to concentrate on managing the construction process and keeping to the target programme. In such cases it is the site quantity surveyor who 'controls the purse strings' and the site manager is kept 'out of the loop' except, perhaps, to be told whether or not the project is achieving its financial targets. Whilst this approach may seem illogical, there is some sense in it because the 'average' site manager will probably not understand the finer aspects of the financial control of projects in general, or the cost value reconciliation (CVR) process in particular, and, in all honesty, will probably not have the time to be interested.

The financial management of a business is dependent upon the quality and accuracy of the information received via its financial reporting system. If the data coming from site are inaccurate then this will distort the 'view' that management has. In smaller firms this information will be distorted because the reporting system is not good enough; in larger companies, misreporting, either accidentally or intentionally, will be equally misleading for management.

From an outsider's point of view, the financial 'goings on' within a particular business will be something of a 'black box' and not only will the available information be limited in detail but it will also be well out of date when it is published. In publicly quoted companies financial information is less outdated because six-monthly results precede the annual accounts and Stock Market regulations require companies to make announcements on specific matters of interest or concern to investors.

The information requirements of outsiders will vary, of course. HM Revenue and Customs (HMRC), for instance, will mainly be interested in how much profit the company has made in any one tax year. On the other hand, existing shareholders and those wishing to invest in the company (i.e. prospective shareholders) will be keen to know whether the financial fundamentals of the business are strong enough to warrant risking, or continuing to risk, their capital.

Satisfying the information needs of management and outsiders depends upon accurate reporting of the true financial position of the business. This is not easy to achieve in practice because information generated on site becomes distorted as it passes through the various levels of the company.

2.7 Financial reports

The running of any business cannot be done in a vacuum and a good deal of reporting of various sorts is required:

- management requires reports in order to control the business;
- lenders (especially banks) often require reports in order to validate banking covenants or to ensure that loans and/or overdrafts are not at risk;
- HM Revenue and Customs requires:
  - annual accounts for Corporation Tax purposes;
  - monthly (or three-monthly) returns of tax and national insurance deducted from employees and subcontractors;
  - monthly (or three-monthly) VAT returns;
- Companies House requires an annual return and the latest accounts (or abbreviated accounts) of the company.

2.7.1 Management reports

The construction industry has a hugely wide ranging representation of firms from the tiny to the mega-big operating in markets that include general contracting, house building, infrastructure, specialist subcontracting, crane and plant hire and so on. In this respect alone, it is difficult to generalise what sort of reports should be expected by management to help them to run the business. Very large firms will have highly detailed and sophisticated reporting systems and small firms will probably lack the staff and expertise to have anything other than simple feedback as to 'what is happening' in the business.

In contracting, the 'core' business is 'contracts' and the main focus will undoubtedly be on reporting how well these contracts are doing from a financial and progress point of view. Notwithstanding this, it is crucial that management does not take its 'eye off the ball' when it comes to cash and liquidity or allow overhead expenditure to creep ever upwards.

Strictly speaking, management should receive reports on the financial performance of the business at least on a monthly basis. Most companies work to a 'reporting date', which will either be the end of the month or a particular date towards the end of the month. Some companies have an annual calendar with these dates highlighted in a box ('box dates'). The reporting date is frequently referred to as the 'cut-off' date.

Management will want to know what was planned, what the actual situation is and what the variance +/- is so that remedial action can be taken if things appear to be going wrong. Management reports, therefore, should include:

- The profit/loss situation on individual contracts determined by matching the true value with the attributable costs at the cut-off date (cost value reconciliation).
- The physical progress on individual contracts in relation to both the target programme (set by the company) and the master programme (determined by the terms of the contract).
- Financial progress in relation to physical progress; in particular is the monthly value generated ahead of or behind or on target with what was expected for a particular cut-off date?
- An overall view of the combined profit/loss situation on all contracts.
- A cash flow forecast updated to the review date showing detailed income from contracts, expenditure on labour, materials, subcontractors, site on-costs and so on, and the cash surplus/deficit.
- A finance budget showing the cash surplus/deficit from the cash flow forecast together with investment income, payments of bank interest, tax and dividend payments and so on.

With the widespread availability of powerful personal computers and software, it is well within the bounds of even very small companies to monitor contract profitability and the cash/liquidity situation. Spreadsheets and accounting packages can easily be used to provide information on the income (value) earned from contracts and the costs associated with those projects. What many managers fail to realise, however, is that comparing income to date with costs to date does not give the true picture.

The value given at any point in time must be the *true* value exclusive of overmeasured items and front-loaded preliminaries and so on, and the costs must be those costs expended in generating the true value. In other words, cost must be attributed to value and this is the purpose of the cost value reconciliation procedures undertaken in larger contracting firms.

The cost value reconciliation is a type of audit that takes place each month in many medium and large contracting companies. Such audits are conducted in closed meetings where the site quantity surveyor will present the CVR to a small group of management staff. Those invited to the meeting may include the managing quantity surveyor, the site manager/agent, the contracts manager/director and the finance director.

The purpose of the CVR meeting will be to ensure that the CVR has been correctly conducted in accordance with company procedures and that the figures correctly represent the true cost-value situation on the project. Of particular concern will be the correct adjustment of overvalued items and amounts earned in advance, the correct inclusion of potential liabilities for subcontractors' claims and for defective work and the possibility that there are hidden contingencies included in the figures.

Managers should be aware that financial reports based on information from sites will be subject to some variability because of inherent inaccuracies in the process of information collection and dissemination. For example, labour allocation sheets are usually filled in at the end of the week by the ganger or foreman, rounded to the nearest half hour and are viewed with considerable suspicion because memories have faded by the time they are completed. Daywork sheets are also problematic, as they are often completed by gangers, foremen or subcontractors who do not understand the principle of 'daywork', and records often lack descriptive and numerical precision.

2.7.2 Lenders and banking covenants

An important part of a contractor's reporting procedures will include the provision of information to lending institutions such as banks. Irrespective of whether the bank is providing an overdraft facility or providing medium-term debt finance, the bank will require information to monitor the contractor's financial situation on a continuing basis, thereby protecting the interests of the bank and its shareholders.

When a sole trader, partnership or incorporated company takes out a bank loan, the bank will draw up an agreement which will form the basis for granting the loan. This agreement or 'covenant' will outline the conditions upon which the loan is granted and the sanctions to be imposed by the bank in the event of a breach of the covenant. Banking covenants impose minimum standards for a borrower's future conduct and performance and, if breached, may result in acceleration of the maturity of the loan (i.e. repaying the loan on demand or within a prescribed time limit) or a reduction of the lending facility.

Large companies which borrow huge sums of money from lending institutions have to be constantly aware of their banking covenants and, should they be in danger of breaching the terms and conditions, they will need to make an announcement to the Stock Exchange and negotiate a solution with the bank in question. A breach of covenant may result in a negotiated restructuring of the debt or in the imposition of additional constraints.

Covenants form an important part of the risk management procedures of banking institutions and, therefore, the nature and extent of the restrictions imposed by the covenant will be proportionate to the risk exposure of the bank and the borrower's financial standing. With strong financial credentials, large companies will perhaps have very few standard covenants attached whereas agreements with riskier borrowers who are perhaps highly geared (leveraged) may attract much more complex and possibly non-standard covenants.

Because they reduce the bank's risk exposure, loan covenants can have the beneficial effect of reducing the borrower's costs of servicing the debt, thereby making the company more competitive and profitable. This is a good arrangement for both the borrower and the lender.

Covenants may be broadly classified as affirmative or negative. **Affirmative covenants** often require a borrower to meet certain standards, such as discharging contractual obligations, paying the bank interest and fees and providing information to the bank at regular intervals. **Negative covenants** restrain the borrower from certain actions, such as spending more than a specified amount on capital expenditure, or they may stipulate that specified measurable financial tests must be satisfied on a regular basis. This might include ensuring that a certain ratio of working capital to debt (i.e. borrowing) is maintained.

Many contracting businesses rely on a bank overdraft as a main source of finance together with whatever owner's funds might be available. This is a high risk strategy because banks usually require security for the overdraft facility and, for smaller companies, this often means the bank taking a legal charge over the family home. Whilst an overdraft is a short-term source of finance, the practical reality is that overdrafts are often regarded as medium- to long-term finance by contractors. In many respects this is fine but when the overdraft limit is reached the reality of operating with insufficient working capital can put the business at risk and the pressure to reduce the borrowing from the bank can be intense.

A contractor is particularly likely to be under the affirmative covenant to supply information to the bank at regular intervals. This might be a monthly or quarterly requirement depending upon the nature and amount of the loans and the contractor's financial stability. The information required by the bank will necessarily have to be simplified because the account manager needs an overview rather than lots of unnecessary detail. Typical information requirements will include:

- a statement of the position on contracts including:
  - a list of all contracts including the client's name, the contract value and the form of contract applicable;
  - the gross valuation to date for each contract, retentions held, the value of work in progress and any outstanding claims and so on, and the amount of previous payments (together giving the current asset value of each contract);
  - the value of contract work remaining and the projected cost to complete;
- a list of debtors (money owed to the company) comprising name of debtor, amount outstanding, date due and age of debt;

- a list of creditors (money owed by the company) comprising name of creditor, amount owing, date due for payment and credit terms;
- a list of all plant and equipment and the current value of each item together with details of items bought on hire purchase – for example, start and end dates of hire-purchase (HP) agreement, monthly payments, balance owing at the end of the agreement;
- an annual cash flow forecast updated monthly.

Table 2.2 illustrates a sample 'schedule of contracts' that might be asked for by a lender showing the contractor's current contract position together with explanatory notes. Lenders will be particularly interested in:

- the total of monies outstanding, including work in progress (£504 500);
- the value of contractual claims (£41 000);
- the projected margin on work to be completed (column N – column O = £100 000);
- the current asset value of contracts (column M), especially if there is a fixed and floating charge over book debts in place which is underwriting a loan or overdraft.

2.7.3 HMRC reports

During any trading year, HM Revenue and Customs requires companies to report information and make payments related to the collection of taxes. As these are statutory requirements, late, incorrect or false declarations can have serious implications for companies, ranging from the imposition of financial penalties to criminal prosecution. HMRC reporting requirements include:

- The submission annually of a **company tax return** accompanied by:
  - ○ a copy of the accounts together with computations showing how the entries in the tax return have been derived from the accounts;
  - ○ a balance sheet as at the last day of the financial year, a profit and loss account, any directors' and auditors' reports and notes to the accounts;
  - ○ Group accounts where appropriate.
- **Employer Annual Return** of income tax and national insurance (NIC) deductions under the PAYE scheme for people employed directly.
- Monthly or quarterly **payments** of income tax and NIC deducted.
- **CIS monthly return** of payments made to subcontractors under the Construction Industry Scheme (CIS).
- Monthly **payment** of tax deducted from subcontractors under the Construction Industry Scheme (CIS).

2.7.4 VAT returns

All VAT-registered companies must submit **VAT returns**. This requires the completion of a standard form in which is stated the VAT charged on sales (output tax), the VAT paid on purchases (input tax) and the amount of VAT due to be paid to or reclaimed from HMRC. Most contractors will have VAT to pay.

Submission of VAT returns and payment of VAT due must be made within a specified time from the end of the VAT period in question. Timings vary depending upon whether submissions are made online (which is now the 'norm') or by post. VAT

Table 2.2   Schedule of contracts.

A+B PLANT LTD

SCHEDULE OF CONTRACTS

	A	B	C	D	E	F	G	H	I	J	K	L	M	N	O
Contract Name	Contract No	Employer	Contract value	Form of Contract	Valuation No	Gross valuation certified	2.5% Main contractor's discount	Retention (5%) held	Work in progress	Previous payments (net)	Value outstanding (F-G+I-J)	Counter claims	Current asset value (K-L)	Value of work to be completed	Cost to complete
DRAINAGE CONTRACT	AB/102	Smith Contractors	355 000	FCEC Blue Form of subcontract	3	210 000	5 250	10 500	17 000	142 000	79 750	11 000	68 750	170 000	190 000
SITE CLEARANCE CONTRACT	AB/103	Anywhere Council	205 000	ICE 7 Minor works contract	2	198 000	0	9 900	7 000	86 000	119 000	0	119 000	0	0
HOUSING SITEWORKS	AB/104	Jones Contractors	610 000	DOM/1 form of subcontract	4	460 000	11 500	23 000	9 000	383 000	74 500	3 000	71 500	160 000	150 000
ROADWORKS CONTRACT	AB/105	Acme Construction	980 000	FCEC Blue Form of subcontract	7	870 000	21 750	43 500	56 000	810 000	94 250	27 000	67 250	290 000	210 000
MUCKSHIFT CONTRACT	AB/106	Water Utilities	270 000	ICE 7 main contract	1	130 000	0	6 500	7 000	0	137 000	0	137 000	140 000	110 000
						1 868 000	38 500	93 400	96 000	1 421 000	504 500	41 000	463 500	760 000	660 000

NOTES

C   The contract value is the original value of the contract before commencement on site

F   The gross valuation is the value of work carried out as per the bills of quantities plus any extra work or variations

G   The main contractors' discount only applies where A+B are sub-contractors

H   Retention monies are not deducted from gross value because the money belongs to A+B (subject to the correction of defects)
Retention monies are held by the employer (or main contractor where A+B are sub-contractors) but the money belongs to A+B

I   Work in progress is the value of work done after the last valuation and before the next valuation

J   Previous payments is the money received by A+B after deducting discount (if applicable) and retention

K   The value outstanding is the gross certified value less discount (where applicable) plus work in progress less previous payments

L   Counterclaims is money claimed against A+B for materials or services supplied by the main contractor or for loss/expense claims (eg where A+B has delayed, disrupted or damaged the work of others)

M   The current asset value is the total of value outstanding less counterclaims (if any) and is represented as stock/work in progress on the balance sheet

N   Work to be completed is the gross certified value less discount (where applicable) plus work in progress less previous payments

O   Cost to complete is an estimate of how much it will cost to complete the remainder of the contract works
The difference between value and cost of work to be completed indicates whether this will be profitable or not.
It could also indicate front-loading of the tender where profit from later items of work is loaded onto earlier items in order to boost cash flow

returns are usually made quarterly and periodic visits will be made by the VAT inspector to check accounting records, books and verify that returns have been made correctly.

A late VAT return and/or a late payment of VAT owed in any 12-month period will invoke a Surcharge Liability Notice. If the offence is repeated a Default Surcharge will be payable on all unpaid VAT amounts. This operates on a sliding scale depending upon the number of defaults made in the 'surcharge period'.

2.7.5 Companies House

All incorporated companies must submit information about the company on an annual basis to the Registrar of Companies at Companies House. There are certain fees to be paid and if the information is submitted late there are penalties as well. The Registrar requires:

- A **company annual return** that provides general information about the company's directors, company secretary (if any), address of registered office, shareholders details and share capital.
- A set of **annual accounts** or abbreviated accounts together with other information similar to that required by HMRC.

2.8 Annual accounts

Limited companies are run on a day to day basis by a board of directors. In a small company this may comprise the principal of the business and perhaps one or more family members. In a larger company, the board of directors will be appointed to certain roles, such as managing director, and there is likely to be a chairman. For reasons of sound corporate governance, the executive directors will supported by non-executive directors who have no responsibility for the day to day running of the business.

The directors of a limited company are responsible to the owners of the company – that is the shareholders – but they also have strict statutory duties as regards governance and financial reporting. However, it is the shareholders who are first entitled to hear the annual report of the management as regards how the business has performed over the trading year and they are also entitled to vote on matters such as the proposed dividend, the composition of the board of directors and the future direction of the company.

The publication of the annual accounts gives shareholders and others the opportunity to gain an insight into the financial picture of what has happened in the business over the previous year's trading. Companies have several months during which to prepare the accounts and they are consequently out of date when they become available.

The level of sophistication of the accounts depends upon the size of the company. In small 'one-man band' companies the accounts will be simple and straightforward and few people will see them apart from the accountant, the owner, maybe the bank and certainly the statutory authorities. Whether the owner will understand or even read them is debatable!

In small to medium sized organisations (SMEs), the accounts will still be relatively prosaic but they will be presented formally at a board meeting and the chairman will undoubtedly have something to say about the previous year's trading – a sort of 'end-of-term report'.

In larger and publicly quoted companies the accounts will take on a greater level of sophistication. The look of the accounts will be 'glossy' with colour printing and there will be photographs of the directors and both completed and current contracts of note. The best place to see these is usually at the company's website where downloads are available and hard copies can usually be ordered.

The information in the annual accounts may seem somewhat alien to non-accountants and those used to dealing with contractors' tender summaries, bills of quantities and interim valuations. The accounts, however, provide a useful and different insight into what is happening in the company once a basic understanding of the various documents contained therein has been gained.

2.8.1 Operating statement

Not all annual accounts have an operating statement – it is usually a feature of the accounts of larger companies. The operating statement is basically a financial summary showing performance data over the current year along with comparative data from previous years (sometimes the preceding year, sometimes taking a five-year view). It is purely an 'overview' of the accounts and will give basic information such as annual revenue (turnover), earnings before interest and tax, operating profit and net cash flow.

There is no set model for the operating statement; the presentation and content vary from company to company. There are usually some histograms or pie charts to be found – even accountants have a sense of artistry!

2.8.2 Directors' report

The directors' report is an important statement about the company's activities and the year's trading figures; it also gives a perspective of what can be expected in the following year and beyond. The requirement for a directors' report is controlled by statute under the Companies Act 2006, which also sets out exemptions and requirements as to the content of the report.

The content of the report varies but there will usually be an overview of the affairs of the company and an explanation of how the board has fulfilled its responsibilities as directors over the year's trading. It may be a fairly brief statement or could run to several pages. There will usually be a brief summary of the annual results, with particular emphasis on profit and shareholders' dividend payments. Reference will also be made to growth levels (both positive or negative) and emphasis will be placed on the economic outlook, the challenges facing the construction market, margins and pricing levels and the board's perspective for the forthcoming year's trading.

The directors will be particularly keen to explain to shareholders how they have responded to the challenges of the market place and how they have compensated for falling markets by undertaking new business opportunities. There will usually be some dialogue concerning the challenges of staffing and skill levels, the forging of partnerships with clients, designers and the supply chain and any acquisitions that the company has made.

Other issues that may be referred to in the report include details of the board and its directors, directors' remuneration and their interests in the company (such as their individual shareholdings), corporate governance and compliance with the Code of Best Practice and a statement of directors' responsibilities for, in particular, preparation of the annual report and accompanying financial statements.

2.8.3 Profit and loss account

The profit and loss (P&L) account is a statement of business performance that records annual sales income, costs and expenses over the accounting reference period (usually a trading year). The P&L account measures how much money has been made in a specific period and, if there is more than one sector to the business, how the money was made. It also states the cost of sales, including the costs of production, overheads and interest charges on borrowings. The annual profit is given by the difference between sales and the cost of sales plus administration costs and interest.

The P&L account records:

● incomings (revenue from sales);
● outgoings (cost of sales plus overheads and expenses);
● and shows whether a profit or loss has been made.

The turnover figure stated is a summation of invoices that have been raised, or sales income that has been generated, together with an estimate of work in progress not yet invoiced (Chapter 8 provides for more detail on turnover). The cost of sales figure includes purchases made from suppliers for raw materials, the labour costs of conversion including the cost of subcontractors and an estimate of materials used but not yet paid for.

It is important to remember that the P&L account does not record whether invoices raised or received have been paid, so it therefore does not indicate the amount of cash the business has. This can be seen in the 'movement of funds' or 'cash flow' statement that also appears in the annual accounts.

Typically, the P&L account is declared annually but, in larger and publicly quoted companies, an interim set of accounts will be published every six months. Some companies produce a P&L account every month as part of the internal control procedures of the business.

2.8.4 Balance sheet

The balance sheet is a statement of the **assets** and **liabilities** of a business. It does not show what the business is worth because the assets are not usually shown at their current value; balance sheet assets may be revalued from time to time to bring them up to date.

The basic principles behind the balance sheet may be best explained by reference to the financial affairs of an individual or married couple as illustrated in Table 2.3, from which it should be noted that:

● The balance sheet balances because it always does!
● The balance sheet balances because every asset has to be financed and this creates a liability.
● The personal investment is highlighted because this is the item that makes the balance sheet balance.

Table 2.3   Personal assets and liabilities.

Assets		Liabilities	
Item	Value	Item	Value
	£		£
House	190 000	Mortgage	100 000
Car	15 000	Car loan	12 000
Computer	500	Credit card	3500
Personal belongings	3000	Loan from parents	30 000
Savings	1700	**Personal investment**	**64 900**
Cash	200		
TOTALS	210 400		210 400

Table 2.4   Contractors' assets and liabilities.

Assets	
Land and buildings	1. Includes development land where the company is a speculative house builder. 2. The company may own its main office(s) or builder's yard.
Plant and equipment	Includes excavators, cranes, wagons, concreting equipment etc.
Debtors	Debtors owe money to the company. This might include clients for whom work has been carried out and invoiced.
Stocks and work in progress	1. Stocks of materials kept in the builder's yard, such as drainage goods, plastic pipes, timber etc. 2. Work carried out but not invoiced including materials on site.
Cash at bank	Liquid cash held in a bank account which can be drawn down immediately.
Liabilities	
Creditors	Creditors are owed money by the company. Includes materials suppliers, plant hire firms, subcontractors and providers of professional services, such as architects and accountants.
Overdraft	A short-term borrowing facility agreed by the bank to provide working capital for the business. Possibly secured on the assets of the business. Interest is paid only on the amount outstanding.
Interest-bearing loans	Medium-to-long-term finance. More permanent than overdraft but interest is payable on the full amount of the loan. May be secured on the assets of the business. Often called 'debt capital'.
Shareholders' claims	This is the 'equity capital' invested in the business by shareholders. Represents long-term capital for which a dividend is paid usually twice per year. Dividends and/or capital growth is the attraction for shareholders but this has to be balanced against the level of risk to which the business is exposed.

- The personal investment is shown as a liability because it represents the amount of personal money invested in buying the assets.
- The personal investment is a claim against the assets in the same way as shareholders have a claim on a company's assets.
- Personal claims take a back-seat in the pecking order of claims against the assets after the mortgage, credit cards and loans.
- The parents' loan is probably unsecured and, therefore, they have no claims against the assets if things go 'pear shaped'.

The same principles apply to the business situation, albeit that the names of the assets and liabilities are somewhat different. The common assets and liabilities normally seen in a contractor's balance sheet are illustrated in Table 2.4.

2.8.5 Movement of funds statement

The movement of funds statement compliments the balance sheet and profit and loss account and shows the sources of money and how it has been spent by the company during a particular trading or accounting reference period. The statement also indicates the short-term viability of the company and its ability to pay its bills as and when they fall due.

Also called the cash flow statement, this mandatory part of the annual accounts is distinct from the balance sheet and profit and loss account in that it does not include future incoming and outgoing cash transactions made on credit. In this respect, cash flow is not the same as net income, which includes sales made on credit such as the work carried out for a client under a building contract. The movement of funds statement is subdivided into three parts:

- **operations** – cash inflows and outflows generated by the core business;
- **investing** – cash movements due to the purchase and sale of assets such as plant and machinery;
- **financing** – changes in debt, loans or dividends.

2.8.6 Auditors' report

The auditor's report is quite different from the other information presented in the annual accounts. It is not an evaluation of what has happened in the year's trading but is simply an opinion given by an independent, qualified professional that the information presented in the accounts is correct and free from material misstatements. There are statutory requirements under the Companies Act 2006 concerning the appointment and functions of auditors, their duties and responsibilities and the content of the auditor's report.

In other words, the auditor's report gives credibility to the accounts, which may help to attract investors or obtain finance and give outsiders confidence in the information presented in the accounts.

The auditors may give:

- An **unqualified opinion** – where the accounts are in accordance with statutory requirements and accepted accounting practice.
- A **qualified opinion** – where the accounts are in order save for one or two instances of non-compliance with accepted accounting practice.

- An **adverse opinion** – where the accounts are materially mis-stated and do not comply with generally accepted accounting practice.
- A **disclaimer** – where the auditor is unable to form an opinion on the accounts as presented.

The preparation of audited accounts is a legal requirement under the Companies Act 2006 provided that the company does not qualify as a 'small' company. The accounts do not have to be audited if the annual turnover does not exceed £6.5 million and total assets are not above £3.26 million (Section 477(2) of the Act). If this is the case, there will be a caveat in the accounts saying that they have been prepared on the basis of information and explanations given to the accountant who prepared them. This exemption will apply to many subcontractors and specialist contractors who operate in the construction industry.

The lack of an audit does not in any way diminish the responsibility of the director(s) for the preparation of the financial statements or for ensuring that the accounts give a true and fair view of the company's affairs, which are both statutory requirements under the Companies Act 2006.

This includes making an assessment of whether the accounting policies adopted are appropriate to the company's circumstances and that they have been consistently applied and adequately disclosed. The auditor must also assess the reasonableness of any significant accounting estimates made by the directors and the overall presentation of the financial statements.

It is obviously unreasonable to expect auditors to visit every office and every construction site but they must be sure that adequate accounting records have been kept and that adequate returns have been made from sites. This involves a lot of checking to ensure that the financial statements made in the annual accounts agree with the accounting records and returns in the company's management information system. The auditor must also delve into the records and ask questions to be able to state that all the information and explanations required for the audit have been given.

The audit is an area of the financial affairs of a business where the project cost value reconciliation interfaces with the annual accounts and where the site quantity surveyor's work may be questioned by an outsider. Whilst management must be made aware of the profitability or otherwise of a continuing contract, Barrett (1992) states that profit must not be declared in the annual accounts unless it can be demonstrated that the profit has been determined by cost value reconciliation procedures which take account of SSAP9, or, in other words, that the contract has been concluded and it has made a profit.

It is only with the assurance that the management reporting system is sufficiently robust in terms of SSAP9 (or the equivalent) that the auditor can give the opinion that the financial statements:

- represent a true and fair view of the state of a company's affairs *and of its profit* for the year ended
- have been properly prepared in accordance with generally accepted accounting practice; and
- have been prepared in accordance with the requirements of the Companies Act.

In some large companies there may be an **audit committee**. This will usually comprise a small number of independent non-executive directors whose remit may include monitoring the integrity of the financial statements of the company, reviewing reports by management on the effectiveness of the systems adopted for internal financial control, financial reporting and risk management, overseeing the process for selecting the external auditors and considering each year whether there is a need for an internal audit function.

2.8.7 Notes to the accounts

The notes to the accounts provide additional detail and explanation to the accounts themselves and clarify the basis upon which the accounts have been prepared and the assumptions made in their preparation.

This an important part of the annual accounts where a breakdown of the figures to be found in the balance sheet, the profit and loss account and so on can be found. The notes enable financial analysts and potential investors to carry out more detailed calculations to assure themselves as to the credibility of the accounts and the strength of the underlying fundamentals of the business.

The notes to the accounts are given a reference number which relates to particular components of the accounts themselves. For instance, stocks and work in progress may be number 12 under the general heading of current assets. This reference number then takes the reader to the relevant note where further detail or background information may be found quickly and easily.

2.8.8 Group accounts

Some larger companies may own the share capital of other firms and, as such, accounts have to be prepared for the group as well as the individual company. These accounts are called **consolidated** or **group** accounts. Where the company only partly owns the share capital of other firms, these are listed in the accounts as **minority interests**.

A group of several companies may be organised into divisions such as construction, housing, infrastructure and fit-out. Within each division there may be one or more companies operating in their own right in their individual market sector or field of expertise.

Being smaller companies, group members may possibly not be big enough to be public limited companies in their own right but, being in a large group, they can still enjoy the benefits of belonging to a larger organisation. Such benefits may include market reputation, access to funding or the provision of guarantees required under certain procurement arrangements (parent company guarantees). There may also be benefits of guaranteed turnover by working for other members of the group. For instance, a fit-out company may undertake work for the development arm of a large group.

In such circumstances, the annual accounts of the parent company will include the consolidated accounts of the group of companies it owns.

References

Barrett, F.R. (1992) *Cost value reconciliation*, 2nd edn. The Chartered Institute of Building, Ascot UK.

ICAEW (1989) Statements of Standard Accounting Practice, No. 9. The Institute of Chartered Accountants in England and Wales (ICAEW), London.

3

Risk and uncertainty

Financial Management in Construction Contracting, First Edition. Andrew Ross and Peter Williams.
© 2013 Andrew Ross and Peter Williams. Published 2013 by John Wiley & Sons, Ltd.

3.1 Definitions

'Risk' and 'uncertainty' are not synonymous but the two words are often used together in common parlance as an expression in the same way as 'delay and disruption' or 'loss and expense'. The words 'risk' and 'uncertainty' may well be linked but they have separate and distinct meanings. Knight (1921), in a seminal work, suggested that *'uncertainty must be taken in a sense radically distinct from the familiar notion of risk, from which it has never been properly separated'*. He also suggested that risk is measurable whereas uncertainty is not. Others authorities differ from this view.

Winch (2010) also distinguishes between uncertainty and risk and suggests that *'uncertainty (is) the absence of information required'* for decision making whereas *'risk is the condition where information is still missing but a probability distribution can be assigned to the occurrence'* of a particular event. Figure 3.1 illustrates Winch's point in relation to the tender invitation to final account period of a project during which the uncertainty is dynamic and shifts according to the release of information.

The ISO Guide 73:2009, *Risk management vocabulary*, which complements the international standard ISO31000:2009 Risk Management – Principles and guidelines, defines risk as the *'effect of uncertainty on objectives'*. In this definition, uncertainties include events (which may or may not happen) and uncertainties caused by a lack of information or ambiguity.

The British Standard on Project Management (EN BS 6079-3:2000) defines risk as *'An uncertainty inherent in plans and the possibility of something happening (i.e. a contingency) that can affect the prospect of achieving business or project goals'*.

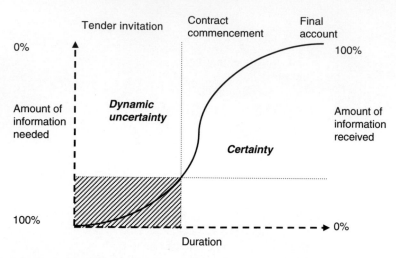

Figure 3.1 Dynamic uncertainty curve (adapted from Winch, 2010).

In the Vocabulary of this British Standard a further definition is that 'risk' is a *'combination of the probability or frequency of occurrence of a defined threat or opportunity and the magnitude of the consequences of the occurrence'*.

Another common definition of risk, and one frequently used, is that it is *'the threat or possibility that an action or event will adversely or beneficially affect an organisation's ability to achieve its objective'*. Alternatively, risk may be considered to be a consequence of the presence of a hazard and represents a measure of the likelihood of the hazard manifesting itself and the severity of its effect if it does.

Donald Rumsfeld, the former US Secretary of Defense, famously said that 'there are known-knowns, known-unknowns and unknown-unknowns' which tends to suggest that:

- there are eventualities that we are not aware of which are therefore unpredictable in magnitude or extent;
- awareness may not be sufficient to be able to ascribe a credible probability to a possible outcome;
- if uncertainty can be recognised it may be sufficiently manageable to allow a project to continue without the benefit of certain knowledge (Friedman *et al.*, 1999);
- the expected return on an investment or a building contract may be less than anticipated but may be predicted within certain confidence limits.

Events following the 2008 banking crisis and the fall-out of the American subprime mortgage lending farrago have led 'thought leaders' to adopt fresh approaches to thinking about risk and uncertainty in both macro and micro aspects of economics and business. Whilst recognising the influence of world-wide 'big events', the focus of this chapter is more on the tangible challenges to making business decisions in construction and how they impact on the reporting regime necessary to ensure legally compliant corporate governance.

3.2 Risk and reward

Exposure to risk almost inevitably leads to financial consequences. Risk outcomes are not always negative, however, and there is an upside to risk as well as a downside. Gains may equally result from taking risks as losses and good business instinct is characterised by knowing when to take a chance (risk) and when not to. When adverse events occur risk is crystallised but profit can only be realised if and when the risk event is eliminated.

Winch (2010) suggests that risk should be distinguished from reward in that **risk** relates to the probability of a detrimental event whereas **reward** identifies the probability of a beneficial event.

Examples of upside risk include:

- Unused risk allowances that contribute to increased profits.
- Programme risk (or float) that can result in time savings, reduced preliminaries and increased profits.
- Taking on 'risky' contracts that may lead to unexpected further work, thereby increasing turnover and profitability.
- Risky work that challenges the competence of both management and workforce and may lead to successful outcomes that may enhance the reputation of the company and lead to new clients, new markets and new opportunities.

Downside risk may be characterised by:

- Loss-making contracts that may lead to loss of profit, reduced dividends and dilution of share value.
- Too many loss-making contracts that may lead to dilution of working capital, reduced asset values and the possibility of instability or even insolvency.
- Programme risk that can result in charges for liquidated and ascertained damages, claims from subcontractors and resourcing problems.
- Accidents on site that may lead to loss of productivity, statutory fines, civil actions for damages, loss of reputation and consequent loss of business.

Figure 3.2 illustrates a qualitative approach to rating risk using a simple 2×2 matrix of risk impact relative to likelihood.

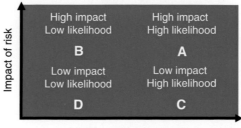

Figure 3.2 Risk rating.

3.2.1 Risk

Each September the Construction Index (http://www.theconstructionindex.co.uk) publishes a league table of the top 100 construction companies in the United Kingdom ranking their performance according to (i) turnover and (ii) profit. Relevant data taken from the most recently filed accounts (2011) for the top five UK construction companies is shown in Table 3.1.

It is notable that the Construction Index Top 100 (2011) indicates that only five of the 100 companies listed reported a pre-tax profit to turnover of more than 10% and more than half (57) of the companies listed reported a margin of 3% or less. The margins of 25 of the Top 100 Construction Companies were in the band 3–5%.

Whilst one year's results should not be interpreted as 'typical' of financial performance trends, historically construction margins have always been low compared to other industry sectors. Consequently, the Top 100 table suggests that the long-held

Table 3.1 Top five UK contractors.

Rank by turnover	Rank by profit	Company	Turnover (£million)	Pre-tax profit (£million)	Margin (pre-tax profit/ turnover) (%)
1	1	Balfour Beatty	10 541	187	1.8
2	2	Carillion	5139	168	3.3
3	20	Laing O'Rourke	3320	26	0.8
4	3	Babcock	2756	115	4.2
5	11	Morgan Sindall	2102	41	1.9

(Table compiled using information sourced from the Construction Index; http://www.theconstructionindex.co.uk/).

Table 3.2 Influencing factors for high risk–low margin industry.

Generic risks	Site-specific risks
1. 'Boom and bust' economic cycles	1. Design changes and variations
2. Intense competition for work	2. Insolvency of subcontractors
3. Too many contractors tendering for the same job	3. Wrong choice of subcontractor
4. Tough contractual arrangements	4. Accidents and health and safety breaches
5. Adversarial attitudes	5. Site access problems
6. Lack of working capital	6. Adverse weather conditions
7. Capital lock-up on projects due to:	7. Unforeseen ground conditions:
• Retentions	• Bad/unstable ground
• Work in progress not invoiced	• Contaminated ground
• Outstanding variations and claims	• High water table
• Claims and disputes with sub-contractors	• Unidentified underground obstructions
• Disputed final accounts	• Unidentified underground services
8. Public liability claims	8. Labour shortages
9. Compensation claims at industrial tribunals	9. Late delivery of materials
10. Fines for accidents and breaches of health and safety legislation	10. Poor workmanship

view that construction is a **high risk–low margin** industry is probably correct. There are any number of possible reasons for this phenomenon; some are illustrated in Table 3.2.

One of the main problems facing many contractors is lack of **working capital** and the industry is beset with firms that are under-capitalised. In other words, such firms have too little capital available to fund their activities and are, thereby, heavily reliant on the industry credit system to make up the shortfall. Classically, short-term costs, such as wages and overheads, are paid for out of the working capital available and materials, subcontractors and plant hire are obtained on credit and paid for out of revenue from contracts.

This results in many firms **over-trading**, meaning that too much work is undertaken without the necessary working capital to fund it. This places a great deal of strain on the contractor's cash flow position. Providing that money is flowing in from projects and other activities everything will no doubt be fine. However, if there are problems on a particular project then this can have serious consequences for a firm's financial stability. Loss-making projects, under-payments, late payments and unsettled claims can quickly have a negative impact on cash flow and, thereby, place undue strain on working capital and the firm's ability to pay its creditors.

3.2.2 Reward

Paradoxically, the construction sector is generally regarded as being attractive to investors, offering potentially high returns on their investments. On the face of it this contradictory statement makes little sense in the context of construction margins and, with such low margins generally, it might be equally prudent to invest in a building society account that carries much less risk. Investors, however, look to other indicators with which to measure their investment decisions. One of these measures is the ratio of **profit to capital employed**.

Thus investors (i.e. shareholders), whether the company is large or small, like to see how well the directors are using the capital invested in the firm and this is perhaps a more reliable way to compare alternative investment vehicles than profit to turnover. The simple example in Box 3.1 illustrates the point:

Box 3.1

A small subcontractor has an annual turnover of £5 million.

The owners of the business (shareholders) have invested £500 000 capital in the business.

Annual pre-tax profit on turnover is £100 000.

$$\text{Pre-tax margin} = \frac{£100\,000}{£5\,000\,000} \times 100 = \mathbf{2\%}$$

$$\text{Return on Investment} = \frac{£100\,000}{£500\,000} \times 100 = \mathbf{20\%}$$

This is a very healthy return that should be reflected in an attractive annual dividend for shareholders.

3.3 Corporate governance

3.3.1 Definition

Corporate governance is the system by which companies are directed and controlled and a company's ability to manage its risk exposure is determined by the quality of its corporate governance regime. United Kingdom standards of corporate governance are partly controlled by voluntary self-regulation and partly by statute.

3.3.2 The Cadbury Report

In the late 1980s, high-profile and infamous cases such as BCCI and Maxwell raised concerns that the standards of financial reporting in the United Kingdom were not as they should be. This was partly the back-drop to the Cadbury Report (Cadbury, 1992), which suggested a Code of Best Practice designed to achieve high standards of corporate behaviour. The Report also provided a framework for corporate governance which has become the basis for the arrangements whereby UK companies now govern themselves.

The Cadbury Committee took the view that compliance with a voluntary code, directed at establishing best practice and coupled with transparency of disclosure to shareholders and outside interests, was likely to prove more effective than a statutory code in achieving high standards of corporate governance.

3.3.3 The Financial Reporting Council

The Financial Reporting Council (http://www.frc.org.uk) is the body responsible for promoting high standards of corporate governance in the United Kingdom. It publishes the UK Corporate Governance Code (formerly the Combined Code) which is regularly updated following a consultation process (FRC, 2010). Non-compliance with the Code is an infringement of Stock Exchange listing requirements and may also lead to an adverse disclosure in the company's annual report.

At the heart of the Code lies **openness, integrity and accountability** in presenting a balanced perspective of the affairs of a company and these principles underpin accounting standards in the United Kingdom.

3.3.4 The Companies Acts

Corporate affairs are also controlled by statute, notably by the Companies Acts, the latest of many manifestations being that of 2006 which was fully implemented by 1 October 2009 (http://www.companieshouse.gov.uk). The Companies Acts are enforced by the Department of Business Innovation and Skills, which has the power to investigate and prosecute a variety of offences, including malpractice by company directors in relation to the keeping and the preservation of company accounting records.

In this context, section 386 of the Companies Act 2006 requires companies to keep adequate accounting records so as to be able to disclose with reasonable accuracy, *at any time*, the financial position of the company *at that time*. Officers of the company who fail in this duty may be subject to imprisonment or to a fine or both. Section 388 of the Act requires accounting records to be kept for prescribed periods and section 393 requires company directors to make sure that the accounts give a true and fair view of the assets, liabilities, financial position and profit or loss.

3.3.5 The Turnbull Report

The Cadbury Committee also recommended that the boards of listed companies should report on the effectiveness of their systems of internal control, and that the auditors should report on this statement.

The later Turnbull Report (ICAEW, 1999) emphasised the need to adopt a **risk-based approach** to creating and reviewing internal control procedures both within companies and within not-for-profit organisations such as local authorities. The risks and uncertainties that a company may be exposed to are usually specified in the Directors' Report. These will include such factors as **credit risk**, **liquidity risk** and **cash flow risk** and it is usual for the directors to identify the measures that the company is taking to mitigate the impact of such risks on the company.

Hazards are events that could result in injury, damage or loss and the presence of hazards gives rise to risk. The assessment of 'market risk' requires the identification of events (hazards) that could give rise to such risk. The Turnbull Report offers guidance in good business practice in the areas of risk management and internal control and emphasises that the system of internal control plays an essential role in the management of those risks that threaten the fulfilment of business objectives. Turnbull suggests that emphasis should be placed on managing the **significant risks** to the organisation and that the 'risk identification overload' syndrome, which results from attempting to identify and deal with **all risks**, should be avoided.

3.4 Market risk

3.4.1 Definition

Strictly speaking, **market risk** may be defined as **the effect on an investment of exposure to market risk factors**. These risk factors concern:

● equity prices
● interest rates
● currency exchange rates
● commodity prices.

3.4.2 Market risk factors

Being a credit-based industry, construction is exposed to market risk factors:

(i) Share (or equity) value is an expression of the value of the business (i.e. what someone is willing to pay for it) and this is determined by such fundamentals as asset values, liquidity and profitability.
(ii) Most construction companies are 'highly geared' and therefore suffer when interest rates rise.
(iii) Many of the materials used in construction work are imported and prices are, therefore, influenced by exchange rates.
(iv) The prices of commodities such as copper, steel and aluminium change according to the market and construction is obviously affected by such fluctuations in commodity prices because of its reliance on such commodities for the manufacture of many of the products used in the construction process.

3.4.3 Governance

Market risk factors must be recognised and managed both in the boardroom and at project level; this is a central function of corporate governance and internal reporting. The monthly cost value reconciliation should equally reflect the true net asset value of a contract, the cost of financing working capital, any increased costs of imported materials or any fluctuations in commodity prices at project level as the balance sheet and profit and loss account do at corporate level.

In its narrow sense, the construction market is considered 'risky' due to the intense level of competition for work, to the chance of client payment default or insolvency, to the 'technical' risks of undertaking projects and to the problem of satisfying the quality and performance expectations of architects, engineers and the like. For the contractor (and subcontractor), such risks have to be balanced with the need to recover costs, make a profit, enhance reputation and be considered for future work.

This means alerting management when things go wrong, and when corrective action is required, and ensuring that outsiders see a true and fair view of the affairs of the business. Developing robust reporting systems helps to ensure that these objectives are achieved. The identification, prioritisation and control of market risk are part of this process.

3.5 Working capital

For a contractor to carry out work before receiving payment, credit needs to be arranged for the supply of materials and for plant hire so that the contractor does not have too much money tied up in the contract. Payments to subcontractors are similarly deferred, which provides an additional source of credit.

All companies, large and small, require working capital in order to operate their business. Working capital is needed to pay wages, to pay bills, to service overhead costs and to pay off any short-term loans that may be maturing. Working capital is needed because bills have to be paid whilst awaiting payment for work invoiced. Work done on a contract but not paid for represents money tied up in a project (**capital lock-up**).

Working capital is a bit like cash in your pocket – it gives you **liquidity**. You may own a house and a car but these items are fixed assets and cannot be sold quickly when bills have to be paid. Without cash (or money in the bank) you cannot settle your bills without resorting to borrowing (**debt**), which means having an overdraft, taking out a loan or using a credit card. The same principle applies to businesses.

3.5.1 Sources of working capital

Classically, businesses obtain finance by borrowing money, by investment by shareholders and by personal investment by the owner(s). Borrowed money is **loan capital** and shareholder and personal investment is **equity** or **risk capital**. In reality, working capital is obtained by other means as well; all the common sources are summarised in Table 3.3.

The relative proportion of the various sources of working capital in a contracting firm is a function of the size and status of the business. Large and powerful contractors have access to funds that are closed to small businesses and some contractors, irrespective of size, may have problems accessing working capital because they may have breached banking covenants or becomes over-exposed to risky markets or simply be too highly geared relative to profitability and balance sheet strength.

Some contracting firms are in the fortunate position of having no debt capital at all and, therefore, all tax paid profits go to the shareholders in the form of dividends or retained profits. Lack of indebtedness is not the sole province of the large contractor and many SMEs have 0% gearing, good profitability and a healthy cash flow. In the main, however, contracting firms are known to be both highly geared and under-capitalised and this condition can lead to the risk of instability or even insolvency when lenders want their money back and payment for contract work in progress becomes problematic.

3.5.2 The annual accounts

Working capital appears on the balance sheet as **current assets**, which consist of **cash**, **debtors** and **stock**:

- Cash
  - Money held in current bank accounts
  - Money on deposit
  - Petty cash.
- Debtors
  - Money owed to suppliers, builders' merchants, plant hire companies and subcontractors
  - Money (paid monthly or quarterly) owed to HM Customs and Revenue in respect of tax and national insurance deductions and VAT.
- Stock
  - Materials on site
  - Materials in the builder's yard
  - Work in progress.

The location of 'working capital' on the balance sheet is illustrated in Table 3.4, which indicates a figure of £153 386. The liquidity position indicated is somewhat worrying as creditors exceed current assets by £25 887 (£179 273 – £153 386) and there is only £12 584 of liquid money at the bank to pay them with. The other current assets shown cannot be considered liquid because they consist of debtors (who may not pay up) and work in progress (which may be difficult to agree and settle). On a positive note, the company is asset rich and may be able to sell some plant or development land to help with the working capital shortfall.

3.5.3 Loan capital

Loan capital (or debt finance) is usually arranged through a bank or other lending institution in return for payment of an agreed rate of interest over the period of the loan. The debt capital may take the form of medium- to long-term loans or bonds. In smaller companies the loan capital may come from family or friends.

Loans are repaid after a fixed period but may be extended by agreement. In order to guard against default, loans may be secured against the assets of the business, which might include land and buildings, plant and machinery or book debts (i.e. money that is owed to the company).

Table 3.3 Sources of working capital.

Source	Explanation	Application
Bank overdraft	• An agreed amount (facility) by which a bank account may be overdrawn • Paid for at an agreed percentage over the Bank Base Lending Rate (e.g. Base Rate + 3%) • Can be an expensive option • The facility can be withdrawn without notice • The amount borrowed is repayable on demand	• Commonly used as the main source of finance by small and medium sized contractors and subcontractors • Considered to be a short-term source of funding • Accepted as 'risky' if the sole source of finance • If the bank supplies contract bonds, the value of the bond is added to the amount borrowed
Equity	• Money invested in the business by shareholders • In small firms the shareholders will usually be the owners/directors and/or family members • In private limited companies there may be up to 50 shareholders • In a public company the shareholders will be financial institutions, insurance companies and the general investing public	• Considered a long-term source of finance
Debt	• Loans made to the company at a specific rate of interest for a specific period • Includes bonds, debentures etc. • Repayable when the period of the loan expires	• Commonly used by larger companies • Considered to be medium-term finance • Large companies are usually subject to 'banking covenants' relating to the allowed proportion of debt to equity (gearing)

Debtors (trade credit)	• Money (i.e. credit) owed to suppliers, subcontractors, plant hire firms etc.
	• Credit can extend from 14 to 90 days
	• Failure to repay on time can lead to:
	  ○ Suspension of supplies
	  ○ Suspension of work by subcontractors
	  ○ Court action for recovery of debt
	  ○ Adjudication in accordance with the terms of the subcontract
	  ○ Loss of reputation
		• Trade credit is a short-term source of finance
		• Usually repaid when contract payments are received
		• Subject to legislation:
		  ○ Housing Grants, Construction and Regeneration Act 1996
		  ○ Late Payment of Commercial Debts (Interest) Act 1998

Hire purchase and leasing	• Commonly used to acquire plant and equipment • The goods may be owned at the end of the HP agreement or any remaining equity used as a deposit for further purchases • Leased goods are never actually owned	• A common source of finance used by small and medium sized contractors and subcontractors • Makes sense from a cash flow point of view • Interest rates tend to be higher than debt finance
Retained profits	• Profits (after tax) retained in the business after paying loan interest and share dividends • The effective rate of interest is the opportunity cost of the money (i.e. what else could have been done with it to earn interest)	• The money 'belongs' to the shareholders • Can be an efficient source of working capital, especially where the rate of return (i.e. profit) on capital employed is good

Debt factoring	• Money owed to the business is sold to a factoring company
	• The debt is sold at a discount
	• Consequently, cash flow is improved but the discount reduces profitability
	• The factoring company has the job of recovering the debt
		• Commonly used by suppliers and merchants
		• A problematic solution for contractors as debts are often difficult to prove due to:
		  ○ Disputed valuations on contracts
		  ○ The complexities of variations and contractual claims
		  ○ The possibility of set-off against subcontractors

Table 3.4 Balance sheet.

Balance sheet (simplified)		
As at 31 March		
	£	
Fixed assets		
Land and buildings	90799	Assets could be sold to release cash
Plant and equipment	270788	Sell plant e.g. excavators, waggons
	361587	
Current assets		
Stocks and work in progress	27081	Could be over-valued
Debtors	113721	Risk of bad debts
Cash at bank	12584	Poor liquidity, insufficient to pay debts
	153386	
Creditors	(179273)	Money owed by company
Net current assets/(liabilities)	(25887)	Money owing to creditors is greater than current assets
Total assets less current liabilities	335700	
Provisions	–	
Net assets	335700	Asset rich/cash poor

In some cases, a loan facility is agreed with the lending institution and funds are drawn down as and when required with interest being paid only on the money borrowed. Borrowings are considered to be short- to medium-term finance.

3.5.4 Risk capital

Equity or **risk capital** is the money invested in the company by shareholders and this is very much viewed as a long-term source of finance. Equity capital carries the greatest risk because shareholders invest without any guarantee that their money will be returned. Shareholders, who may be the director(s) of a small company or the investing public in a public limited company (plc), are effectively the owners of the business and they are repaid in the form of a dividend on their investment, usually twice yearly in public companies. It is the role of the directors to decide what the dividend per share will be based on the profits made by the company in the trading year.

Exposure to market risk directly impacts on the investment that the shareholders have made and the value of the shares will go up or down accordingly. Investors in both private and public limited companies hold shares and, whilst shares in private limited companies are not traded on the Stock Market or to the general public, their value is nevertheless similarly affected by exposure to market risk.

The difference between shareholders and lenders is that shareholders risk losing their entire investment if things go wrong whereas investors are entitled to be repaid and may also have a charge over the assets of the company.

Sole traders are a bit different in that they are companies owned by an individual or by partners but because they are not limited companies there are no directors as such. Such businesses have unlimited liability for their debts. It is difficult for a start-up company to obtain permanent loans because there will be normally be no business assets with which to secure the loan. Consequently, working capital may come from individual savings or may be borrowed from family or friends. It is likely that a bank overdraft will be arranged as well but overdrafts will normally be secured against personal assets, which in most cases will be property, commonly the family home.

3.5.5 Overdrafts

Many contracting companies view their **bank overdraft** facility as a source of working capital. This is a mistaken view, however, as an overdraft is a temporary facility repayable on demand.

Overdrafts can be 'called in' at any time - usually the wrong time! - and, if this happens, the company could be in big trouble. When a sole trader or limited company is trading at its overdraft limit, the bank has to decide whether to call the overdraft in or allow the situation to continue for a while. Calling in the overdraft may result in the company becoming insolvent but giving more time may allow the company to trade on and reduce the amount borrowed.

3.5.6 Trade credit

The construction industry as a whole relies on the credit afforded by suppliers, builders' merchants and plant hire companies. Employers (clients) enjoy the same privilege from main contractors who, in turn, are granted credit by their subcontractors.

Many small contractors and subcontractors fund the majority of their working capital requirements by taking advantage of trade credit but, where a company relies too much on short-term debt, it is vulnerable to cash flow pressures. This can happen when there is a delay in an expected contract payment or less money than expected is received.

A reasonable 'rule of thumb' is that a firm should have working capital of about one-eighth of turnover to be comfortable and enable trade creditors to be paid and avoid suspension of supplies or legal action for recovery.

3.5.7 PAYE, NIC and VAT

As discussed in Chapter 2, contractors are required to make returns of various sorts to HM Revenue and Customs and this usually involves making payments for:

- Income tax and National Insurance Contributions (employer and employee) in respect of directly employed operatives and staff.
- Income tax and NIC deducted from subcontractors who do not have the correct tax status to be paid in full.
- VAT collected on sales less payments made on purchases.

These payments are made 'in arrears' and as such represent a temporary source of working capital for the contractor.

It is a very dangerous 'game' - played by some in all honesty - to assume that this money is anything but belonging to the public revenue. Late payments are subject to penalties and persistent lateness or failure to pay will result in recovery action being taken and the possibility of a winding up order being sought by the tax authorities.

3.5.8 Capital gearing

The relationship between loan capital and equity is called the **gearing ratio**. This is the ratio of money borrowed compared to equity expressed as a percentage:

$$\frac{\text{Total borrowings}}{\text{Share holders' funds}} \times 100$$

Some contractors have no borrowings at all and all working capital requirements come from shareholders' funds (equity). Other contractors will be in the 80-100% gearing 'comfort zone' whilst still more will be 'highly geared', where borrowings exceed equity to a significant extent. An example of the gearing ratio calculation is illustrated in Table 3.5.

Highly-geared companies are exposed to the risk that interest rates may rise significantly. Provided that profits are adequate to 'cover' the interest payments all will be well but if interest rates rise and profits fall at the same time then there may be problems in servicing the debt.

Table 3.5   Gearing ratio.

Loan capital = £700 000 Equity = £500 000 Gearing ratio $= \dfrac{£700\,000}{£500\,000} \times 100 = \textbf{140\%}$	• Borrowings exceed equity by 40% • A company with a ratio over 100% is highly geared

3.5.9 Working capital ratios

In accounting terms, working capital is defined as current assets less current liabilities. The **current ratio** (or working capital ratio) compares the current assets and current liabilities of the firm and this gives a clue as to liquidity. This ratio should be about 2:1 in a healthy company.

Current assets, however, are not all liquid (i.e. they cannot be readily converted into cash) and, therefore, a company may have a shortfall of working capital even though the current ratio looks healthy. This is a particular problem in contracting because stocks and long-term contracts (i.e. work in progress) can be overstated in the accounts or the debt owed may never be fully recovered. Subcontractors often suffer in this respect and may be forced to 'do a deal' with the main contractor at the end of a contract simply to get hold of some cash. The downside to this is that the sum recovered may be significantly less than that expected or stated in the accounts.

Because current assets contain items that are not readily convertible into cash (e.g. stocks of materials and work in progress), a better measure of liquidity (i.e. the availability of working capital) is the **acid** or '**quick' ratio**. This represents the relationship between liquid assets and current liabilities.

'Liquid' assets are defined as debtors + cash and, conventionally, the liquid ratio should be 1:1. The liquid asset value is given by the calculation current assets (including debtors) less stocks and long-term contracts (i.e. monies recoverable on contracts but not certified or, in other words, work in progress).

3.5.10 Liquidity

The theory behind the acid ratio is that debtors (i.e. those who owe the company money) can be converted into cash by factoring to a specialist factoring company at a discount or by making them subject to a specific **floating charge** by a lender (such as the bank) who could either collect or sell the debt. 'Debtors' would, therefore, be regarded as cash to all intents and purposes.

This is fine in theory but factoring companies will only take on debts if there is no question as to what is owed and by whom, and this is often not clear in contracting. 'Apparent' debts may be problematic due to:

- disputed measurements;
- verbal instructions not confirmed;
- unagreed claims;
- unvalued variations;
- counter-claims for all sorts of (sometimes spurious) reasons.

Charging assets to the bank can also be problematic. 'Debtors' are difficult to value and they also change regularly in value as bills are settled and new debtors arise. This is why 'debtors' are normally subject to a floating charge rather than a fixed charge which operates a bit like a mortgage on a house and is less flexible. 'Debtors' would probably be heavily discounted by the bank as an asset, meaning that any loan based on the value of debtors would only be granted at something like 60% of the net book value of the debt.

Consequently, the real test of a company's liquidity is 'cash' and anything owned that can be sold quickly, such as plant and equipment, motor vehicles and so on.

3.5.11 Cash flow

Cash flow may be regarded as the oil that lubricates the engine of a company. It represents the movement of money in and out of a firm and is inextricably linked to the industry credit system.

Cash flow is also linked closely with working capital and work in progress, as it is the working capital that provides the funds to pay bills when they fall due and it is the payment for work in progress that replenishes the working capital.

If the work in progress is undervalued, or the contractor's payment is less than expected, the contractor may run out of working capital and this, in turn, will result in a cash flow problem. Such problems may mean that the contractor becomes over-stretched at the bank where he may be at the limit of his overdraft or borrowing facility. The next problem will be difficulties in paying the wages or settling the accounts of suppliers or subcontractors, which, in turn, could lead to the withdrawal of credit facilities or the suspension of work by subcontractors.

The deduction of retention from the contractor's payment, usually at 3–5% of the sum due, adds to these problems because it represents a negative cash flow influence on the project. This is because the money is effectively 'locked up' capital which the contractor cannot get access to until the contract is completed.

'Cash is king' is an often heard expression in construction and many seemingly profitable contractors have 'gone to the wall' due to insufficient working capital.

3.6   Competition

3.6.1   Definition

The construction industry is characterised by a highly competitive marketplace, especially in contracting. Construction clients seek value for money whether this is achieved through traditional competitive tendering, e-tendering, framework agreements (such as ProCure21+) or other procurement arrangements.

The construction industry has always had a culture of 'lowest price wins' and this is still largely the case. More informed clients to the industry, however, take a different view and look for 'value' in the form of:

- Price
- Time
- Quality
  - Health and safety provision
  - Environmental responsibility.

It is now common practice to have 'two-envelope' bidding, whereby tenderers for a contract submit a price bid in one envelope and a quality/time bid in a second envelope. It is usual to open the quality bid first and, if this is acceptable, the price bid second. The price/quality bids are usually weighted (say 60/40) to decide who wins the contract.

Competition extends down the supply chain from main contractors to subcontractors, suppliers, manufacturers and service providers and so on but with less sophistication and more emphasis on price and reliability of performance.

3.6.2   Procurement

Despite the increased use of negotiation and two-stage tendering and the decline in use of bills of quantities reported by the RICS (2011), it is also reported that the majority of contracts in the United Kingdom are let on the basis of traditional procurement. This would suggest that competitive tendering is still in widespread use in the industry both at main contractor and subcontractor levels.

Whilst selective competitive tendering is preferable (i.e. a limited competition) anecdotal evidence would suggest that this is far from common practice and that forms of 'open tendering', either with a limited but lengthy list of tenderers or no limit at all, are still commonly found in use in the industry.

3.6.3   Subcontractors

Subcontractors particularly suffer from the pressures of open tendering where there may be a large number and wide variety of firms permitted to tender for a contract. Some forms of e-bidding allow any number of subcontractors to tender with the obvious motive of driving down prices to the lowest possible level. The type of reverse auctions that are becoming popular in construction can lead to ridiculously low pricing – well below an economically viable level.

The downside risk for main contractors is that rock bottom prices may result in claims, defective work, subcontractor default or insolvency with the consequent cost and time implications. More often than not replacement subcontractors, which may be unwilling to take on the work at such low prices, have to be found.

The cost value reconciliation may need to make provision for potential subcontractor liabilities in such circumstances.

3.7 Profitability

3.7.1 Definitions

Profit is usually compared to turnover. This is the amount of work the company has done in a trading period.

In construction, profit margins are historically low and quite commonly can be as little as 2 or 3% of turnover. However, when profit is compared to capital employed, the percentage is usually much higher because many contractors are under-capitalised in relation to their workload. This makes construction a highly profitable (but risky) investment for owners and shareholders.

Profitable companies are not always stable because the profits may not be earned quickly enough to avoid a cash flow crisis. Consequently, the ability of a firm to settle its debts (which is a test of solvency under the Insolvency Act 1986) is perhaps a better test of stability. Profitable companies can, and have, become insolvent and unprofitable companies are not necessarily unstable or likely to become insolvent, at least in the short to medium term, simply because they are able to find ways to manage their cash flows.

Consequently, the normal measure of profitability – profit on turnover – is not a reliable indicator of business health. A better indicator is profit before interest and tax (PBIT) compared with interest payable on loans.

3.7.2 Income gearing

PBIT appears on the profit and loss account as 'operating profit', as illustrated in Table 3.6. Interest payments are effectively a business 'overhead' that is paid on

Table 3.6   Profit and loss account.

Profit and loss account		
For the year ended 31 March	£	
Turnover	1640471	i.e. sales
Cost of sales	(1498760)	
Gross profit	**141711**	i.e. profit + overheads
Administration expenses	31294	i.e. overheads
Operating profit	110417	Represents profit before deduction of interest charges and tax
Net interest (payable)/receivable	(41568)	Represents interest payable on bank loans and other finance charges less any interest receivable on bank deposits and investments
Profit on ordinary activities before tax	**68849**	
Tax on profit on ordinary activities	(19924)	
Profit for the financial year	**48925**	

Table 3.7   Income gearing.

Operating profit = £110 417 Interest payable = £41 568 Income gearing = $\dfrac{£110\,417}{£41\,568}$	Interest cover is **2.66:1** **NB:** This is a healthy ratio because if profit were to halve for some reason interest payments would still be covered

moneys borrowed by the company to support its activities. Because the **capital gearing ratio** is the relationship between borrowings and equity (or shareholders' funds), the more highly geared a company is the more interest it will have to pay. The risk issue here is that where a company is highly geared the greater is the proportion of profit which must be paid out in interest payments.

Consequently, if profits fall, there may not be sufficient monies to pay the interest, let alone the tax and dividend to shareholders. This clearly places the company at a greater risk than one where the gearing ratio is lower. The problem may be managed to some extent by deferring the shareholders' dividend but interest payments usually cannot be deferred. The relationship between PBIT and interest is called **interest cover** or the **income gearing ratio** and it is a good measure of a company's exposure to financial risk.

Table 3.7 shows an example of how the income gearing ratio is calculated and indicates that operating profit covers interest payments by a healthy factor of 2.66 – not bad!

3.8   Work in progress

The perennial problem in construction is payment for work in progress. This represents work carried out on site which has not been valued or certified for payment. Contractors and subcontractors habitually overvalue the work they have done, expect to be paid the amount claimed and are surprised, disappointed and angry when their expectations are not realised.

Contractors and subcontractors spend money on materials, labour, plant and overheads in the expectation of getting paid for their work on a regular basis. If actual revenue is less than anticipated this can cause cash flow problems which, in the worst case scenario, can lead to difficulties paying wages, suppliers and subcontractors and, possibly, insolvency. Work in progress becomes a problem when the contractor has not been recompensed through the contract and this situation might arise for a number of reasons:

- Work carried out on site is under-certified by the architect or contract administrator due to error, omission or disagreement with the contractor's valuation.
- Money has been withheld because work has not been carried out to an acceptable standard under the terms of the contract (e.g. not in accordance with the specification or contract drawings).
- The architect or contract administrator has issued instructions for a variation to the contract but this has not been included in the quantity surveyor's valuation.

- The contractor has carried out work in good faith without formal instructions from the architect or contract administrator or on the basis of verbal instructions and this work has not been valued.
- The contractor has or intends to submit a claim for loss or expense under the contract but this has yet to be agreed by the architect or valued by the quantity surveyor.

Lenders (e.g. banks) are keen for contractors to control cash flow closely and work in progress (debtors) is frequently offered as security for lending (such as a loan or overdraft). Although work in progress is technically a debtor, and as such is viewed as a liquid asset – that is one that can be readily converted into cash – these so-called 'liquid' assets may not be as liquid as they seem on paper!

Debtors may well owe money but this is not so easily collected and there are invariably arguments over how much is owed on construction contracts. Main contractors are regularly underpaid because instructions have not been confirmed or variations orders not issued and any subcontractor will tell you that they are rarely paid the amount invoiced to the main contractor for a wide variety of reasons. Being first in the payment 'queue' nevertheless gives main contractors a big advantage and they are much more likely to be paid close to expectations, and on time, than the more vulnerable subcontractors who regularly have to battle for every penny. There are always problems and arguments over money in contracting.

There is also the possibility that debtors may go out of business before they pay what is owed. This can literally happen overnight and it is a shock to the system when it does. Frequently, in such cases, the likelihood of receiving any dividend at all from the liquidators is very slim and is unlikely to even approach, never mind exceed, 30p for every pound owed. Administration can offer a more realistic prospect of getting some money back but this can take time. Once debts have gone 'bad' it usually best to take a deep breath, tighten the belt and move on!

3.8.1 Payment in arrears

Contractors work for clients who pay for their work in arrears and contractors obtain supplies of labour, plant, materials and subcontractors on credit. Monthly payment in arrears is the way that most construction contracts work and when this money comes in the bills can be paid. Working capital is needed to pay wages because these are normally paid weekly, that is before the contract revenues come in.

The contract work done is usually paid for by valuing the quantity of work carried out or, alternatively, by reference to stages of completion determined either by the construction sequence or the programme. Sometimes work is valued on the basis of ogive curves (S-curves) which predict value to time on a graph.

Although payment problems do arise from time to time, main contractors are usually paid a fair value and on time. Therefore, cash flow problems are more likely to be the result of a lack of working capital than delayed payment on contracts.

Subcontractors similarly extend credit to main contractors but their problems are somewhat different. They are further down the 'food chain' to begin with and many cannot afford to dispute payment issues with the main contractor. Disputes are costly, adjudication under the 'Construction Act' may be considered 'an expensive form of rough justice' and disputes mean that further business with the main contractor may not be forthcoming.

There is no problem with the payment in arrears way of working, and indeed there are many benefits to it, but the system relies on everyone in the chain paying their bills as and when due. Therein lies the difficulty!

3.8.2 Valuations and payments

Small subcontractors are particularly at risk and are often kept waiting for payment for inordinately long periods, despite the provisions of the Construction Act. On occasion, they may be owed money by a larger contractor and may owe money themselves for supplies to a subsidiary company of the same group – one refusing to pay and the other demanding settlement and threatening to cut off further credit. This is the construction equivalent of being 'between a rock and a hard place'!

Being an industry which is easy to enter, with little or no test of competence or financial standing required, and being largely reliant on credit in order to conduct its business, the prospect of insolvency is a major risk issue in construction. Insolvency is difficult to predict and complex to manage and, consequently, all construction practitioners need to be alive to the possibility that clients, main contractors and subcontractors alike may fail financially. There may be warning signs – companies in financial difficulties may find it hard to maintain their quality of work, to obtain supplies of materials and to resource the project – and the industry 'grapevine' is often a reliable indicator of problems ahead.

Two of the reasons why construction firms are vulnerable to insolvency are, firstly, because of cash flow pressures brought about by lack of working capital and, secondly, due to problems with work in progress. These issues are discussed next.

3.9 Insolvency risk

In the 12-month period ending fourth-quarter 2011, a total of 2688 construction companies became insolvent and over 5000 companies had been lost to the industry since the beginning of 2010. The expectation is that this will not improve quickly (PWC, 2012).

Key risks in the current economic climate are falling demand, increases in uncontrollable costs and poor cash flow management, which is critical to firms finding themselves in difficult circumstances. In construction terms all of this translates to reduced turnover, increased costs of materials and overheads and poor control over debtors (money coming in). This comes as no surprise to those who work in the industry.

Any insolvency, whether an individual or company, can have serious and far reaching consequences. Unpaid bills, unhappy creditors, homes repossessed, assets sold at knock-down prices and, not least, damaged self-esteem and anger are just a few of the outcomes. In construction, the consequences of insolvency can be particularly complex and emotive resulting in:

- delay and difficulty completing projects;
- financial loss to employers (clients), consultants, contractors, subcontractors and the entire supply chain;
- complex financial and legal implications;
- angry creditors who are unlikely to get their money back.

Emotions can run high, especially as the United Kingdom limited liability system can seemingly, on occasion, allow companies to go out of business one day and then start up the next day bigger and better than ever leaving a long line of creditors to foot the bill.

3.9.1 Industry structure

Construction has, historically, been an industry which is vulnerable to the risk of financial problems and insolvency and this, to a certain extent, is a feature of the make-up of the firms which operate in the industry. The industry is populated with a great number of very small firms and a small number of relatively large firms which carry out the bulk of construction work. Bearing in mind that there are something like 200 000 firms in construction, most of which employ less than five people, it is not surprising than many do not stay in business for too long.

The large companies operating in the construction industry tend to be publicly quoted companies in the main, although there are still some large companies in private hands. Private limited companies tend to be the medium sized and smaller contractors, whilst many of the very small businesses are 'sole traders' or partnerships – that is individuals working on their own account without the protection of limited liability.

3.9.2 Sole traders

Sole traders and partnerships are in an even riskier position than limited companies when things go wrong because they are unincorporated and, therefore, have unlimited liability for the debts of the company. Many of the smaller subcontractors in the industry are in this position and risk losing everything if they have financial problems that cannot be resolved.

The principal reason why sole traders risk everything is that their family home is often charged to the bank as security for an overdraft or loan facility. This means that an overdraft can be 'called in' and the property sold to pay off the debt. Constantly working at or near the overdraft limit is highly risky and any bad debt or disputed or delayed payment can tip the balance the wrong way. The pressures on individuals concerned can be enormous!

When main contractors 'put the squeeze' on small subcontractors they perhaps do not realise (or maybe do not care) what is at risk and what the consequences for the subcontractor and their family will be. It is a sad fact of life, a poor reflection on the industry and one of several reasons why there is a skills shortage in construction (personal view!).

3.9.3 Limited liability

A limited liability (or incorporated) company is a legal entity that may sue or be sued in the courts just like an individual person. However, the limited liability 'bit' does not refer to the company's liability for its debts or actions but to the liability of those individuals who have invested in the company, such as the owners and shareholders. Consequently, each investor's liability is limited to the amount that they have invested in the business. Therefore, if you buy 1000 shares in Acme Construction Ltd at £3.75 per share, the most money that you can lose should the company become insolvent is £3750 (i.e. 1000 shares at £3.75) plus any dividends that might have been due.

The problem for the owners of many small and medium sized contracting firms is that they may be required to give personal guarantees or provide security for their bank loans and overdrafts. This often means that the debts of the company are secured by a charge on personal assets – this normally means the family home and so, despite the protection of limited liability, if a firm gets into trouble the bank will want to protect itself and the asset will be repossessed, meaning that the owner of a small firm may lose his/her house. The eventual result will probably be the insolvency of the business and the personal insolvency (bankruptcy) of the individual.

This can and does happen, sometimes through simple ignorance of business matters, sometimes through bad pricing of contracts, sometimes because the subcontractor does not understand the payment mechanisms or expects too much and sometimes because main contractors either do not pay on time or will not pay the just entitlement.

3.9.4 Large firms

Large firms are not insulated from the harsh realities of financial problems or insolvency and 2010 saw the demise of some well-known names in the industry.[1] This is not surprising considering the low level of margins in general contracting and the extent of contract risk undertaken by contractors. Where risk is misjudged or contracts go wrong (for a multiplicity of reasons) available margins do not leave a great deal of room for error.

The severe weather conditions in the latter part of 2010 played a key part in the failure of some construction companies where a halt in activity over a number of weeks was enough to 'tip them over the edge' (Lowery, 2011).

3.10 Instability

Table 3.8 illustrates the three possible financial conditions which a firm may be in:

- Stability
- Instability
- Insolvency

It is probably the case (but difficult to prove) that many contractors and subcontractors are in the 'unstable' zone and, therefore, vulnerable to insolvency without ever becoming insolvent. This is probably more evident in recessionary economies when lack of working capital and the vagaries of the industry credit system take their toll.

Many contractors and subcontractors find ways of managing their cash flow in such difficult times but some, and some large companies included, do fall by the wayside. Managing cash flow is a constant juggling act because payments from contracts do not always arrive when expected, the amounts received may well be less than expected and yet wages have to be paid and suppliers' and subcontractors' invoices have to be settled.

---

[1] Administrators were appointed to **Connaught plc** and **Connaught Partnerships Ltd** on 8 September 2010 and building services company **Rok** announced that it would go into administration on 8 November 2010.

Table 3.8 Financial conditions of a company.[a]

Solvent	Unstable	Insolvent

• Sufficient liquidity to fund day-to-day operations
• Able to pay debts as and when they fall due
• Ratio of Current Assets to Current Liabilities of at least 2:1
	• Over-reliant on short-term debt e.g.:
	  ○ Overdraft
	  ○ Trade credit
	• Persistent cash flow problems:
	  ○ Difficulties funding weekly wages
	  ○ Always chasing outstanding payments
	  ○ Overdraft persistently at or approaching the agreed limit
	• Persistently unable to pay suppliers and subcontractors on time within credit terms
	• Frequent disputes over payment
		Informal insolvency:
		Going concern test
		• Unable to pay debts as and when they fall due[b]
		• Able to pay debts on time but a balance sheet deficit of assets to liabilities
		Balance sheet test
		• Assets insufficient to meet liabilities[c]
		
		Formal insolvency
		i.e.: entering a formal insolvency procedure such as:
		• Administration
		• Liquidation
		• Receivership

a. This is a simplified diagram.
b. Refusal to pay should not be confused with inability to pay. Refusal to pay may be due to a disputed payment for defective work or materials supplied.
c. The balance sheet test is complicated due to:
   • Difficulty of valuing long-term contracts
   • Off-balance sheet finance
   • The treatment of contingent liabilities e.g.:
     ○ Liability to the client for defective work carried out
     ○ Subcontractors claims for variations and loss and expense.

When a company is unstable it is a fine judgement as to whether the company is insolvent or not. Companies frequently fail for cash flow reasons – they simply run out of cash – and many profitable companies have gone into liquidation for this reason. On the other hand, some companies have sufficient cash to pay their bills but fail the solvency test because they have insufficient assets to cover their liabilities – that is they are technically insolvent.

3.10.1 Living with instability

Due to the nature of construction, the boundary between instability and insolvency may be difficult to define and can be extremely complex. For instance, a contractor may have several – maybe dozens – of contracts at various stages of completion and it is often extremely difficult to determine the true asset value of these contracts. These contracts may be:

• signed and about to start;
• just started;
• part-way through;
• nearly complete;
• complete but the final account not submitted.

Coupled with this is the added complication that contracts may be:

- short-term (i.e. started and completed in the same financial year);
- short-term but spanning one or more financial years;
- long-term taking, perhaps, several years to complete.

Further considerations to take into account include:

- the value of the work in progress taking into account any defective work;
- debtors (work done and invoiced but not paid);
- suppliers and subcontractors who are owed money (creditors);
- some subcontractors may have claims against the contractor;
- the contractor may have outstanding claims against the client;
- there may be problems with subcontractors accounts such as defective work or work not carried out;
- the contractor may have supplied work or services to subcontractors (such as the provision of scaffolding, removal of waste, provision of materials or plant etc.) which will have to be charged to the subcontractors account ('contra-charged');
- some contracts may have below ground problems with claims pending;
- other contracts may have been subject to lots of variations and architect's/engineer's instructions, some of which may have not been confirmed in writing.

When a business is 'unstable' all these issues (and more) add to the pressures on owners, managers and directors who have to find enough money to pay the wages every week, find ways of staving off creditors until revenues from contracts have been received, pull 'rabbits out of the hat' to find cash to reduce the bank overdraft, win enough work to keep the workforce employed and deal with 101 other problems as well.

3.10.2 Indicators of instability

When companies are in trouble there are usually warning signs such as:

- slow progress on contracts;
- under-resourcing of labour and plant;
- difficulty obtaining materials;
- disputes with subcontractors over payment;
- increased reliance on contractual claims;
- pushing for payment and requests for payment in advance.

In addition to the 'formal' tests of insolvency prescribed by the Insolvency Act 1986, there are a number of 'informal' indicators of business health which may assist managers to better cope with the financial management of their business. One of these is described in the next section.

3.10.3 Multiple discriminant analysis

A predictive model of business health which is favoured by some accountants and financial analysts is the **Altman-Z Score** test.

This is a calculation using *multiple discriminant analysis* which seeks to place a company somewhere along a scale of 'healthy to insolvent' by using multiple

Table 3.9 Ratios used in calculations.

X1	Liquidity	working capital[1] : total assets
X2	Age/profitability of firm	retained earnings[2] : total assets
X3	Profitability	earnings before interest and tax (PBIT) : total assets
X4	Financial structure	market capitalisation[3] : total liabilities
X5	Capital turnover	sales (turnover) : total assets

Notes:
[1] Current assets less current liabilities.
[2] Profits retained in the business after tax and shareholder's dividends.
[3] The value of the company as represented by the number of shares issued x the market price.

predictors of business failure. Altman's five-variable model was found to be a very reliable predictor of business failure but the technique relies on up-to-date data being available. The five ratios used in the calculation are shown in Table 3.9.

The discriminant function found to be most effective was:

$$Z = 0.012X_1 + 0.014X_2 + 0.033X_3 + 0.006X_4 + 0.010X_5$$

The Altman model recognises that there is an 'overlap area' between health and insolvency. This lies between a Z-score of 1.8 (indicating potential failure) and 3.0 (indicating relative safety).

3.11 Credit control

Credit control is a vital feature of the construction process. Suppliers and subcontractors extend credit to main contractors and they, in turn, extend credit to employers who pay for the work done in arrears.

Suppliers like to think that they will be paid 30 days following the end of the month in which their invoice is issued and subcontractors 'dream' of payment 14 days after their payment application. Dream on!

The industry is notorious for late payment and most contractors and subcontractors like to hold on to their money for as long as possible before paying their debts. Suppliers, in particular, exercise strict control over credit but good customers are invariably allowed some leeway in settling their account and 45–60 days credit is fairly normal for reliable payers. Those who push it to 90 days or more are likely to have supplies suspended or credit facilities withdrawn entirely.

For contractors, the control of debts is fundamental to a healthy cash flow position. Debtors are those who owe money to the firm and, in contracting, this will largely consist of monies owing on contracts which have been certified or agreed for payment but not yet received.

For the main contractor, debtors will be clients (employers) and for subcontractors it will be main contractors who owe money to the company. Debts arise because most construction work is done on credit and payment is deferred for a period (usually one month).

Further debts owed will be retention monies deducted on contracts which are partly collected upon practical completion of the contract and partly at the end of the defects correction period. Retention monies are considered as longer term debts because defects periods can be 12 months or even longer in some cases.

3.11.1 Debtor days

An indication of the extent to which credit control is being managed effectively is given by the relationship between **debtors** and **sales** (i.e. turnover). This information is contained in the company balance sheet (debtors) and profit and loss account (sales). The **debtor days** derived shows the average length of credit extended to customers in days.

This can be compared to the contractual credit periods under the standard forms of construction contract, as indicated in the following example which takes its debtors and sales information from Tables 3.4 and 3.6:

$$\frac{\textbf{Debtors}}{\textbf{Sales}} = \frac{£113\,721}{£1\,640\,471} \times 365 = \textbf{25.3 days}$$

- Debtors are paying the company on average within 25.3 days.
- Compare this to the settlement period of 14 days from issue of the interim certificate in the JCT Standard Building Contract.

3.11.2 Creditor days

By comparing **creditors** to **sales**, the average number of days taken to settle debts is indicated. Creditors are those to whom money is owed and would include materials suppliers, builders' merchants, plant hire companies and the like. The trick here is to hold on to other people's money and pay out only when need be which improves the positive cash position. This practice is now subject to some degree of statutory control under *inter alia* the 'Construction Act'.

Taking the sales (turnover) figure from Table 3.6 and creditors from Table 3.4, the **creditor days** calculation is as shown below:

$$\frac{\textbf{Creditors}}{\textbf{Sales}} = \frac{£179\,273}{£1\,640\,471} \times 365 = \textbf{39.9 days}$$

- The company is settling its debts after 39.9 days (on average).
- Compare this to the normal settlement periods of 14 days for subcontractors and 30 days for suppliers and merchants.
- Some contractors do not settle debts for 60 and sometimes 90 days (if they can get away with it!!).

Managing debtors and creditors is something of a balancing act and represents the 'sharp-end' of the many areas of risk and uncertainty in contracting.

References

Cadbury, A. (1992) Financial Aspects of Corporate Governance. The Committee on the Financial Aspects of Corporate Governance and Gee and Co. Ltd (a division of Professional Publishing Ltd), London.

FRC (2010) The UK Corporate Governance Code. Financial Reporting Council (FRC), London.

Friedman, S.M., Dunwoody, S. and Rogers, C.L. (1999) *Communicating Uncertainty*. Lawrence Erlbaum Associates, Mahwah, NJ.

ICAEW (1999) The Turnbull Working Party Report. Institute of Chartered Accountants in England and Wales (ICAEW), London.

Knight, F.H. (1921) *Risk, Uncertainty and Profit*. The Riverside Press.

Lowery, D. (2011) Company failures jump 20% in fourth quarter of 2010. *Building*, 11 February 2011 (http://www.building.co.uk/news/company-failures-jump-20-in-fourth-quarter-of-2010/5012992.article).

PWC (Price Waterhouse Cooper) (2012) UK suffers nearly 10,000 construction & manufacturing insolvencies in two years. 23 January 2012 (http://www.ukmediacentre.pwc.com).

RICS (2011) *Contracts in Use, A survey of building contracts in use*. The Royal Institution of Chartered Surveyors (RICS), London.

Winch, G.M. (2010) *Managing Construction Projects*, 2nd edn. John Wiley & Sons Ltd, Chichester.

4

Contracts and documentation

4.1 Types of contract

Keating suggests that a contractor's right to payment for construction work carried out depends upon the terms of the contract entered into by the contracting parties (Ramsey and Furst, 2012). Consequently it is good practice, though not always the case in the real world, for there to be express provision within the contract to determine how and when the contractor will be paid for his work.

This is an important consideration in the context of this book as the valuation of work in progress, the valuation of variations and extra work and the timing of

Financial Management in Construction Contracting, First Edition. Andrew Ross and Peter Williams.
© 2013 Andrew Ross and Peter Williams. Published 2013 by John Wiley & Sons, Ltd.

payments impact significantly on the financial reporting of contracts and the cash flow of the business.

Keating (Ramsey and Furst, 2012) further suggests that a claim for payment may arise under three separate circumstances:

1. A lump sum contract – where payment is made either on completion of the work or by instalments.
2. An express contract other than for a lump sum – where the work is either measured and valued according to a schedule of prices (measure and value contract) or is paid for on the basis of actual cost plus a percentage or lump sum fee (cost plus contract).
3. A claim for a reasonable sum or quantum meruit – where there is no express agreement to pay a particular sum or where no fixed price has been agreed.

In practice, there are three types of contract common in construction – **lump sum**, **measure and value (**or **re-measurement)** and **cost reimbursement** (i.e. cost plus). The distinction between the three types of contract is important, especially in respect to the valuation of work in progress and settlement of the final account. The type of contract is instrumental in the decision as to which form of contract to choose (e.g. JCT, NEC, ICC or bespoke form) in order to ensure that the valuation and payment provisions are appropriate to the contractual arrangements desired.

4.1.1 Form of tender

The contractor's offer to carry out the works is usually stated in the **form of tender** that accompanies the bid. The form of tender is a document that is prepared by the employer, sent out with the tender documents, completed and signed by the contractor and submitted to the employer, either on its own or with the priced bills of quantities.

The Joint Contracts Tribunal (JCT) and New Engineering Contract (NEC) contracts do not have a standard form of tender but the Infrastructure Conditions of Contract (ICC) – Measurement Version and ICC – Design and Construct Version both have a form of tender provided in the bound contract document. At tender stage it is, nevertheless, usual to provide a form of tender as a separate sheet along with all the other tender documents and an envelope in which to return the offer.

The form of tender is important because it clarifies that the contractors bid is either for a single fixed lump sum (a lump sum contract) or for such sum as may be ascertained according to the contract (a measure and value contract). JCT lump sum contracts refer to a **contract sum** and ICC re-measurement contracts refer to a **tender total** in order to make the distinction clear.

NEC3 ECC contracts are less clear. Eggleston (2006) confirms that there is no definition of the contract price in the NEC contract and no statement that the contractor will provide the works for a contract price. Consequently, Primary Option A (Priced contract with activity schedule) and Primary Option B (Priced contract with bill of quantities) could be used either as lump sum or measure and value contracts. Eggleston suggests that the form of tender should make it clear where the contractor's offer is for a single lump sum to avoid any confusion.

4.1.2 Lump sum contracts

A contractor's tender of a stated price for a specific job of work, when accepted by the employer, is a lump sum contract. A lump sum contract is intended to remain fixed but may be subject to change for ordered variations, price fluctuations and claims if permitted by the form of contract used. Further adjustment may be made in respect of quantities marked 'approximate' and prime cost and provisional sums, if included in the tender.

The RICS Survey of Contracts in Use (2008) found that 89.1% of all building contracts let in 2007 (64% by value) were lump sum contracts and that, despite the trend away from traditional 'with quantities' contracts towards design and build, measured bills of quantities are still popular, especially to support contractors' bids for lump sum design and build contracts.

Turner (1990) suggests that lump sum contracts (whether traditional or design and build) give the construction client greater price certainty but, to achieve this aim, Morledge (2006) advises that, for traditionally procured projects, the design should be complete before construction commences. The RICS survey suggests that just under half of lump sum contracts are let on a drawings and specification basis and that there is more or less an even balance between traditional bills of quantities and design and build for the remainder of contracts.

The most popular standard form of contract for lump sum building contracts (by number of contracts awarded) is the JCT Minor Works form whereas the JCT Design and Build Form is most dominant when measured against the value of contracts awarded. However, traditional lump sum contracts of a reasonable size would

---

Box 4.1   Lump sum contracts

Example 1

A painting and decorating contractor is asked to quote for redecorating a detached house for a domestic client. The price quoted is £3000. The client agrees to the price and the work is carried out.

This is a **simple lump sum contract** where, unless otherwise agreed, payment is made upon completion of the work. It is unlikely that any form of contract will be used although the JCT does publish standard contracts for **home owners/occupiers** whether or not a consultant has been appointed to oversee the work.

Example 2

The owner of a small engineering company requires an extension to his premises and has engaged an architect to design a scheme. The client's budget is £250 000 including all fees and other ancillary costs. A quantity surveyor has prepared bills of quantities from the architectural drawings and the form of contract is to be the JCT Intermediate Building Contract. Tenders have been invited from six suitable contractors. The lowest bid received is £215 996 and has been accepted.

This is a **lump sum contract**, where the contractor will be paid £215 996 for the job – payable in monthly instalments according to the amount of work carried out. The form of contract gives the architect the right to vary the contract by adding or omitting work as required and this may well affect both the interim payments made to the contractor and the final bill at the end of the job.

normally be awarded using the JCT Standard Form or the JCT Intermediate Contract, which are still popular. The ICC – Design and Construct Version is a lump sum contract. When the NEC3 ECC is used, Eggleston (2006) considers that Option A is a lump sum contract but, nevertheless, counsels care in drafting the form of tender to avoid confusion (Box 4.1).

4.1.3   Measure and value contracts

Keating (Ramsey and Furst, 2012) defines a measure and value contract as one *'where the amount of work when completed is to be measured and valued'* and is to be contrasted to a lump sum contract. A measure and value (re-measurement) contract differs from a lump sum contract because there is no fixed and agreed total price for the job – no contract sum. The contractor may indeed price a bill of quantities and a total sum may be derived from the prices quoted and the quantities stated but this is only a **tender total** where the ensuing contract intends that the work shall be re-measured on completion. Under the JCT Standard Building Contract With Approximate Quantities (SBC/AQ) 2011, the contractor's price is based on the tender figure which is converted to an Ascertained Final Sum following re-measurement.

Measure and value is a common arrangement on civil engineering contracts where the extent of the work involved cannot be precisely determined beforehand and the parties agree to calculate the accurate quantities when the work has been carried out. Consequently, the contractor's rates/prices may remain fixed but the contract sum is not determined until the contract is complete. Other circumstances where a measure and value contract may be used includes repair and maintenance work where the true extent of the work involved cannot be predicted until the project is underway.

Measure and value contracts are usually measured as the job goes along and the value of the work is determined by applying the contractor's rates/prices to the quantities calculated, normally using a combination of drawings and the physical work done on site. This is not to say that there cannot be variations, contractual claims and so on on such a contract. The meaning of 'variation' will be defined in the contract but this will not include changes in the quantities, unless there is a dramatic difference compared to the amount of work originally envisaged. 'Normal' entitlements to loss and expense and the disruptive effect of variations will usually be provided for in the contract.

The normal pricing document for measure and value contracts is a bill of approximate quantities but some projects – repair and maintenance for example – will use a schedule of rates with no indicative quantities at all. The ICC – Measurement Version, the JCT Standard Building Contract With Approximate Quantities (SBC/AQ) 2011 and the JCT Measured Term Contract (MTC) 2011 are examples of standard measure and value contracts. Eggleston (2006) considers that NEC3 ECC Option B is a re-measurement contract but care drafting the form of tender is advised (Box 4.2).

4.1.4   Cost reimbursement contracts

The basis of cost reimbursement contracts is that the contractor will be paid for the actual cost of carrying out the work and a supplementary fee. The fee may be a fixed fee or a percentage of the actual cost and is meant to provide for the

Box 4.2   Re-measurement contracts

A petrochemical works has a large portfolio of buildings that require regular maintenance and repair. This involves a wide variety of work, such as replacing defective rainwater goods, re-glazing windows, re-pointing brickwork, roofing repairs, internal and external painting and so on. It is proposed to let a three-year maintenance contract to a suitable contractor. The extent of the work is uncertain but is likely to amount to over £1 million over the three-year contract. A schedule of typical work items without quantities has been prepared by the client.

This is a **measure and value (re-measurement) contract**, where the work carried out by the successful contractor is physically measured on site, valued according to the rates priced in the schedule and agreed for payment by the client's representative. A JCT Measured Term Contract could be used as the form of contract for such a situation.

Box 4.3   Cost reimbursement contracts

A private property owner has been notified by the local authority that one of his properties is in a dangerous condition and must be made safe immediately. The property owner contacts one of the contractors on his approved list and asks for the necessary work to be carried out immediately. This is confirmed in a facsimile message which also quotes the contractor's fee at 15%. The contractor sends out a crew and all necessary plant and equipment and makes the building safe. The contractor later sends its bill to the client.

This is a **cost reimbursement contract** where the contractor's invoice will consist of the actual cost of labour, materials, plant and equipment used plus 15% to cover overheads and profit.

contractor's overheads and profit. The fee may also include the contractor's site overheads (preliminaries).

In many standard lump sum or measure and value contracts there is a provision for valuing variations where their value cannot be ascertained by measuring the work or by using the bills of quantities rates as a basis for valuation. This provision is called 'daywork'. The basis of **daywork** is payment of the prime cost of labour, materials and plant with the addition of percentages to cover on-costs, overheads and profit. Daywork is discussed in more detail in Chapters 12 and 17 but suffice to say that this is essentially a form of cost reimbursement despite being commonly applied in lump sum or measure and value contracts.

The JCT Prime Cost Building Contract (PCC) 2011 is an example of a cost reimbursement contract and the ICC – Minor Works Version has the option to be used in this way. Cost reimbursement is available under the NEC3 ECC by choosing Option E (Box 4.3).

4.2 Financial implications of contracts

Contractors and subcontractors tendering for construction work should carefully read the 'preliminaries' bill included with the tender enquiry in order to identify

- the conditions of contract that will apply if the tender is successful;
- any amendments to the standard form;
- any non-standard or onerous clauses;
- the period of interim certificates;
- the sums included for liquidated and ascertained damages;
- the retention percentage that will apply;
- the defects correction period (normally 6-12 months, sometimes longer);
- the construction period or possession and completion dates;
- special insurance requirements;
- whether a performance (or other) bond is required.

Careful attention to detail at the tender stage will help to identify risk, flag up potential cash flow issues and avoid surprises during the contract when it will be too late to do anything about it. Risk can be allowed for in the tender. A judgement can be made to mitigate the effect of onerous contract conditions, allowances for possible liquidated and ascertained damages (LADs) can be included in the tender and the cost of capital lock-up (including retention monies) can be calculated and added to the bid.

There are, however, some less obvious areas of risk and reward that contractors and subcontractors should consider that are concerned, not with the form of contract, but with the type of contract that will apply. Some of these issues are considered in Sections 4.2.1-4.2.3.

4.2.1 Lump sum contracts

Lump sum contracts are attractive financially to contractors because there is a degree of certainty in the amount of money that they will eventually be paid. This is because the quantities are fixed and only subject to change for legitimate variations. Such variations will be fairly, if not overgenerously, valued according to the contract provisions and further recompense may be forthcoming through extensions of time with costs and loss/expense claims where additional profit may be generated. Expenditure of provisional and prime cost sums is also fertile ground for making money.

Latham (1994), however, expressed the view that choosing a lump sum contract would be a recipe for disaster if site work was to progress without a complete design, whilst Arain and Low (2011) consider variation orders to be an unwanted but inevitable reality of every construction project that increase the possibility of contractual disputes. Many contractors would no doubt argue that variations cause a great deal of extra administrative work and inconvenience and that they never recover the true cost of variations despite the valuation provisions in standard contracts.

There is another possible risk issue for contractors to consider in that some of the quantities in a lump sum contract may be approximate and, therefore, subject to re-measurement. This happens when the architect cannot define precisely the extent of the work required. Quantity surveyors often inflate approximate quantities so as to provide a contingency allowance, and therefore the contractor could lose turnover, overhead contribution and profit if the re-measured quantities are less than those stated in the contract bills.

When the main contract is a lump sum contract, subcontractors should be careful to ensure that their subcontract conditions are 'back-to-back' with the main contract. A lump sum contract offers some degree of security for a subcontractor because the quantities are largely fixed. Where the main contract is a lump sum and the subcontract is re-measurement, subcontractors could be at the mercy of an unscrupulous contractor who may well 're-measure' the subcontract works in a less favourable way than the employer's quantity surveyor would.

There is potential for 'downside' risk for subcontractors under a lump sum contract and care should be taken to confirm verbal instructions from the main contractor and 'keep an eye on' potential variations to the contract that need to be recorded, measured and valued. Approximate quantities in the bills should re-measured and agreed at the time that the work is done and not left to be argued about later on when valuable evidence has been covered up and memories have faded.

4.2.2 Measure and value contracts

Measure and value (re-measurement) contracts offer risks and opportunities to both contractors and subcontractors from a financial point of view. Where the contract includes quantities these will be 'approximate' (i.e. not actual or correct) and subject to admeasurement when the work has been carried out. Consequently, the re-measured quantities could go up or down. If the contract is based on a schedule of rates without quantities being given then the work will be measured 'from scratch' and the appropriate contract rates will be applied for payment purposes.

Where approximate quantities are given, increases or decreases in the quantity of work stated in the bills of quantities is not a variation and does not require a written instruction. If, however, the quantities misrepresent or are not indicative of the nature or extent of the work involved then the contractor may be able to argue that the contract rates do not apply and should be 're-rated' to allow for the different circumstances. ICC – Measurement Version confers powers on the engineer to fix reasonable rates in these circumstances.

If the contract rates for work that is not a variation are, nevertheless, affected by another variation, the contract administrator may have the power to fix reasonable rates for the work affected. The ICC – Measurement Version provides an example of this and the contractor has the right to claim a higher rate if he thinks this is appropriate.

There is no rule which establishes the degree of deviation between the approximate quantities and the actual quantities, and this can be a problem for contractors. Quantity surveyors are notorious for 'hiding' contingencies in bills of quantities so that money is available to the designer if it needs to be spent on unexpected changes. Bills of approximate quantities can contain such hidden contingencies and the contractor can never be sure as to the real value of the contract until the works have been re-measured. This can have a considerable financial impact when there is an eventual reduction in quantities and may lead to a reduction in turnover (i.e. the contract value) and, consequently, to under-recovery of overheads and profit.

Table 4.1 illustrates this point, where a contract subject to re-measurement and, valued initially at £1.2 million, suffers a 10% reduction in the quantity of measured work. On the face of it the contractor will recover less than the anticipated margin for overheads and profit but, in reality, it is profit that will suffer because overheads

Table 4.1 Implications of a re-measurement contract.

			£
A	Contract sum		1200000
B	Overheads (O/H) and profit (P) at 12%*	$£1200000 - \dfrac{£1\,200\,000}{1.12}$ =	128571
	Anticipated O/H recovery =	$^{7}/_{12} \times £128\,571$ = £75000	
	Anticipated Profit =	$^{5}/_{12} \times £128\,571$ = £53571	
C	Net total of measured work		1071429
D	Reduction in measured quantities	Say 10%	107143
E	Net value of re-measured work		964286
F	Add O/H and P	12%	115714
	Less O/H recovery	as per tender	75000
	Profit	=	40714
G	Loss of profit	**£53571** less £40714 =	12857
H	**Reduction in profit**	$\dfrac{£12857 \times 100}{£53571}$ =	**24%**

* assuming 7% overheads + 5% profit.

are a cost of turnover and must be recovered before a profit can be made. Overheads, therefore, have a greater priority than profit. It is only when there is no profit that overhead may be under-recovered. In this example, the contractor will lose 24% of the profit anticipated at tender stage due to the reduction in the measure. If this is extrapolated into the contractor's annual accounts, it is the shareholders who will bear the loss in terms of loss of retained profits and, possibly, loss of dividend.

4.2.3 Cost reimbursement contracts

A contract which pays the contractor on the basis of his costs and a fee would seem to be the ideal contractual arrangement. The contractor is in a 'no-lose' situation and, provided that his book-keeping is well managed, there should be no reason why the contractor cannot make a reasonable profit at the end of the contract.

However, one of the attractions of contracting is the opportunity to make profits that are better than 'reasonable' by:

- making the most of opportunities in the contract documents by:
  - front-loading rates
  - tactical pricing of undermeasured items
  - capitalising on mistakes in item descriptions or in the application of measurement rules;
- profiting from:
  - variations and extra work
  - expenditure of provisional sums
  - contractual claims and loss and expense entitlement
  - extensions of time with costs;

- making money by:
  - squeezing subcontractors' prices
  - shopping around for the best discounts from suppliers
  - 'playing' the industry credit system to best advantage.

The chance to do any or all of the foregoing is limited with a 'cost-plus' contract, where the employer's quantity surveyor is scrutinising 'the books' and making sure that only legitimate costs and expenses are charged to the contract. Depending upon the degree of scrutiny exercised, the contractor may be able to be 'creative' with time sheets and goods received documentation and there is always the possibility that the employer's quantity surveyor will 'take his eye off the ball' from time to time!

Making money from a cost reimbursement arrangement on a lump sum or measurement contract is a different story. Daywork can be a 'money spinner' for contractors and subcontractors provided that the rates have been pitched correctly and that daywork sheets are not scrutinised too closely (Chapters 12 and 17).

4.3 Project documentation

4.3.1 Definitions

The basis for administering any construction project is the documentation which makes up the contract. This is agreed by the contracting parties – that is the employer (client) and the contractor – as representing the agreement they have entered into. The type and level of detail of the documentation will vary according to the nature of the project and the procurement arrangements chosen by the client.

Whatever the project, the conditions of contract identify the risks, responsibilities and obligations of the parties to it and, in conjunction with other documents, these comprise the **contract documents**. Where the JCT Standard Building Contract is used, the 'Contract Documents' are defined. Therefore, under a **traditional building contract**, say for a new school, the contract documents would comprise:

- the contract drawings;
- the contract bills;
- the agreement and conditions;
  - the articles of agreement
  - the conditions or contract clauses
  - the appendix to the conditions;
- the specification and, where there is a portion of the works to be designed by the contractor:
  - the employer's requirements
  - the contractor's proposals
  - the CDP (contractor's designed portion) analysis.

If the same school were to be built under a **design and build** arrangement, a bill of quantities will not be provided with the tender documents, which would normally comprise:

- the employer's requirements
  - scope drawings
  - written brief or specification.

In order for the contractor to offer the employer a price for building the job the contractor will inevitably prepare his own quantities based on the contractor's design. If the tender is successful, a contract will be drawn up which will comprise:

- the employer's requirements;
- the contractor's proposals;
  - scheme drawings (plans, elevations, typical sections)
  - specification;
- the contract sum analysis
  - a breakdown of the contractor's price.

The contract sum analysis is normally presented either as a series of lump sums, possibly linked to activities on the contractor's programme, or a list of prices based on the main construction stages of the project (e.g. substructure, envelope, finishes, services etc.). It is unlikely that the contract sum analysis will be in a bill of quantities format.

For smaller projects, or projects involving repairs or alterations to existing property, the contract documentation would be somewhat different. Under such circumstances it may not be possible to measure the work required and so the bills of quantities would be replaced by schedules. There are two main types of schedule:

- schedule of rates;
- schedule of works.

4.3.2   Priority of documents

The **priority of documents** under common law is that the written word prevails over the printed word. In a JCT contract the opposite is normally the case. JCT SBC/Q 2011 states that *nothing contained in the Contract Bills ... shall override or modify the Agreement or these Conditions*. Therefore, the printed word (i.e. the printed contract signed by the parties) has priority over the written word (e.g. the bills of quantities) as representing the true intentions of the parties to the contract. Beyond this, it is the architect or contract administrator who has the duty to explain any further ambiguity between the documents.

Under the ICC – Measurement Version, the contract documents are *mutually explanatory* and the engineer shall explain any ambiguities or discrepancies and shall issue appropriate instructions to the Contractor. NEC3 ECC has a similar provision, in that the project manager shall issue instructions regarding any ambiguity or inconsistency in the contract documents. Eggleston (2006) expresses concern, however, that NEC3 ECC provisions may be interpreted as having a much wider meaning than authorising changes to the works information.

4.3.3   Drawings

For the vast majority of construction projects, the drawings are crucial to the physical and financial management of the job. Drawings normally portray a two-dimensional model of the project and convey to third parties the intentions of the designer(s) as well as helping other designers (e.g. structural engineer, mechanical engineer, electrical engineer etc.) and contractors to do their work. Drawings for a building contract usually comprise:

- site plan;
- floor plans;
- cross-sections;
- external works and drainage;
- elevations;
- a variety of descriptive schedules showing:
  - door and window details
  - ironmongery
  - painting and decoration requirements
  - drainage details.

Drawings enable the client's quantity surveyor to present a model of the project in quantitative and descriptive terms to tendering contractors and, later on, to those carrying out the construction work. Such models include:

- bills of quantities;
- schedules of rates;
- schedules of work.

Additionally, drawings provide invaluable assistance to the contractor engaged to carry out the construction work, such as:

- choosing appropriate methods of construction;
- ensuring safe access and egress from places of work;
- deciding on the order and sequence of the work;
- selecting appropriate scaffolding and other temporary works;
- deciding how to provide suitable supports for deep excavations;
- making use of available working space and circulation for plant and vehicles.

During most contracts the client, or designer, will wish to make changes to the design or circumstances may arise whereby the design has to be altered (e.g. due to unforeseen ground conditions, to revise specifications or to add or omit work). In such cases, revised drawings are usually issued and it is from these drawings that instructions are issued to the contractor to vary the work. The contractor should keep a drawing register and ensure that drawing revisions are logged and cross-referenced to variation orders and other instructions to make sure that nothing is overlooked from a payment point of view.

4.3.4 Specification

The specification is a document that describes the quality and standard of materials and workmanship expected of the contractor. It sets the tone for the contractor's pricing of a contract which, if misjudged at tender stage, can prove costly when on site. The contractor's work will be judged by reference to the specification and, if the work carried out is not up to standard, it may have to be removed and done again. Specifications are difficult to write with any precision and, therefore, the common sense test of 'reasonableness' may be written into the contract to prevent the architect/contract administrator imposing excessively strict standards (e.g. JCT SBC/Q 2011).

The specification is essentially a written explanation of the design so that, when read in conjunction with the drawings and the bills of quantities (if used), the contractor has a complete 'picture' of what is expected from him on site. The specification should be written specifically for each individual contract but, because it is a time consuming and costly document to prepare, previous specifications are often regurgitated. This practice, whilst understandable, can cause confusion when the document has not been edited properly and the contractor may be faced with a lengthy specification for structural steelwork when the drawings show a precast concrete frame! Poorly drafted specifications can be a fruitful area for contractors' claims and conflicts between the specification and other documents may lead to variations from which the contractor could profit.

There are two types of specification:

- prescriptive
- performance.

A **prescriptive specification** is written in prose and may follow a non-standard format or be based on a standard specification, such as the National Building Specification (NBS). British Standards, Codes of Practice or Board of Agrément Standards may be used. Alternatively, reference may be made to manufacturers' literature or proprietary products, although this may fall foul of competition legislation.

Traditionally, the specification is laid out in sections that follow the trade sections of the bills of quantities or may be arranged according to the Common Arrangement of Work Sections (CAWS) of Standard Method of Measurement 7 (SMM7) (RICS, 1998). Standard civil engineering specifications include the Specification for Highway Works and the Civil Engineering Specification for the Water Industry. Table 4.2 shows a typical extract from a prescriptive specification for blockwork.

Table 4.2 Types of specification.

Ref	Specification
Prescriptive specification (blockwork)	
2.1	**Types of Block** The Contract Administrator will specify either on the drawings, in the bills of quantities or in the specification, the type, size and sometimes the makes of blocks to be used. In any case blocks shall conform to BS6073.
2.2	**Blocklaying** Concrete blocks shall not be wetted immediately before or during laying. Where necessary the consistency of the mortar shall be adjusted to suit the suction of the blocks.
Performance specification (window system)	
18.4	**Performance** The proposed window system must have an expected lifespan of 60 years with regular maintenance.
18.5	**Strengths** The window system, glazing and fixings must be designed with full consideration to the anticipated imposed loads based upon the following: Window loads in accordance with CP3: Chapter 5: Part 2: (1972) for a basic wind speed, $V = [\ ]$ in category [ ].

Performance specifications avoid the imprecision of prescriptive standards by placing the emphasis on the final product or service required. They may be used for defining end-product requirements, such as the load carrying characteristics of piling, structural steelwork or precast concrete decking, curtain walling or raised access floors. Alternatively, performance specifications may define the required outcomes for services installations such as lifts, escalators or heating and air conditioning. Table 4.2 gives an extract from a performance specification for a window system.

Performance specifications encourage innovation and cost effective methods of work, reduce administration costs, increase efficiency and give the designer the opportunity to confine his attention to how well the system works in the completed building rather than policing the 'nuts and bolts' of how the system was put together on site. From the contractor's viewpoint, performance specifications are more exacting and, therefore, more costly because the test is one of fitness for purpose rather than compliance to a defined standard.

A prescriptive specification for carpeting a new hotel would be described in terms of how the product was manufactured and laid whereas a performance specification would focus on outcomes such appearance retention, dimensional stability and colour fastness. Both types of specification have practical limitations, including the difficulty of defining subjective requirements, such as 'appearance', and whether the specified standards can be achieved at an acceptable cost. The two types of specification are often combined in practice.

Where a specification is used without bills of quantities then the specification would normally become a contract document along with the drawings and so on. However, where bills of quantities are provided, the specification is not usually a contract document unless it is bound within the bills of quantities. In civil engineering work, standard specifications are commonly used and these will be included as contract documents, by specific reference, for the project in question.

Specifications may be combined with the drawings to provide a convenient means of arriving at a lump sum contract. This type of contract is where the contractor quotes a lump sum on the basis of these documents with no quantities given. The contractor produces the quantities and takes the risk for any mistakes or omissions that might be made. Typically, this arrangement might be used for a domestic client who wishes to have an extension built but it can also be used for substantial contracts as well. Where **'drawings and spec'** is used, it is sensible to ask the contractor for some documentation (such as a schedule of prices) for the purpose of valuing and agreeing variations.

The RICS (2008) survey indicates that some 47% of the number of building and repair and maintenance contracts let in 2007 were on a 'drawings and spec' basis (18% by value) and 14.7% (by number) and 9.2% (by value) of all JCT contracts used in 2007 were 'without quantities'.

4.3.5 Bills of quantities

The purpose of bills of quantities is to convey the scope and extent of the work involved in a construction project so as to enable contractors to submit tenders based upon a uniform and reliable representation of the quantity and quality of work required. In traditional competitive tendering, it is usual practice for the employer's quantity surveyor to prepare the bills of quantities and reliability in the quantities is provided by the use of a recognised standard method of measurement, such as SMM7 or CESMM4 (ICE, 2012).

The basic idea is that the contractor prices the items in the bill of quantities and the total price for the job is given by multiplying the various quantities and rates together; these are then summarised and totalled. The RICS survey (2008) reveals, however, that the use of bills of quantities has been in decline for some years and that bills of quantities were used in only one in five contracts included in the survey. Considering the growth in popularity of design and build procurement and the increased use of contracts based on drawings and specifications this conclusion is not surprising. Bills of quantities are commonly used for civil engineering work.

A distinction needs to be made between bills of quantities prepared as a tender document (formal bills of quantities) and those prepared internally by contractors as a means of establishing the quantity of work required in a project so that a price can be calculated (informal bills of quantities). The key difference between the two is that **formal bills of quantities** are based on standardised measurement rules and **informal bills of quantities** are based on so called 'builders' quantities' determined by each of the individual tendering contractors. Tenders based on builders' quantities introduce an aspect of risk and variability of tender prices that the late nineteenth century proponents of a single set of quantities, based on standardised measurement, sought to avoid!

Whether or not the bills of quantities are prepared formally, the structure of the document will usually follow established practice and convention:

- a preliminaries bill;
- a number of sections containing sets of quantities or 'measured items' for each trade or work package;
- a list of provisional sums for items of work where detailed quantities cannot be determined but a financial allowance needs to be included in the price.

In formal bills of quantities the bill of 'preliminaries' provides an important function, both as a means of conveying key information and as part of the pricing of the project; it includes:

- details of the employer, architect, quantity surveyor and CDM (Construction (Design and Management) Regulations 2007) coordinator and so on;
- a description of the project and any constraints regarding time for completion, phasing of handover, access restrictions and so on;
- a list detailing the conditions of contract to be used with relevant clause numbers;
- a place where the contractor can price his site overheads (preliminaries).

A 'preliminaries bill' is not needed in informal bills of quantities as contractors normally have their own internal pro forma forms for pricing such items.

The measured items in bills of quantities are distinguished by a long established conventional layout that reads from left to right and comprises:

- a reference number or letter which may be:
  - given uniqueness by prefixing the relevant page number
  - linked to a work breakdown structure (WBS)
  - cross-referenced to the estimator's pricing notes;
- an item description that identifies the work required based either on formal SMM rules or the contractor's own conventions;
- the quantity as measured from the drawings which may be:

Table 4.3   Bills of quantities.

Ref	Description	Quant	Unit	Rate	£
A	*Excavating* To reduce levels 1m maximum depth	5290	m³	*7.50*	*39 675.00*
B	For pile caps and ground beams between piles 1m maximum depth; commencing from reduced level	1049	m³	*18.95*	*19 878.55*
	Page total to summary				*59 553.55*

NB: Prices inserted by contractor at tender stage.

- ○ a firm quantity subject to change only by a variation instruction
- ○ an approximate quantity (marked as such in a lump sum contract) and thus subject to re-measurement;
- a unit denoting that the measurement is linear (m), superficial (m²), cubic (m³), unit (number) or by weight (kg/tonne);
- a place for the estimator to insert a unit rate which, in standard methods of measurement, is deemed to include labour, materials, plant, waste, cutting and overheads and profit;
- a column for the price which consists of the quantity multiplied by the rate.

It should be noted that formal bills of quantities may be fully re-measureable where a re-measurement contract is used and that builders' quantities are firm and at the contractor's risk as regards accuracy and completeness when informal bills of quantities are used.

A sample of a typical layout for a page of measured items is shown in Table 4.3.

4.3.6   Schedule of rates

Schedules of rates are commonly used for repair and maintenance work and for measured term maintenance contracts where the contractor is appointed for a fixed period (say three years) to carry out maintenance work on a portfolio of buildings such as social housing or government buildings. They are used where the nature of the work required is known but not the extent.

Few drawings would be available (perhaps a site plan and general arrangement only) but the work would be described in detail in the specification and the schedule of rates. The type of contract would be a **measure and value contract** (re-measurement) and the form of contract might be a standard form for term contracts or a 'without quantities' form (e.g. JCT SBC/XQ 2011).

There are three basic types of schedules of rates:

- An 'in-house' or 'ad hoc' bill of quantities style document, in SMM format, without any quantities where the contractor enters his rates for the various items of work described.
- An 'in-house' or 'ad hoc' schedule of work items, not necessarily described in accordance with a standard method of measurement, where either:

Table 4.4   Schedule of rates.

Ref	Description	Unit	Rate	£
A	Form door opening in existing 250 mm cavity wall size 1600×2100 mm. Include for closing cavity and providing and fixing new 250×225×1900 mm long precast concrete lintel.	No	*976.30*	Work done is measured on completion to determine total value
B	Prepare and apply one mist coat and two full coats of matt emulsion paint to plasterboard surfaces	m²	*7.56*	

NB: Prices inserted by contractor at tender stage.

- ○ the contractors enters his rates or
- ○ rates are already entered and the contractor quotes a plus/minus percentage on the schedule.
- ● A published schedule for various types of work (e.g. building, electrical, mechanical, painting and decorating etc.) with either:
  - ○ rates pre-entered, where the contractor quotes a plus/minus percentage on the schedule or
  - ○ items left blank for the contractor to enter his rates.

Examples of published schedules include:

- ● The PSA Schedules of Rates (Carrilion).
- ● The M3NHF (National Housing Federation) Schedule of Rates (NHF).
- ● The National Schedules of Rates (rates are broken down into materials, labour and plant) (NSR).

At tender stage the contractor would simply give a price per unit for the items of work listed on the basis of a general understanding of the scope and scale of the work as described in the tender documents or, alternatively, a percentage deduction or addition to the rates quoted in the schedule. The eventual total price for the work carried out would be given by the quantity of work actually done multiplied by the rate in the schedule.

Table 4.4 shows a typical schedule of rates layout with the rates priced by the tendering contractor.

4.3.7   Schedule of works

A schedule of works describes the work required, not in accordance with a particular method of measurement, but in the form of 'composite' items of work. A **composite item** includes several work items that would normally be separated and might, typically, describe the forming of a new door opening including the provision of a lintel, making good the opening and the supply and installation of the door and all its component parts. A composite item for the construction of a new concrete floor

Table 4.5   Schedule of works.

Ref	Description	£	
3.8	Build up new 100 mm blockwork partitions as shown on drawing No RF/171/B. Blockwork to be bedded on to top of concrete slab. Allow for forming openings for 2 No doors and 1 No enquiry window including providing and fixing 'Catnic' or other approved lintels. Both sides of partitions to be finished with 13-mm two coat lightweight plaster.	*1759.57*	• Work done is not measured as such • Lump sums may be adjusted to account for variations
3.9	Provide and fix 3 No softwood windows size 1050×1215 mm complete with 25 mm softwood window board and SAA furniture.	*976.45*	
	Page total to summary	**2736.02**	

NB: Prices inserted by contractor at tender stage.

might include the excavation, hardcore, damp-proof membrane, any reinforcement and formwork and the concrete itself.

In a bill of quantities, all these items would be individually measured and priced but in a schedule of works the contractor would give a price for everything listed in the item, as indicated in the example in Table 4.5.

The schedule of works places more risk on the contractor's pricing with respect to ensuring that all the necessary costs have been included and it is much less precise than either bills of quantities or schedules of rates from the point of view of valuing variations. There is no standard format for schedules of work but some guidance in formulating item descriptions may be found in SMM7 Class C10 (Demolishing structures) and Class C30 (Spot items).

References

Arain, F.M. and Low S.P. (2011) Effective Management of Contract Variations using a Knowledge Based Decision Support System. CEBE Working Paper No. 10, Centre for Education in the Built Environment (CEBE), Cardiff University, UK.

Carrilion (published annually) The PSA Schedules of Rates. The Stationery Office Ltd, London (http://www.psa-sor.com/psa/services/services_rate.asp).

Eggleston, B. (2006) *NEC3 Engineering and Construction Contract: A Commentary*. Blackwell Publishing, Oxford.

ICE (2012) CESMM4 The Civil Engineering Standard Method of Measurement. Institution of Civil Engineers (ICE), London.

Latham, M. (1994) Constructing the team: Joint review of the procurement and contractual arrangements in the UK Construction Industry. HMSO, London.

Morledge, R., Smith, A. and Kashiwagi, D. (2006) *Building Procurement*. Blackwell Publishing, Oxford.

NHF (published annually) The NHF Schedule of Rates. M3 Housing Ltd, Mitcham, UK (http://www.m3h.co.uk).

NSR (published annually) The National Schedule of Rates. NSR Management, Aylesbury, UK (http://www.nsrm.co.uk/schedules).

Ramsey, V. and Furst, S. (2012) *Keating on Construction Contracts*, 9th edn. Sweet and Maxwell.

RICS (1998) *SMM7 – Standard Method of Measurement for Building Works*, 7th edn. The Royal Institution of Chartered Surveyors (RICS), London.

RICS (2008) *Contracts in Use, A survey of building contracts in use during 2007*. The Royal Institution of Chartered Surveyors (RICS), London.

Turner, A.E. (1997) *Building Procurement*, 2nd edn. Macmillan Press, London.

5

Payments in construction

Financial Management in Construction Contracting, First Edition. Andrew Ross and Peter Williams.
© 2013 Andrew Ross and Peter Williams. Published 2013 by John Wiley & Sons, Ltd.

5.1 Industry credit system

The vast majority of construction work is carried out on credit and most construction contracts provide for the contractor to be paid in arrears. In many cases this means that the contractor will have to carry out up to a month's work before applying to the employer for payment and will then have to wait for a further period of between two and four weeks before actually receiving payment.

To make this system work, a system of credit for the purchase of materials, for plant hire and for subcontractor procurement has developed over the years. The labour market also extends credit, albeit for the limited period of one week for site operatives.

The usual process by which contractors are paid is that the work in progress is **valued** and **certified** and the certificate is later honoured by cheque or electronic bank transfer (such as BACS, CHAPS or FPS). The payment timescale will depend upon the contract conditions but receipt of the payment will depend on the method chosen – cheques take six days to clear while electronic transfers can be as fast as 'same day'. A similar system of payments applies to subcontract services whereas plant hire and materials suppliers are not paid until an agreed credit period has expired.

On occasion the contractor may be paid 'up front'. This usually happens on very large projects where the contractor may be paid advance fees for mobilising the job, that is setting up the site and perhaps carrying out enabling works. Some standard contracts provide for an advance payment, such as under the JCT SBC/Q 2011 or the JCT Design and Build Contract 2011 (DB) but, in such cases, it may be necessary for the contractor to provide an advance payment bond to protect the employer.

A delicate balance exists in the supply chain between income and expenditure. Contractors and subcontractors must pay money out on wages and overhead costs before they get paid themselves by which time payment has to be made for materials, plant hire and so on. If there is insufficient money coming in to counterbalance expenditure, then the working capital will dry up and the firm will face the possibility of insolvency.

Deferred payment is a feature of all the standard forms of contract and subcontract but, when one party or the other fails to pay on time, cash flow problems may arise for those firms with inadequate working capital to suffer the shortfall in income. The control of cash flow is crucial to survival; this is why Lord Denning referred to it a 'the lifeblood of the building trade' (*Gilbert Ash (Northern) Ltd v Modern Engineering (Bristol) Ltd*, 1973).

5.1.1 Labour and wages

Labour may be directly employed by the contractor or may, alternatively, be supplied via the industry's controversial 'labour only' system, whereby subcontractors or self-employed people are engaged to supply labour for a particular project. This system is controversial because of its long history of income tax evasion.

Where the contractor has his own labour force, minimum wages are paid depending upon the skill level of the operative in accordance with the Working Rule Agreement (WRA) published by the Construction Industry Joint Council (CIJC). The CIJC is made up of representatives from the construction unions and the employers (contractors).

The WRA establishes basic rates of pay for general operatives, operatives with particular skill levels and craft workers such as joiners and bricklayers. Additional allowances need to be made by employers for non-productive overtime, travel, subsistence and sick pay, as well as employers' national insurance, holidays with pay and any bonus payments, to make up the total costs of employment. Such additions to the basic pay scale are called **labour on-costs**.

Subcontractors and self-employed operatives are paid on a different basis due to legislation operated by HM Revenue and Customs (HMRC). Under the Construction Industry Scheme (CIS), such organisations may only be paid gross (that is without deduction of tax) if they hold a Subcontractors' Tax Certificate. On the other hand, if they only hold a Registration Card they must be paid net by making deductions in respect of income tax and national insurance contribution liability from any payments due. The rates of pay for self-employed operatives are outside the scope of the Working Rule Agreement and are agreed locally with individual employers.

Contractors who pay for labour in this way must be aware that they will be made liable for the tax and national insurance contributions of firms or individuals who are paid gross when they should have been paid net.

Wages in the construction industry are usually paid on a weekly basis with directly employed workers normally having to work a week in hand, that is they will be paid on Thursday or Friday for the previous week's work. Labour-only subcontractors are normally paid at the end of the week for the work done in that week.

5.1.2 Materials

Contractors can only obtain materials on credit (i.e. without paying for them immediately) by opening accounts with suppliers. Suppliers may be specialist suppliers providing materials such as cladding panels, precast concrete flooring units and partitioning systems or they may be builders' merchants which supply a wide range of construction products such as cement, drainage goods, bricks and blocks, plasterboard and so on.

Before an account can be opened, it is usual for the contractor to be checked for credit worthiness, following which the contractor will be allowed credit on purchases up to an agreed monthly value. Most suppliers have a 'credit control department' which deals with such matters.

Suppliers' credit terms are usually monthly-in-arrears, which means that payment is expected 30 days or one month following the end of the month in which the materials are delivered. Payment is based on the supplier's invoice which is prepared on the basis of the contractor's order and the delivery ticket which is signed by the contractor on receipt of the materials on site.

5.1.3 Subcontractors

Despite the prevalence in the industry of labour-only subcontractors, the usual understanding of the term 'subcontractor' is that of a specialist contractor who carries out distinct parts or packages of the main contract works and in doing so provides the necessary labour, materials, plant and supervision under a formal subcontract agreement. Examples of the sort of work undertaken by such specialists include:

* groundworks
* piling

- formwork and concrete
- roofing and cladding
- plastering
- electrics and plumbing
- heating, ventilating and air conditioning
- painting and decorating.

Subcontractors are normally paid in arrears, monthly or sometimes fortnightly being most usual. Payment will mostly be based on a valuation of work carried out, either by measurement or a percentage assessment of work done. Sometimes stage payments will be made in accordance with the 'stage' of construction reached or by reference to completed items on the subcontractor's programme of work.

Payment terms for subcontractors vary in accordance with the prevailing form of subcontract, which might be the main contractor's own bespoke form of contract or a standard form such as those issued by the JCT or NEC.

5.1.4   Plant hire

Over the last twenty or thirty years it has become increasingly less attractive for contractors to have their own plant department and, consequently they 'hire in' items from specialist plant hire firms such as:

- excavators
- cranes
- materials handling equipment
- scaffolding and patented formwork systems
- small items of plant such as abrasive cutters and wood working power tools.

Plant is generally hired on a daily or weekly basis and is paid for according to the length of the hire period and the agreed hire rate. There may be additional costs for the contractor to bear, such as transport of plant to and from site, fuel and consumables such as saw discs and blades. Some plant hire firms offer specialist lifting services and this may include the provision of an appointed person to manage the lifting operation.

Important items of plant such as excavators and cranes are hired under the Contractor's Plant Association (CPA) Conditions, which specify the duties and obligations of the parties to the hire agreement.

Unlike materials, plant is generally invoiced once the item of plant is off-hired and payment terms are usually 30 days from invoice. Where the item of plant is hired on a long-term basis, interim payments, based on the time on site and the agreed hire rate, are usually paid.

5.1.5   Credit terms

Credit terms under standard forms of contract and subcontract can vary between 14 and 42 days whilst suppliers will usually extend credit for 30 days from the end of the month in which delivery takes place. Collecting the money, however, is not so easy. Subcontractors, for instance, are often paid only when the main contractor receives payment from the client. This is known as 'pay-when-paid' – a practice supposedly made illegal by the Housing Grants, Construction and Regeneration Act 1996.

Suppliers, on the other hand, are commonly kept waiting for 60 or 90 days and may be forced to suspend supplies or even withdraw credit facilities from persistent offenders. Court recovery action may well ensue and a visit from the bailiff may follow the court judgment!

Payment to subcontractors is often subject to deduction for set-off or for contra charges or, alternatively, the subcontractor may be asked to accept a lower sum in exchange for faster payment. There is nothing wrong with the credit system in the industry provided everyone pays on time. This does not happen in the real world, however, and consequently problems of mistrust, ill feelings, debt collection, litigation and insolvency are common.

5.1.6 Discounts

It is common in the construction industry for contractors to expect a discount either for prompt payment (cash discount) or as a special dispensation for being 'in the trade'.

Trade discounts relate to price reductions from the normal retail list price of materials and are available exclusively for the 'trade'. Trade discounts may vary according to the scale and regularity of the order. Once agreed, trade discounts are always allowed by the supplier whether or not the contractor settles the account on time.

Cash discounts, on the other hand, should only be allowed if the contractor pays promptly for the materials or subcontract services, although this 'rule of thumb' is often overlooked in practice.

In order to provide a cash discount of, say, of 5%, the supplier needs to add the correct sum to the price of the materials so that 5% can be deducted if the contractor pays the supplier on time. Table 5.1 illustrates an example of the necessary calculation to provide a 5% discount.

Table 5.1   Providing a 5% discount.

	£	
Price of facing bricks per 1000	570.00	
Add cash discount $1/_{19}$	30.00	
Supplier's quote to contractor	600.00	per 1000 less 5% cash
Consequently £600 less 5% = **£570.00**		

It is common practice for subcontractors to offer discounts to the main contractor and some standard and in-house forms of subcontract include express terms requiring the subcontractor to offer a discount. Such discounts are intended as a reward for prompt payment and, consequently, if payment is made by the main contractor in accordance with the terms of the subcontract, a discount of (say) $2^{1}/_{2}$% may be deducted from the subcontractor's payment. This is a **cash discount**.

To allow for this discount, the subcontractor must make an appropriate addition to his price if he is not to lose part of his profit margin. Table 5.2 shows an example of how a $2^{1}/_{2}$% cash discount may be provided for in a subcontractor's quotation.

The idea of adding money to a quotation or price so as to provide a discount sometimes causes confusion. The reason for the system is to provide an incentive for the payer to pay on time and, at the same time, provide the supplier with a 'cushion'

Table 5.2   Providing a 2½% discount.

	£	
Total of subcontract price	780 000.00	
Add cash discount ¹/₃₉	20 000.00	
Total subcontract price	800 000.00	less 2½% monthly account
Consequently £800 000 less 2½% = £780 000		

to cover the cost of borrowing should the payment be late. The discount has to be added because any discount given would otherwise have to come from the supplier's profit margin on his sales revenue. When you think about it the system makes sense!

It should be noted, however, that, as far as subcontractor's discounts are concerned, many main contractors take their cash discount even when payment is made late. Such discounts are frequently regarded as a reduction of the subcontract price and main contractors claim the discount whether or not the subcontractor's account is paid on time. This sometimes happens because subcontractors do not understand how discounts work and sometimes because subcontractors are in a poor bargaining position. A subcontractor cannot force a main contractor to 'pay up' unless the issue is taken to adjudication or litigation and the main contractor knows that most subcontractors will not bother to do this, and possibly risk losing future work, for a relatively trifling amount of money.

Trade and cash discounts are not normal practice in the plant hire business.

5.2   Payment problems

Payment problems and unfair payment practices are widespread in the construction industry and there is a well-known culture of delaying payment to subcontractors and suppliers. Over the years such practices have become almost a science and most subcontractors and suppliers will have heard all the excuses:

- there is no-one in to sign the cheque*;
- the cheque is in the post*;
- the directors are in a meeting;
- you have missed this week's computer run;
- there is a problem with the measure;
- we have not been paid by the client;
- the QS is on holiday.

Not all main contractors are bad payers by any means, however, and arguments over money often arise through ignorance or lack of experience. Subcontractors in particular frequently:

- misjudge the work that they tendered for;
- fail to read the tender enquiry carefully;
- do not understand the contract conditions to which they will be bound;
- do not appreciate the workings of the standard method of measurement;

*when cheques are eventually phased out, creditors can at least look forward to fewer fairy stories!

- ignore potential risks in the hope of winning a valuable contract;
- carry out work without written instructions;
- fail to confirm verbal instructions;
- fail to prepare their valuation correctly;
- overvalue their work in progress (i.e. work they have done) which may be incomplete or defective.

It is understandable that subcontractors sometimes get 'hot under the collar' when they are paid less than expected and sometimes their frustration is justified. Often, however, they do not appreciate the basics of how construction work is measured and valued. For instance:

- many smaller subcontractors do not understand the difference between a lump sum and a re-measurement contract;
- nor do they appreciate the importance of records and written instructions;
- they do not understand how variations are valued;
- nor do they appreciate that a signed daywork sheet will not necessarily result in payment on a daywork basis.

A subcontractor's disappointment is often compounded by the pressure of cash flow problems in the business and the frustration of a contract that is stacked in favour of the main contractor. There is no sentiment in business!

Such is the nature of the construction industry that nothing is 'typical'. Main contractors come in all 'shapes and sizes', as do subcontractors. There are some small firms in the industry that are extremely well run and efficient and there are larger organisations that perform poorly and mistreat their subcontractors shamefully. Some main contractors have money problems as do subcontractors. Shortage of work, tight or non-existent profit margins, pressure from banks, recessionary pressures, slow payments from clients, outstanding retention monies and poor credit control are just some of the reasons for the money problems found in the construction industry.

The provisions of the Construction Act 1996 were intended to help reduce the industry's problems but some contractors still play 'hard ball' with their subcontractors and defer payment for as long as possible purely to help their own cash flow. Keeping hold of other peoples' money is a simple way to both improve cash flow and make money at the same time by reducing the cost of interest on borrowings. Money has a value and a cost and **value minus cost = profit**. It's simple economics really!

5.2.1 Trust and money

In the context of a chapter on payments, it is important to revisit the 1994 Latham Report, 'Constructing the Team', referred to in Chapter 1, which describes the cultures and methodologies of the construction industry and identifies the factors which militate against the successful outcome of projects. The report was commissioned to find ways to 'reduce conflict and litigation and encourage the industry's productivity and competitiveness' but said little about the industry's payment problems.

However, Latham's interim report, 'Trust and Money' – also referred to in Chapter 1 – did highlight payment issues and clearly pointed out that the participants in the construction process did not trust each other. The interim report also emphasised that withholding of payments, the abuse of dominant positions (e.g. contractors over subcontractors) and 'pay-when-paid' terms, were rife in the industry.

Thanks largely to Latham, legislation, made specifically for the construction industry in the Housing Grants, Construction and Regeneration Act 1996 (the 'Construction
Act'), was introduced to reduce conflict by improving payment practices and facilitat-
ing 'fast track' dispute resolution via a statutory right to adjudication.

The provisions of the Construction Act, and the subsequent changes thereto made
in 2009, are key features of what should be the 'gold standard' of current payment
practices and procedures in the industry. However, it seems that 'the jury is out' on
the extent to which the provisions of the Act are generally applied in practice.

5.2.2 The Construction Act 1996

Fellows and Liu (2011) make the point that the interim report of the 'Latham' review
of the United Kingdom construction industry – 'Trust and Money' (Latham, 1993) –
emphasised that there was not enough of either 'trust' or 'money' between the
participants. They also state that the full Latham Report – Constructing the Team
(Latham, 1994) – obscured much of the message regarding trust and the various
aspects of money and further beg the question that little has changed in the industry since either the interim or final report.

Anecdotal evidence would seem to concur with Fellows and Liu that 'leopards do
not change their spots' and it could be argued that attitudes and practices have not
changed fundamentally since Latham despite the increased emphasis on supply
chain integration, frameworks and partnering. Notwithstanding Latham's call for
standardisation of contracts and fairer contract terms, it is not uncommon to find
contract clauses in practice that are less than favourable to either main contractors
or subcontractors, especially regarding payment clauses and those concerning dispute resolution and contractual redress for non-payment.

The Construction Act 1996 resulted from Latham and this legislation heralded a
number of important changes to payment practices in the industry as well as
introducing the statutory right to adjudication and the right to suspend work for
non-payment. One of the features of the Act was to outlaw the practice of 'pay-
when-paid' but subsequently, and not untypically of the industry, other similarly
unfair practices have emerged. These practices include 'pay-when-certified' and
'pay-when-notified'. Underlining this, Fellows and Liu (2011) conclude *that the
industry operates opportunistically for its self-protection*' and for the '*survival of
individual firms*' and that '*the culture of the industry is deeply ingrained and very
difficult to change*'.

To deal with some of these unfair practices, the Construction Act 1996 was
amended by means of Part 8 sections 138–145 of the Local Democracy, Economic
Development and Construction Act 2009 (HMSO, 2009) resulting in extensive
changes ('new rules') to existing provisions regarding payment, suspension and
adjudication in particular. The provisions of the Construction Act concerning these
issues, together with the 'new rules' introduced in 2011, are considered below. It
remains to be seen how effective these changes will be in the future.

5.2.3 The Construction Act – scope and application

Largely as a result of the Latham Report (1994), problems regarding cash flow and
the lack of trust in the industry were tackled by the inclusion of construction specific
provisions in Part II sections 104–117 of the Housing Grants, Construction and
Regeneration Act 1996.

This important and ground-breaking legislation thereby made all construction contracts, whether standard or otherwise, subject to statutory control. The scope and application of the Act is set out in sections 104 and 105, which, respectively, define 'construction contract' and 'construction operations'. As far as this book is concerned, 'construction contract' includes both construction work and the supply of professional services, such as design and contract administration, and 'construction operations' can be taken to be those normally considered as construction work.

Part II sections 104–117 of the Housing Grants, Construction and Regeneration Act 1996 does not apply to contracts with a 'residential occupier' (e.g. construction work carried out for the occupier of a dwelling house).

5.3 The scheme for construction contracts

JCT, NEC, ICC and other standard contracts have been amended to comply with the Construction Act but those contracts that are not compliant are, nonetheless, subject to statutory requirements as regards adjudication of disputes, payment provisions and the prohibition of conditional payments. This arises by virtue of section 114 of the Construction Act, which makes provision for regulations to be made to create **The Scheme for Construction Contracts** (HMSO, 1998).

The aim of the Scheme is to replace terms in contracts that might be missing, thereby helping to improve cash flow and provide a means for the swift resolution of disputes. Where the Scheme applies, its terms are given legal effect in the contract in question as 'implied terms' by virtue of section 114(4) of the Housing Grants, Construction and Regeneration Act 1996.

In common with the Construction Act 1996, the Scheme for Construction Contracts was amended as a result of the introduction of the Local Democracy, Economic Development and Construction Act 2009.

5.4 Payment under the Construction Act

The improvement of payment practices is at the heart of the Construction Act 1996 and the payment provisions of contracts, whether written or oral, are determined by sections 109–113 of the Act. Where the contract does not comply with the Act, the provisions of The Scheme for Construction Contracts will apply. From 1 October 2011, a number of 'loopholes' – especially those concerning conditional payments – have been closed, thereby necessitating changes to standard conditions of contract (described in the following sections). The 'new rules' were introduced by virtue of the Local Democracy, Economic Development and Construction Act 2009.

5.4.1 Payment period

The provisions of the Construction Act as regards payment must be considered against the background of the construction industry supply chain and the payment 'pecking order'. For instance, it is usual industry practice that the further down the payment chain one goes the longer the payment period will be (e.g. the main contractor is paid first and the subcontractors wait a bit longer).

This 'fact of life' is recognised in section 110(1) of the Construction Act which leaves the parties free to agree the time period between the payment due date and the final date for payment. In standard contracts, payment periods are fair and reasonable in recognition of the importance of cash flow to the industry.

Where non-standard conditions are used, however, contractors and subcontractors should be especially alert to the payment provisions stated in the tender enquiry which could be much longer than normally expected.

5.4.2 Periodic payments

Section 109(1) of the Construction Act 1996 confers a statutory right to **periodic payment** or **payment by instalments** or **payment in stages** unless the contract specifies, or it is agreed by the parties, that the duration of the work is estimated to be less than 45 days duration.

Most standard forms of contract allow for interim payments and most provide for a maximum period of interim certification of one month. As a consequence, main contractors and subcontractors can look forward to regular payments throughout their contract, thereby aiding cash flow and enabling payments to flow down the supply chain to sub-subcontractors, suppliers and so on.

It must be remembered, however, that standard contracts are not exclusively used in construction and, therefore, unless the contract specifies otherwise, the statutory 45 days will apply before the right to instalments or stage payments is triggered. Consequently, if the contract or subcontract period is less than 45 days, it may be necessary to 'cash flow' the job for seven weeks (or more if there are delays or extra work) before an application for final payment can be made. There will then be a further delay of some weeks before payment is actually received. Subcontractors (especially) beware!

The Construction Act makes a distinction between **payments that are due** and **the final date for payment** of any sum which becomes due. This provision of the Act recognises normal industry practice whereby work in progress is **valued** at regular intervals (usually monthly), and becomes due when **certified**. It also acknowledges the 'breathing space' provided in most contracts between the date of the certificate and the date when the amount certified has to be paid.

5.5 Payment notification under the Construction Act

5.5.1 Contractual provisions

Payments to contractors (and thereby to subcontractors) are commonly triggered by the issue of interim certificates. Under JCT SBC/Q 2011 conditions, for example, the period of interim certificates is stated in the Contract Particulars: Part 1. This is where the date for issue of the first certificate is stated which must be no more than one month following the date for possession of the site. Subsequent certificates are issued at monthly intervals thereafter. These provisions set the payment timetable for the contract.

JCT SBC/Q 2011 requires that the interim certificate should state:

- **the amount due** to the contractor - that is, the value of the work carried out less retention and less any previous payments;
- **what the amount relates to** - that is, the work done to which the payment relates;
- **the basis upon which the amount was calculated** - that is, details of the quantity of work done, the value of any variations and details of any further sums due to the contractor (such as payments for loss and expense).

JCT SBC/Q 2011 states that the final date for payment shall be 14 days from the date when the certificate was issued.

Notwithstanding such contract conditions, the Construction Act 1996 makes specific provision for the issue of **payment notices** (section 110A Payment notices: contractual requirements refers) and these must be issued in addition to the normal interim certificate.

5.5.2 Payment notice

The Construction Act 1996 section 110A(1)(a) requires a **payment notice** to be issued by the 'payer' (e.g. employer) to the 'payee' (e.g. the contractor) in order that the 'payee' knows how much he will be paid and the basis of that payment. Alternatively, the notice may be issued by a 'specified person', who could be a contract administrator, such as an architect or engineer. This is more likely to happen in practice. The payment notice must be issued no later than five days after the **payment due date**, which is signalled by the issue of the interim certificate. It is from this date that the allowed time for issuing a payment notice under section 110A of the Construction Act 1996 runs. Similar payment notice provisions are included in the Scheme for Construction Contracts.

There is no sanction if the notice is not issued, and therefore the Section 110 (Payment) Notice is simply an early warning as to the proposed payment to be made. However, if the contractor does not receive a payment notice from the employer or the contract administrator, the contractor's payment entitlement would be that which is due under the contract, that is the amount certified.

To comply with section 110A(2) of the Construction Act, the payment notice must state both the amount considered to be due and the basis upon which the amount was calculated. The effect of such notices depends upon the express terms of the contract. For example, if the contractor makes an application for interim payment and the employer fails to issue a payment notice within the prescribed period, then the contractor's valuation must be paid in full. The case of *Watkin Jones & Son Limited v Lidl UK GmbH* (27 December 2001) refers.

Section 110A(1)(b) of the Construction Act provides for the possibility that the 'payee' (e.g. the contractor) may provide the payment notice. This may happen where the employer fails to give a payment notice to the contractor who then issues a **default payment notice**.

5.5.3 Default payment notice

Section 110B(2) entitles the 'payee' (e.g. contractor) to issue a payment notice to the employer where either the employer or contract administrator fails to do so. The effect of this will be to delay the final date for payment whilst the default payment notice is issued and also to enable the contractor to state in the notice the sum that he considers is due. This sum may well differ from the sum that otherwise would have been paid had the correct payment notice been issued. Section 110(B) of the Construction Act refers.

5.5.4 Withholding (or pay-less) notice

Section 111(1) of the Construction Act requires the 'payer' (e.g. employer) to pay the notified sum on or before the final date for payment irrespective as to whether the notice has been issued by the employer, by a 'specified person' (e.g. the contract administrator), by the 'payee' (e.g. contractor) or by the 'payee' (e.g. contractor) under a default notice.

However, the 'payer' or 'specified person' may issue another notice of the payer's intention to pay less than the notified sum (a **withholding** or **pay-less notice**). This notice must be issued no later than a 'prescribed period' before the **final date for payment**. This period may be agreed by the parties – the JCT SBC/Q 2011 states a period of five days before the final date for payment – or, failing such agreement, the statutory default period is seven days before the final date for payment in accordance with paragraph 10 of the Scheme for Construction Contracts.

A withholding (or pay-less) notice is a response to a specific application for payment and, as such, must specify how much of the sum that is applied for is to be withheld, together with the ground(s) for withholding payment.

Withholding money from payment applications is common practice in construction and both contractors and subcontractors are frequently underpaid for all sorts of valid (and sometimes invalid) reasons. Most construction contracts are complicated and frequently there are arguments concerning the valuation of measured work, the valuation of variations and extra work, the authorisation and payment of daywork, the value of materials on site and so on.

If a contractor or subcontractor is overpaid, and subsequently becomes insolvent, this money may not be recoverable by the 'payer'. Consequently, there is a fine balance to be struck and it is human nature to err on the side of caution. Try explaining that to an irate subcontractor!

5.6 Conditional payments

Under traditional style contractual arrangements, payments normally flow from the employer to the main contractor and then to the subcontractors and thence to any sub-subcontractors and so on. It should be noted that the payment terms between the employer and a main contractor may well be different to those between the main contractor and his subcontractors and, therefore, subcontractors may have to wait much longer to get paid than the main contractor. Despite this, the payment system works well enough under normal circumstances.

However, if there is a delay in payment between the employer and the main contractor, the subcontractors may be kept waiting even longer for their money with the consequent cash flow problems that this might cause. In the past this has led to conditional payment practices in the industry where one party (e.g. main contractor) makes payment to another party (e.g. subcontractor) subject to receipt of payment from a third party (e.g. employer).

This practice was made illegal by the Construction Act 1996 and the prohibitions on delayed payments have been further strengthened by the 'new rules' introduced from 1 October 2011. A variety of conditional payment procedures are controlled by the Act as described in the following sections.

5.6.1 Pay-when-paid

This is a form of conditional payment whereby payment by one party (e.g. main contractor) to another party (e.g. subcontractor) is conditional upon the main contractor receiving payment from a third party (e.g. employer). Pay-when-paid clauses are 'bad news' for subcontractors as they may be kept waiting for payment for reasons known only to the employer and the contractor and, therefore, beyond the subcontractor's control or influence.

Pay-when-paid was prohibited by section 113 of the Construction Act 1996 and remains unchanged under the 'new rules'. However, section 113(1) provides that this form of conditional payment is allowable when the third party (i.e. the employer) is insolvent (see also Chapter 15).

5.6.2 Pay-when-certified

Originally used to avoid the pay-when-paid prohibition of the 1996 Act, this form of conditional payment makes payment to one party (e.g. subcontractor) contingent upon certification of payment to another party (e.g. main contractor) by a third party (e.g. employer).

This practice is now illegal by virtue of the 'new rules' contained in section 110(1)(A) of the Construction Act. There are some exclusions to this prohibition where, for example, there is a management contract and the management contractor merely manages and supervises the work and the subcontractor's payment is subject to certification by the employer. There are also exclusions which relate to PFI contracts.

5.6.3 Pay-when-notified

Pay-when-notified arises when there is a contract condition (e.g. between the employer and main contractor) which permits the employer to decide on the payment due date. A notice as to what payments are due under the contract is then issued to the contractor. This mechanism allows the employer to enjoy a 'flexible' payment due date but, conversely, leaves the contractor in the position of not knowing exactly when he will be paid. There is not necessarily any suggestion that the contractor will not be paid at all but, nevertheless, exactly when payment will happen is at the whim of the employer.

Section 110(1)(D) of the 'new rules' to the Construction Act prohibits this form of conditional payment.

5.7 Late payments

5.7.1 Legislation

In addition to the provisions of the Construction Act, the corporate late payment culture has been recognised at European Union level as a problem affecting all industries and businesses. The EU **Late Payment Directive**, which will be effective in 2013 and transposed into United Kingdom law sometime after that, establishes 30 days as the standard payment term and identifies terms longer than 60 days as 'grossly unfair'.

To be fair to the construction industry at large, it is unlikely that there is anything other than a relatively small minority of firms that will fall into the 'grossly unfair' bracket. Those that do will probably not be in business for very long in any case, as they are likely to be financially unstable and a step away from insolvency or, if not, likely to be facing the distinct probability of having supplies suspended or credit withdrawn altogether. A bad name for excessively late payment circulates very quickly on the construction 'grapevine' and no-one is desperate enough for business knowing that payment for the goods and services rendered is going to be a lengthy and, possibly, losing battle.

Late payment is clearly a concern at Government level as well because the 'Late Payment Working Group' has been reformed with the brief to tackle the problem of late payment, which is in danger of becoming endemic, and to review corporate payment performance and legislation.

5.7.2 Interest

Where payment is withheld from the contractor without reason, standard contracts provide for the employer to pay interest on the outstanding amount. For example, under the JCT SBC/Q 2011, the contractor is entitled to simple interest on the amount outstanding at the rate stated in clause 1.1 – Definitions, that is 5% above the prevailing Bank of England rate. The contract stipulates that any interest owing would be a debt owed by the employer to the contractor which could be pursued through the civil courts without having to establish a breach of contract.

Where no contractual provision exists, statutory rights to interest are provided for under the Late Payment of Commercial Debts (Interest) Act 1998.

5.8 Suspension of performance

Latham (1993) emphasised that the construction industry is renowned for money problems and lack of trust but, to most people in the industry, this was 'old news'. Disputes over money, slow payment, underpayment and, sometimes, no payment at all have been common features of the industry for many years. Such problems can happen to main contractors as well as subcontractors with damaging effects on profitability and, crucially, cash flow. The question arises as to what can be done to resolve matters.

The Construction Act 1996 has gone a long way to helping redress such difficulties but, even so, this legislation has required some fairly extensive 'fine tuning' with 'new rules' introduced by virtue of Part 8 sections 138–145 of the Local Democracy, Economic Development and Construction Act 2009.

An important feature of the original Construction Act was the statutory right to suspend performance for non-payment (in full) of a sum due under a construction contract in circumstances where no effective withholding (or pay-less) notice has been given. Prior to this legislation, the only recourse available to the injured party under most contracts was a civil action for damages for breach of contract (although some subcontracts provided an express term for the right to suspend). Clearly such action was lengthy and costly and failed to address the essential problem, that is improving cash flow.

The Construction Act's 'new rules', introduced after 1 October 2011, retain and extend the statutory right to suspend performance. Section 112 of the Act gives this right where a notified sum is not paid on or before the final date for payment. Furthermore, the right to suspend applies to *any or all* of a contractor's (or subcontractor's) obligations under the contract. This means that *part* of the works may be stopped without the need (or expense) to demobilise the whole of the contract works as was previously the case. In any event, the costs and expenses of any legitimate suspension is payable to the injured party under section 112(3A) of the Act.

Section 112(2) requires that suspension of performance of the contract cannot take place before giving a minimum of seven days notice together with a statement of the grounds for suspension; the right to suspend ceases when payment in full has been made. In addition to the statutory right to recover the costs of suspending the work, section 112(4) of the Act recognises that the time lost as a result of the suspension will be disregarded in computing the time for completing the contract works. This right extends to an injured party (e.g. contractor) or to a third party (e.g. subcontractor). There may well be practical difficulties in agreeing the time computation, however, especially where there has been a partial suspension.

Notwithstanding the provisions of the Construction Act, there is a laudable culture in the industry that the project should not suffer and it is rare that a contractor or subcontractor will 'stop the job' despite arguments over money. On the other hand, where a subcontractor does suspend performance, it is not unknown for the main contractor to engage another subcontractor to finish the work and argue over the 'rights and wrongs' of such action at another time. Some contractors are happy to play 'call-my-bluff' and effectively dare the injured subcontractor to risk adjudication or arbitration with the consequent cost and time implications.

5.9 Adjudication

For many years the construction industry relied upon arbitration and litigation as the sole means of resolving disputes. Arbitration agreements were common in construction contracts, and still exist today, but were considered by some people as 'an expensive form of rough justice'. Litigation is expensive and lengthy and neither process served the industry well in terms of resolving disputes quickly and effectively during the contract. From a cash flow point of view this was the crucial issue.

Latham (1994, Chapter 9) pointed out the folly of the litigious nature of the industry and called for a statutory right to adjudication – a short and cost efficient means of resolving disputes during the contract – whilst at the same time preserving the right of the parties to enter arbitration or litigation proceedings at a later stage following practical completion of the works.

As a consequence, the Construction Act 1996 included a statutory provision for the adjudication of disputes in building contracts (section 108 refers). This provides a summary procedure for reference of a dispute (usually over money!) to an independent third party with a decision to be forthcoming within a timetable of 4–6 weeks. The adjudicator's decision (with some exceptions) is binding until the matter is resolved through litigation or, where the contract so provides, by arbitration or both.

Construction contracts are, therefore, required to include a suitable provision, in writing, to enable a party to refer a dispute to adjudication. Failing such compliance with the Construction Act, the Scheme for Construction Contracts provides a 'fallback' procedure enabling disputes to be settled using the adjudication process.

Since the inception of the Construction Act, an adjudication 'cottage industry' has developed and there is a catalogue of disputes that have reached adjudication. Additionally, a body of law has developed surrounding the rights of the parties and the duties and powers of the adjudicator such that the process is extremely procedural and 'legalistic'. Arguments abound as to 'when is a dispute a dispute' and whether proper notices have been served and when. Legal representation is 'a must' for the parties involved and, for small subcontractors especially, dipping a toe into any legal process is daunting and inevitably costly. No party is insulated from the possibility that a sound case may be lost or that a shaky case may win!

The whole point of adjudication is to arrive at a quick, practical and sensible solution to a problem so as to unblock money and improve cash flow in the industry. Tate (various dates) raises the question that the process may be too truncated to fully consider the issues and that this results in 'rough justice'. A more cynical point of view might be that the real winners are usually the legal people who are not shy at charging 'eye-watering' fees!

5.10   Value Added Tax

5.10.1   VAT in construction

VAT rules are complicated and there is no substitute for reading HM Revenue and Customs notices, which prescribe the types of supply which attract tax and which rates of tax apply in different circumstances. An added complication is that rates of VAT and associated rules are subject to change. However, as a general rule, all construction work is standard rated with the exception of certain residential 'new build' and renovation/alteration work, which is either zero rated or subject to a reduced rate.

Standard JCT contracts make reference to VAT in the main conditions and in the contract appendix, which states whether or not supplemental conditions apply to the contract. The main clause states that the contract sum excludes VAT and also provides for the contractor to be reimbursed should the works or part thereof become exempt from VAT (because input tax is not recoverable where the output is an exempt supply). This avoids any complications with the contract should the VAT rules change during the life of the contract.

The supplemental conditions set out procedural rules for dealing with VAT. These include arrangements for the contractor to notify the employer as to the rate(s) of VAT which apply to the work carried out and for the contractor to provide a written statement of the VAT liability in respect of the certificate being issued.

There are three rates of VAT that might apply to construction projects:

● standard rate (20%)
● reduced rate (5%)
● zero rate (0%)

5.10.2   How VAT works

Irrespective of who prepares the interim valuation, payment of the interim certificate requires an invoice from the contractor stating the amount of the payment and the amount of any VAT applicable. This is called a VAT invoice. The same principle applies to subcontractors who must submit a VAT invoice to the main contractor for payment to proceed.

The contractor's (or subcontractor's) VAT invoice includes a VAT number which corresponds to the company's VAT registration with HM Revenue and Customs and HMRC officials pay regular visits in order to check the VAT returns and ensure that the system is being administered properly in accordance with the rules.

Box 5.1   Example VAT calculations

Example 1
Where a builder who is VAT registered (i.e. VAT taxable turnover >£73 000*) buys £100 worth of materials and carries out work which costs £500 in labour and plant and the work is **standard rated**, the VAT liability is:

A	Input Tax on materials	= £100×20%	= £20.00
B	Output Tax on labour+materials	= £(500+100)×20%	= £120.00
C	VAT Payable C=B−A	= £120.00 − £20.00	= £100.00

Example 2
Where a builder who is VAT registered (i.e. VAT taxable turnover >£73 000*) buys £100 worth of materials and carries out work which costs £500 in labour and plant and the work is **zero rated**, the VAT liability is:

A	Input Tax on materials	= £100×20%	= £20.00
B	Output Tax on labour+materials	= £(500+100)×0%	= £0.00
C	VAT Re-payable C=B− A	= £0.00 − £20.00	= −£20.00

*The VAT registration threshold varies according to the Chancellor of the Exchequer's Budget.

VAT works on the basis that:

Output Tax − Input Tax = VAT Payable (to HMRC) or, if a negative figure,

VAT Re-payable by HMRC

Two example calculations are shown in Box 5.1.

Some supplies may be classed as 'Exempt', which means that there is no output tax chargeable. However, there may be an input tax but this cannot normally be re-claimed.

References

Fellows, R.F. and Liu, A.M.M. (2011) Trust and Money: 20 Years of (No) Progress? In: Proceedings of the CIB International Conference Management and Innovation for a Sustainable Built Environment (MISBE2011), 20-23 June 2011, Amsterdam, The Netherlands.

HMSO (1998) The Scheme for Construction Contracts (England and Wales) Regulations 1998. HMSO, London.

HMSO (2009) Local Democracy, Economic Development and Construction Act 2009, Part 8 Construction Contracts. HMSO, London.

Latham, M. (1993) Trust and Money: Interim report of the joint government industry review of procurement and contractual arrangements in the United Kingdom Construction Industry. HMSO, London.

Latham, M. (1994) Constructing the team: Joint review of the procurement and contractual arrangements in the UK Construction Industry. HMSO, London.

Tate, R. (various dates) Legal Beagle series of articles, Dispute resolution. http://www.publicarchitecture.co.uk.

6

Managing the supply chain

Financial Management in Construction Contracting, First Edition. Andrew Ross and Peter Williams.
© 2013 Andrew Ross and Peter Williams. Published 2013 by John Wiley & Sons, Ltd.

6.1 Supply chain management

6.1.1 Definitions

Constructing Excellence (2004) suggests that the term 'supply chain' describes the linkage between those companies that convert a series of basic materials, products or services into a finished product for the construction client. It is further suggested that all parties to the construction process – client, consultant, main contractor, subcontractor, or supplier – are part of a supply chain.

Winch (2010) explains that 'supply chain' refers to a chain of firms linked through a series of contracts. He also distinguishes between the relationships developed at the 'first tier' of **principal suppliers**, who are in direct contract with the client, and the second and subsequent tiers of **subcontractors** which also make up the supply chain.

The supply chain concept presupposes that each company in the chain has a client to whom goods and/or services are supplied but that an **integrated supply chain** will be focussed on understanding the objectives of the end user or 'project client' and will work with his interests uppermost. Because of the project-based nature of construction and the procurement methods used, supply chains are not homogeneous, and therefore individual members of one supply chain are usually members of different supply chains on different projects.

6.1.2 Integrated supply chains

Integrated supply chains replace the traditional approach to procurement whose emphasis is on forming relationships linked only by contracts that have been procured on the basis of lowest price against a fixed set of requirements specified in the drawings, specification and bills of quantities and where there is no motivation to work in the client's best interests.

Pryke (2009) suggests that focussing on the supply chain, in contrast with traditional procurement, provides more effective ways of *'creating value, innovation, continuous improvement, integration of systems and improved profitability'*. He also says that the term 'supply chain' implies a linear process but explains that *'clusters of suppliers come together in a series of dyadic* (Google it!) *exchanges that may be observed as dynamic networks of relationships'*.

Consequently, Pryke feels that *'clients, consultants, contractors and suppliers are positioned as nodes connected by linkages comprising knowledge transfer, information exchange, directions and financial and contractual relationships'*. These linkages represent *'transitory relationships and iterative flows'* and the nodes are continually linking and disconnecting.

A central feature of integrated supply chain management is that benefits accrue to all companies in the supply chain; these benefits include:

- the incentive to manage out waste from the process resulting in:
  - reduced real costs
  - greater certainty of out-turn costs;
- delivery of better value to the client leading to:
  - repeat business;
- greater confidence in budgeting future turnover; and, most importantly
- ring-fenced profit margins ensuring a 'win-win' project out-turn for all participants.

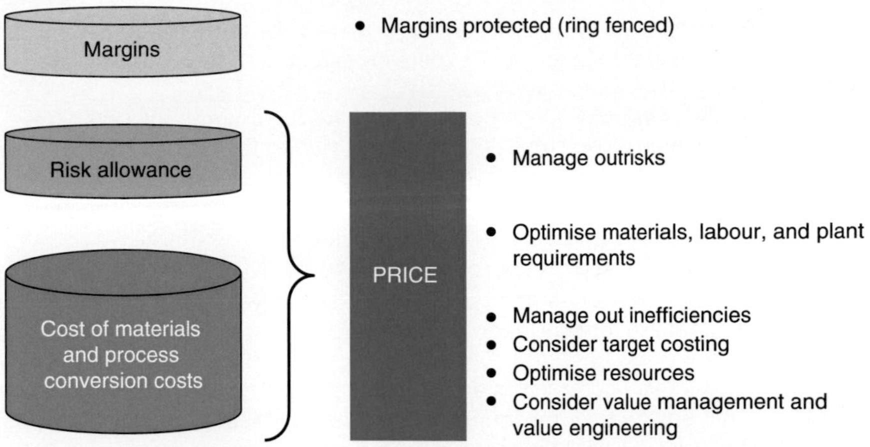

Figure 6.1   Supply chain cost management (adapted from Constructing Excellence, 2004).

6.1.3   Managing cost and profit

A central feature of successful collaboration with supply chain partners is the approach to cost management. Figure 6.1 illustrates the approach suggested by Constructing Excellence in which profit margins are protected or 'ring fenced' in order that a more enlightened approach may be made to cost and risk management. The idea is that 'price' is given by purchase cost + costs of conversion + an allowance for risk with a suitable agreed margin for profit set aside and assured.

This approach makes sense in many ways, because concentration on cost rather than profit facilitates greater pro-activity in cost and waste reduction without the worry that reduced cost will lead to reduced margins. A positive attitude to risk reduction is also engendered, which helps to deliver value without endangering profitability on the contract.

Most contractors and subcontractors will be guarded when it comes to discussions about profit but the Constructing Excellence approach takes the heat out of negotiations and places the focus on obtaining best value. If supply chain margins are offered some protection this will allow supply chain members to deliver quality and value without undue emphasis on self-interest and protectionism. This is easier said than done in practice and a high level of trust and transparency is required to make this work. Contractors and subcontractors are notoriously coy when it comes to revealing how much money they make!

6.1.4   Practical applications

Not surprisingly, the principle of integrated supply chain management has not yet been widely accepted in the construction industry, which is notoriously reactionary, but there is evidence that a number of client organisations and major contractors have successfully applied its principles to their projects. British Airports Authority, SEGRO plc (formerly Slough Estates) and the Ministry of Defence are committed to the precepts of sustainable procurement and members of the UK Contractors Group (UKCG), such as Balfour Beatty plc and BAM Construct UK, have sustainable supply chain strategies in place.

BAM Construct UK has a highly developed approach to supply chain management aimed at delivering added value to clients and ensuring that its subcontractors possess the right blend of skills at the right price in order to ensure quality and project delivery of the highest standard. The company categorises its subcontractors into four tiers:

1. **Category one** subcontractors work frequently with the Company.
2. **Category two** subcontractors have started to work with the Company on a regular basis and good relationships and trust are developing.
3. **Category three** subcontractors have worked with the Company but close working relationships are still being developed.
4. **Category four** is subcontractors which have applied for consideration for tenders, have undergone assessment and have been included on the Company's database for consideration for future tender invitations.

Not all main contractors adopt this approach to subcontractor categorisation, but most ensure that subcontractors on their database are capable of doing the job if 'the price is right'. Selection criteria include:

● evidence of compliance with statutory health and safety legislation and best practice standards;
● evidence that the company has qualified for the Construction Health and Safety Assessment Scheme (CHAS) (or equivalent) and that the registration is current and up to date;
● evidence that all site operatives hold a current Construction Skills Certification Scheme (CSCS) card or the equivalent;
● the requirement that all site supervisors have undergone the Construction Skills (or equivalent) two day health and safety training as a minimum standard;
● evidence of a responsible approach to waste management and environmental awareness;
● compliance with appropriate standards of employment policy and procedures to ensure a competent labour force.

6.1.5 Context

Supply chain management is a big topic – Pryke (2009) has written an entire book on the subject! Consequently, we take a somewhat narrower, possibly more traditional, view and concentrates on the relationships between the first tier 'main contractor' and the second tier 'subcontractors'.

The focus of this chapter is necessarily confined to the ambit of the book – cost, value and financial reporting and, whilst the authors recognise the benefits that truly integrated supply chains bring to the industry and its clients, they also acknowledge that the 'real world' supply chain relationships are still largely rooted in the traditional practices and attitudes alluded to by Latham (2003).

Attention will, therefore, be paid to 'subcontractors' within the normally accepted definition of the word.

6.2 Subcontractors

6.2.1 The growth of subcontracting

Specialist contractors have long been a feature of the construction industry and their use dates back to the Middle Ages. Since the late eighteenth century and up to the early 1970s however, the 'general contractor' was the dominant presence in the industry. Many of these contractors had both their own directly employed workforce and 'specialist' departments for joinery, plumbing, plastering, painting, floor and wall tiling and so on. Such companies commonly carried a large fleet of plant and offered most of the traditional construction trades 'in-house'. Relatively little work was sublet.

The mid-1970s saw a big change in the industry with the advent of the then much maligned 'labour-only' subcontractor. These were individuals, often working in gangs, who carried out work directly for main contractors who supplied the permanent and temporary materials required for the job. Direct payments were made to these subcontractors and tax and national insurance contribution evasion was rife in the industry.

Successive Governments have tried to prevent this tax evasion and various registration schemes were introduced over the years albeit with patchy results. More recent construction industry specific tax legislation, however, coupled with significant improvements in health and safety legislation and standards, have helped to polarise labour-only subcontractors into company structures, many of which have been established and are run by former tradesmen blessed with entrepreneurial and organisational abilities.

The 'labour-only' subcontractor still exists, but mainly under the umbrella of a company which obtains, organises and allocates contract work to individual tradesmen, deals with the main contractor and takes on the associated business and employment risks.

As a result, the major contractors carry out very little of their turnover using a directly employed workforce and specialist companies abound. The 'general contractor' is not dead, though, and several well run medium sized companies offer such a service but they are the last of the breed of contractor that followed in the footsteps of the greats such as William Joliffe, Sir Edward Banks and Thomas Cubitt.

6.2.2 Types of subcontractors

Subcontractors are frequently referred to as 'package' or 'work package' contractors but they are, nevertheless, tied contractually to a main contractor through a subcontract agreement of some sort. An exception to this rule is that of 'construction management' procurement, where the 'work packages' are contracted directly with the client or employer. For contractors, subletting work is not only a means of buying in specialist services or expertise but, in a strategic context, provides a way of hiving off non-core business in order to spread commercial risk, to offer clients a total service beyond a contractor's core business and to better manage resources resulting from variations in workload.

Some conditions of contract (e.g. ICC conditions) allow the appointment of a 'nominated' subcontractor and others (e.g. JCT conditions) have the facility of 'listing' subcontractors in the bills of quantities. Both methods provide a means for the contract administrator to influence the main contractor's choice of subcontractor but **nomination** requires a specific subcontractor to be chosen whereas **listing** gives the main contractor a degree of choice. In most conditions of contract, approval is needed to sublet any part of the contract works and, in some conditions, specific approval of preferred subcontractors is required.

The increased use of subcontractors has, to some extent, contributed to fragmentation of the industry and might be considered a contributory factor in the generally poor safety record in construction. Counterbalancing this, however, subcontractors offer specialisation and knowledge which is often outside the scope of a 'general' contractor and they can often be more flexible in terms of programming and response time. Frequently, decisions can be made more quickly and, as there are usually less tiers of management, communications can be more efficient.

Subcontractors come in many shapes and sizes. Some are very big companies in their own right and are sophisticated and powerful. The big ones may be independent companies or subsidiaries of large contracting organisations and they may employ 3000 people or more. Some offer highly specialised services and may be bigger than some of the main contractors they work for. The few 'mega-large' subcontractors are in the £500+ million turnover bracket.

Subcontractors in the 'large' bracket may have a turnover in the £100–500 million range and are clearly substantial businesses. Medium sized subcontractors might be considered to be those in the £10–100 million turnover class. The vast majority of subcontractors are in the 'small' bracket with a turnover of less than £10 million. A significant proportion of the subcontractors in this grouping will have an annual turnover of less than £1 million and the majority of these will be 'very small' with an annual workload of less than £500 000.

6.2.3 Construction Industry Scheme

Mainly as a consequence of the growth of labour-only subcontracting, problems of tax evasion and non-payment of national insurance contributions have plagued the industry. The latest, and arguably most successful, of successive Government schemes designed to deal with such problems, is the Construction Industry Scheme (CIS), which was introduced by the 2004 Finance Act and became effective from the 6 April 2007.

The legislation requires:

● registration by subcontractors;
● verification by contractors to confirm whether subcontractors should be paid net or gross;
● payment statements;
● monthly returns (CIS300) requiring a declaration that the contractor has considered the employment status of his subcontractors to ensure that those claiming to be self-employed are not operating as employees and are, therefore, not subject to the PAYE system.

The Scheme applies to payments made by 'contractors' to 'subcontractors' for work involving construction, repairs, installation, cleaning, painting and decorating, demolition and so on. 'Contractors' are regarded as those who employ subcontractors and this definition applies to many of the industry's clients, to main contractors who sublet work and to subcontractors who engage self-employed workers or otherwise sublet down the chain.

Before making payments to subcontractors, contractors must make sure that they are registered for CIS and, if so, establish whether the subcontractor should be paid gross or with tax deducted at 20%. 'Gross status' is subject to business, turnover and tax compliance tests. If any subcontractors are not registered with HMRC at all, tax must be deducted from payments to them at the rate of 30%.

Late and persistently late monthly returns are subject to a scale of penalties and a further penalty may be incurred if the return is incomplete or incorrect. Penalties may also be incurred for failure to produce records when required and for failure to provide subcontractors 'under deduction' with a statement for each payment or for each month, whichever is applicable. Under CIS, subcontractors must be demonstrably self-employed. If, however, they are to all intents and purposes employees, the PAYE and national insurance contributions that should have been deducted will be payable by the contractor and HMRC may pursue such liabilities dating back six years and impose interest charges and penalties if appropriate.

6.2.4   Trade and other references

At one time, prospective subcontractors would be asked to provide three 'trade references' which the main contractor could take up in order to gain some idea of the subcontractor's financial stability. The same idea was also employed by suppliers needing to be sure that their bills would be paid on time. The industry 'grapevine' was also used to derive 'off-the-record' information and contractors could ask friendly builders' merchants, suppliers and plant hire companies about the credit worthiness of a particular subcontractor. A subcontractor would also be asked for a bank reference and, although couched in bankers' language, a favourable reference would reassure suppliers and main contractors that a subcontractor was stable enough to do business with.

It would appear that, from a limited survey conducted by Cooke and Williams (2009) and from anecdotal evidence, those days are gone and that such sensitive information may now come under the scrutiny of the Data Protection Act 1998.

Other methods that could be employed include:

- Method 1
  - Conduct a company search at Companies House
  - Ask the subcontractor how much he is paying his bank or insurance company to provide performance bonds (the higher the premium paid the greater the perceived risk).
- Method 2 (Interview)
  - Ask questions about the cost of insurances
  - Ask for bank details
  - Ask about the company's Tax/NIC and VAT record.
- Method 3
  - Ask how long the company has been trading
  - Ask for monthly turnover and credit figures
  - Ask for evidence that creditors are being paid to terms.

6.2.5   Bonds

A good test of a subcontractor's liquidity is to ask for a performance or 'default' bond to be provided. This is often mandatory for main contractors on public sector contracts and some private clients regularly ask for bonds and warranties. Such requirements may be 'stepped-down' to subcontractors. Bonds are provided by a 'surety' – normally a bank or insurance company – for a fee or premium. Default bonds are normally for 10% of the contract/tender sum.

A bond is a legal agreement involving three parties:

- surety
- subcontractor
- main contractor.

Essentially, the bank (say) as surety enters into a legal agreement (deed) with the main contractor in which the surety guarantees to pay a sum up to the total value of the bond to the main contractor if the subcontractor defaults on his contract with the main contractor. The deed is signed by the surety, the subcontractor and the main contractor. To protect the surety, the subcontractor will counter-indemnify the surety against financial loss by:

- counting the surety's total exposure as a secured loan or as part of the overdraft facility;
- granting the surety a fixed and/or floating charge over assets.

The tricky bit for subcontractors (especially small ones) is that the bond remains in place until satisfactory closure of the defects correction period; this is geared to the main contract period and not that of the subcontract. Most subcontracts are of a relatively short duration and, therefore, a subcontractor could have several bonds running in any one trading period until a saturation point is reached. This means that the working capital available to the subcontractor will reduce in direct proportion to the value of the bonds in place and could reach the point where no more work requiring bonds can be taken on. This is why the availability of bonds is a good test of liquidity.

6.3 Subcontract tenders

6.3.1 The decision to sublet

The decision to subcontract work is a strategic decision and should be considered as part of the main contractor's overall approach to a project. The scope of subcontract packages warrants careful consideration and planning. The contractual issues in making subcontract appointments are also important as, to a large degree, they determine the nature of the relationships on the project. Contracts with subcontractors should be consistent with the main contract with appropriate distinction between the rights and obligations of the parties involved and the contractual and commercial risks to be borne by each of them.

The legal effect of subletting is to transfer the benefit of part of the contract to a third party but the burden of performance always rests with the main contractor. This can give rise to liabilities (e.g. for defective work) which the main contractor must allow for in his financial reporting of the contract.

Statutory obligations under health and safety and other legislation need to be made clear regarding who is to do what in terms of health and safety on site, construction waste and environmental protection. Many main contractors now require subcontractors to be pre-qualified according to an accredited industry specific health and safety standard before being considered for inclusion on their list of approved subcontractors.

6.3.2 Tender enquiries/send outs

An important part of the estimating process is the 'tender enquiries' stage, where requests for prices are made to subcontractors (and suppliers). These enquiries are sometimes referred to as 'send outs'.

The 'enquiries' process begins with the formal bills of quantities (or the informal bills if prepared 'in house') and involves:

- deciding which parts of the project to sublet;
- identifying which bills of quantities items are applicable to each package of work;
- identifying any work for the same subcontract package that may be scattered about in various parts of the bills;
- identifying relevant pages from the specification;
- collating bill of quantities and specifications pages for each subcontract package;
- adding relevant extracts from the preliminaries bill so that the project description, project details, important constraints, conditions of contract applicable and so on are clear in the enquiry;
- sifting through the tender drawings for those relevant to the subcontract package;
- adding a cover sheet with a summary of the enquiry details, any other relevant information, the enquiry return date, any main contractor imposed terms and conditions, site visiting arrangements, main contractor contact details and so on;
- deciding which approved subcontractors to include in the enquiry;
- arranging for the bills of quantities pages, specifications pages and drawings to be copied or converted into electronic files;
- sending out enquiries;
- chasing up late quotations.

6.3.3 Scoping of work packages

One of the key decisions facing the main contractor is how to 'package' the work in the subcontract enquiry. The bills of quantities will normally have been prepared using one of the standard methods of measurement used in the industry (or an approximation thereof) but the bills are not necessarily arranged conveniently for subcontract packaging (or 'parcelling' as it is sometimes called).

If work is packaged by following the bills of quantities work sections, enquiries may contain work activities which are outside the scope of the work normally carried out by a particular subcontractor. Table 6.1 compares two examples of common trades according to how SMM7 classifies the work and what are generally accepted

Table 6.1 Work classification.

SMM7 Work Section	Includes	Normal 'Trade' Classification	Includes
Groundwork	• Ground investigation • Soil stabilisation • Site dewatering • Excavation and filling • Piling • Diaphragm walling • Underpinning	Groundworks	• Bulk excavation • Filling • Foundations and concrete slabs • Drainage • External pavings
• Structural/Carcassing metal/timber • Cladding/Covering	• Battens • Roofing felt • Roof tiling and slating	Roofing	• Felt and battens • Tiles/slates

'trade' classifications for the work. Any tender enquiry should be specific enough to obtain an unequivocal quotation and avoid the need for a subcontractor to sublet the package to 'sub-sub' or even 'sub-sub-sub' contractors.

Common sense indicates that it is best to parcel work using normal 'trade' demarcations in the industry by abstracting out relevant parts of the bills of quantities and thus ensuring that each subcontractor will receive an enquiry for work with which it is familiar. In this regard, the RICS New Rules of Measurement should be of great assistance when fully adopted in the industry.

6.3.4 Subcontractor selection

Once the 'scoping' of the various packages has been established, there should be cross-checks between packages in order to avoid overlap and duplication. It is common practice for quantity surveyors to be heavily involved in the selection of subcontractors but this should be a 'team' effort involving site managers and health and safety advisers as well. Subcontractor vetting and approval includes consideration of competence and resources and thorough checks should be made. Further paperwork includes consideration of subcontractor risk assessments, method statements and so on.

It is vital to follow a common set of procedures for determining the extent or 'scope' of work packages and the Construction Industry Board (CIB) publication, **Code of Practice for the Selection of Subcontractors** (CIB, 1997), sets out a recommended selection process:

● qualification
● compilation of tender list
● tender invitation and submission
● tender assessment
● tender acceptance.

The Code identifies a number of key principles and procedures that have been adopted by many contractors for the selection and appointment of work package contractors and these procedures include consideration of:

● scope of works including attendance, support and site services;
● lead times and package duration;
● resource levels and productivity;
● concurrent packages;
● form of contract and bonds;
● design responsibility and liability;
● commercial considerations including discount requirements.

6.3.5 Pre-subcontract stage

Prior to formally placing a subcontract, it is often beneficial to interview subcontractors whose tenders are of interest in order to clarify what has been included in the submission. This provides an opportunity to test whether the subcontractors have a real understanding of the project and the key issues to be dealt with.

Matters agreed should be put in writing and, if appropriate, included in the subcontract documentation. Arrangements may also be put in hand to visit a subcontractor's current sites in cases where the proposed work is high risk or there are particularly difficult hazards to deal with. In these circumstances there may well be implications for the management of the package, and perhaps its price, in assessing and managing major risks attached to the work and any interfaces arising.

It is important at the tender stage to make it absolutely clear to any proposed subcontractors exactly what arrangements the principal contractor has made for health and safety management on site. This will include what facilities are to be provided by the principal contractor and others and what the subcontractor is expected to contribute.

6.4 Subcontract stage

6.4.1 Placing the subcontract

The JCT publishes a bewildering array of subcontracts, both generic and designed, to fit with standard main contracts, which correspond with pretty much every conceivable procurement route. There are also standard subcontracts that are designed to fit the NEC and ICC Infrastructure Conditions too. Without listing all the subcontracts exhaustively, the conditions that fit with the common main contracts are shown in Table 6.2.

Simply because a form of subcontract is available that fits with a standard main contract does not necessarily mean that it will be used by the main contractor. Some main contractors prefer to use their own bespoke subcontracts, irrespective of the main contract conditions prevailing, and some use DOM/1 or other standard forms of subcontract, possibly with 'in-house' amendments.

The essential question is whether the main contractor wishes the subcontract to be 'back-to-back' with the main contract. Strictly speaking, 'back-to-back' agreements require a subcontract that enables the main contract conditions to be stepped down to the subcontract. The NEC3 main and subcontract forms do this precisely. However, Eggleston (2006) advises that there is much more to a 'back-to-back'

Table 6.2 Standard main contracts and subcontracts.

Conditions	Main contract	Subcontract
JCT 2011	Major Project Construction Contract	Major Project Subcontract
	Standard Building Contract	Standard Building Subcontract
	Intermediate Building Contract	• Intermediate Subcontract Conditions
		• Intermediate Named Subcontract Conditions
	Minor Works Building Contract	Generic Short Form of Subcontract
NEC3	NEC3 Engineering and Construction Contract	NEC3 Engineering and Construction Subcontract
	NEC3 Engineering and Construction Short Contract	NEC3 Engineering and Construction Short Subcontract
Infrastructure Conditions of Contract (ICC)	Measurement Version	CECA Form of subcontract (Blue Form)

arrangement than the conditions of contract and it is also necessary for the works information detail to match as well.

True 'back-to-back' arrangements do not suit all main contractors and many prefer the subcontractor to be tied to a lump sum price irrespective of the main contract arrangements with the employer. In such circumstances, 'back-to-back' refers to the subcontractor's rates and prices rather than the subcontract conditions.

When contractors tender for a contract, they usually use one of the subcontract prices received as a basis for the rates that they price into the bills of quantities with a margin for overheads and profit. Later on, the preferred subcontractor may well be a different company to that used at tender stage and the rates quoted may well vary considerably from the original choice. This puts the main contractor's bills of quantities rates 'at risk' should there be variations to the contract, for instance, and therefore the main contractor would much prefer the preferred subcontractor's rates to 'line up' with those used at tender stage. The agreement of suitable rates would be a matter for pre-contract negotiation prior to awarding the subcontract.

6.4.2 'Battle of the forms'

Many small subcontractors do not understand that the tenders they submit to main contractors, even if acceptable in principle, may not be incorporated into the eventual subcontract in the way that they may suppose. Nor do they realise the importance of correspondence and the minutes of pre-contract meetings which may also become part of the subcontract. Such documents may well amount to a counter-offer by the main contractor and, whilst the subcontractor may assume that the work will be carried out on his own terms and conditions, it may well be that the main contractor may accept the subcontractor's offer but impose his own terms and conditions instead. This is a very common occurrence in dealings between main contractors and subcontractors.

Thus, if the basis of the main contractor's acceptance is on terms materially different from those of the subcontractor, then this would constitute a counter-offer which has the effect of setting aside the subcontractor's original offer. The 'to-ing and fro-ing' of offers and counter-offers, often on the standard conditions of each party, results in the so-called '**battle of the forms**'. Legally speaking, the terms and conditions of the party that 'fires the last salvo' will prevail. It is often the case that a subcontractor will enter into a formal agreement without objecting to the main contractor's terms and then be faced with a problem later on. In the 'battle of the forms' the usual winner is the main contractor!

6.4.3 The discount 'spiral'

As explained in Chapter 5, discounts are a familiar feature of the construction industry. The theory of the discount culture is quite benign in that it sets out to reward those who settle their accounts on time (cash discounts) and recognises that regular 'trade' business should be accorded preferential prices (trade discounts) compared to 'one-off' retail customers.

Unfortunately, the discount culture has a more sinister side and subcontractors in particular are often caught up in a 'discount spiral' in the chase to obtain work which then becomes unstoppable during the pre-contract and post-contract stages. Figure 6.2 illustrates how the spiral works and suggests reasons why discounts may 'spiral' at an exponential rate from tender stage to final account settlement.

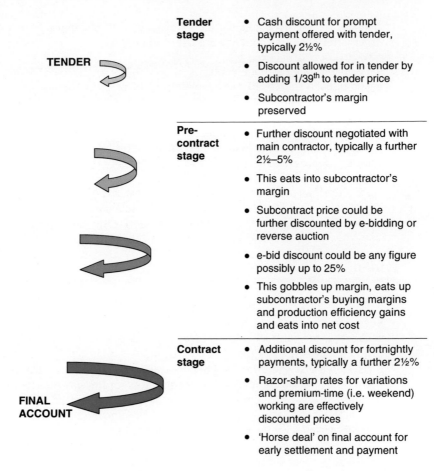

Tender stage

- Cash discount for prompt payment offered with tender, typically 2½%
- Discount allowed for in tender by adding 1/39th to tender price
- Subcontractor's margin preserved

Pre-contract stage

- Further discount negotiated with main contractor, typically a further 2½–5%
- This eats into subcontractor's margin
- Subcontract price could be further discounted by e-bidding or reverse auction
- e-bid discount could be any figure possibly up to 25%
- This gobbles up margin, eats up subcontractor's buying margins and production efficiency gains and eats into net cost

Contract stage

- Additional discount for fortnightly payments, typically a further 2½%
- Razor-sharp rates for variations and premium-time (i.e. weekend) working are effectively discounted prices
- 'Horse deal' on final account for early settlement and payment

Figure 6.2 **The discount spiral.**

Figure 6.2 is hopefully self-explanatory but it must be emphasised that:

- the cash discount offered at tender stage is allowed for in the subcontractor's price by adding 1/39th to the total to give a 2½% discount;
- further discounts erode the subcontractor's margin and, possibly, net cost;
- the reverse auction (or e-bid), which is arguably a synonym for open tendering, may lead to severe price discounting well below an economically viable level.

Unfortunately for subcontractors, most main contractors still take their cash discounts even when they pay late and there is little that subcontractors can do about it. The colloquial expression 'adding insult to injury' springs to mind!

6.4.4 Partnering

Whether or not a formal partnering agreement is in place, and irrespective of whether an integrated supply chain relationship has been established or not, good subcontractors frequently partner with main contractors as a matter of course and as part of the service they provide.

In this regard, good subcontractors endeavour to be helpful to main contractors as a matter of course and they are happy to provide:

● flexible working arrangements;
● additional resources to suit the main contractor's programme;
● early warning of problems on site;
● suggestions for specification savings or more efficient working methods;
● return visits at no extra charge;
● free short-term plant 'loan' or free use of materials;
● best endeavours to accelerate the programme at no extra cost.

More often than not, main contractors forget this part of a subcontractor's service, or do not value it sufficiently highly, but the good subcontractors are the ones they turn to when the poorer ones fail or cannot do the work at the silly prices they have quoted.

6.5   Payment

6.5.1   Terms of payment

Subcontract payment terms will be written into the subcontract, or contained in a schedule to the contract, and will expressly state when a subcontract payment application will be settled by the main contractor.

As stated elsewhere in this book, not all main contractors are bad payers and there is no intention to malign main contractors for the way that they conduct their business. It is also true, however – ask any small subcontractor – that some main contractors habitually abuse the terms of the subcontract and settle accounts on different terms once the subcontract is underway.

Admittedly, subcontractors can benefit from *ex-gratia* payment terms from main contractors, meaning that payments can often be arranged to be on a fortnightly basis, despite the express terms of the subcontract for monthly payments, in order to help with cash flow. Such arrangements, however, usually come at a cost and this may contribute to the 'discount spiral' referred to in Figure 6.2.

6.5.2   Retention and defects correction

In common with main contracts, subcontracts are subject to retention provisions at rates determined by the specific terms of the subcontract. Consequently, retention is deducted from subcontract payments applications at the contract rate until either:

● the retention limit (if any) is reached or
● practical completion of the subcontract is achieved.

Practical completion – or substantial completion in civil engineering contracts – signals the release of one moiety (half) of the accumulated retention fund with the remainder being paid over, subject to satisfactory rectification of defective work, once the final certificate is issued.

The downside of all this for subcontractors is that:

● the retention release 'trigger' dates are those that apply to the main contract and not the subcontract;

- subcontractors do not know what these dates are because they are not privy to the contract administrator's certificates;
- main contractors do not volunteer the information;
- chasing up outstanding retentions is time consuming and frustrating for the subcontractor;
- valuable working capital is frequently locked up for longer than is strictly necessary.

As a result, retention monies deducted from the subcontractor may remain outstanding for a significant period after the point when the subcontractor completes his work and this may well give rise to potential problems.

A subcontractor who has finished his package of work – for example plastering – is no longer able to protect his work from damage by others and it is frequently the case that things do go wrong and have to be put right. The question is 'who pays for it?' Here begins a complex web of claim and counter-claim in order to endeavour to find an answer. The usual process is:

- the subcontractor pays a return visit to site to put the work right;
- the work done is recorded, often on daywork sheets (time and materials);
- the subcontractor asks the main contractor to sign the sheets, not always successfully!
- the subcontractor submits a payment application to include the re-mobilisation costs and the daywork account;
- this creates a liability in the main contractor's contract accounts;
- the subcontractor's account is contra-charged to the responsible party (usually another subcontractor);
- the contra-charge is disputed;
- the main contractor disputes the plastering subcontractor's account and may either refuse to pay or seek to settle at a lower figure;
- some sort of deal is done or the subcontractor takes a 'hit' or (depending upon the amount of money at stake) the matter goes to adjudication.

6.5.3 Valuations and applications for payment

Despite the 'raw deal' that subcontractors sometimes get from main contractors, the obverse side of the coin is that payment applications are frequently disputed by the main contractor for the right reasons.

Subcontractors often start at a disadvantage when it comes to valuing their work:

- they usually have incomplete documentation;
- any drawings they may have will be well out of date;
- they may not have a copy of the standard method of measurement;
- if they had a copy they would probably not understand the measurement rules;
- they may not have a full copy of the form of subcontract and they will almost certainly not have a copy of the main contract form;
- consequently, they may not appreciate the rules for valuing variations or how the principle of valuing variations on a daywork basis works;
- they will probably have incomplete site records and most of the site instructions received from the main contractor will be verbal;
- the subcontractor is not sharp enough to ensure that all such instructions are confirmed in writing.

The large subcontractors are pretty sophisticated and can give main contractors a hard time on some of these issues but the smaller ones (the bulk of subcontractors) usually come off worst in the exchanges.

An important issue, largely misunderstood by subcontractors, is the difference between lump sum and re-measurement contracts and some subcontractors would not be able to say what type of contract they were working under. This topic is explained in Chapter 4 but the point to make here is that the distinction is crucial to the correct valuation of work in progress and to submitting an accurate payment application.

Most standard subcontracts, except notably the NEC3 subcontract, require the subcontractor to value work in progress independently and to submit an application for payment to the main contractor. A major problem for subcontractors is how to deal with varied work.

If the subcontract is a lump sum contract there is no point measuring the work physically on site because the bills of quantities is the starting point for determining value on such a contract. The subcontractor, however, will probably not have received copies of the variation instructions that the main contractor has received from the contract administrator and will, therefore, be in a dilemma as to how to value the work in progress. Any 'change orders' or instructions that the subcontractor has received will probably lack proper authentication and so the choice will be:

- re-measure the whole job – this will not be accepted by the main contractor;
- submit daywork sheets for 'out of scope' work and hope they will be signed and paid (some hope!);
- make an assessment of obvious changes to the scope of works using 'builder's quantities' – doomed to failure!

If the valuation is overvalued and overoptimistic, the main contractor will cut it to ribbons and if it is undervalued the subcontractor will be underpaid and may not recover his costs on the contract, even if he knows what they are.

The subcontractor may fare better with a re-measurement contract but it depends who does the re-measurement. If it is the Professional Quantity Surveyor (PQS), and the information is passed on by the main contractor, fair enough, but if the main contractor does the re-measure there is likely to be a less favourable outcome.

6.5.4 Liabilities, claims and accruals

Many subcontractors do not realise that any justifiable (i.e. not spurious) payment applications they make to the main contractor above and beyond the original subcontract value must be treated seriously even if, in discussions with the subcontractor, the main contractor dismisses the application out of hand. Such applications might be for variations carried out, for disruption to the subcontractor's programme of work, for contractual entitlement claims that the subcontractor feels are justified or for the costs of rectifying damage to the subcontractor's work by others and so on.

The main contractor is bound to report the true value of his contracts within his reporting system and must do so on a prudent basis in keeping with the main contractor's accounting policy and with normally accepted standards of accounting practice and corporate governance. Accordingly, the main contractor must identify the value of any such subcontract 'claim' as this creates a subcontract **liability**, which will need to be included in the main contractor's overall liability allowance for that particular subcontractor in the contract accounts. Subcontract claims could lead to adjudication, arbitration or litigation and should be taken seriously by the main contractor. As explained in Chapter 18, the main contractor's total subcontract liability is the sum of his potential financial exposure to each and every subcontractor.

A subcontract liability is also created when a 'potential' subcontract claim is in the offing, but before a formal application for payment is made, and this needs to be recognised on the basis of prudent accounting.

The difference between the total liability for a subcontractor and any payments made on account is treated as an **accrual** in the main contractor's accounts because the balance has not yet been agreed or paid.

References

CIB (Construction Industry Board) (1997) *Code of Practice for the Selection of Subcontractors*. ICE Publishing, London.

Constructing Excellence (2004) Supply chain management. Constructing Excellence, London.

Cooke, B. and Williams, P. (2009) *Construction Planning, Programming and Control*, 3rd edn. John Wiley & Sons Ltd, Chichester.

Eggleston, B. (2006) *NEC3 Engineering and Construction Contract: A Commentary*. Blackwell Publishing, Oxford.

Latham, M. (1993) Trust and Money, Interim report of the joint government/industry review of procurement and contractual management in the UK construction industry. HMSO, London.

Pryke, S. (2009) *Construction Supply Chain Management*. John Wiley & Sons Ltd, Chichester.

Winch, G.M. (2010) *Managing Construction Projects*, 2nd edn. John Wiley & Sons Ltd, Chichester.

7

Getting work

7.1 Business development

Most contractors have a marketing strategy which aims to develop a close working relationship with potential clients. This helps reduce speculative bidding which can often have little chance of success. Business development personnel network extensively with designers, clients and business groups; they scan planning applications and the local press and have a good knowledge of who is likely to build. The main aim of the business development staff is to put the contractor is a position to negotiate for a future project, be asked to complete a pre-qualification questionnaire or be added to an approved tender list.

The business development staff have a good knowledge of the budgeting cycle of their potential clients – in the public sector, such as the NHS or local authorities, this commences in April – and will use their contacts to find out which future projects are being procured and how. Once a potential client and project have been identified, the contractor may assemble an expression of interest, which will contain exemplar case studies of projects of a similar nature. Business development can be a costly process; for example, one organisation with a turnover of £12 m employs two full-time

Financial Management in Construction Contracting, First Edition. Andrew Ross and Peter Williams.

business development staff whose target is to bring in £5 m worth of tender enquiries every month.

Other sources of information are business information publications. A national company, ABI, has a staff which gathers together all the information submitted to the planning offices and sells it on to subscribers in the form of a look-up databases which can be searched by region, project size or sector.

Construction firms have a preference for particular market sectors and tend also to pursue projects which fall into a size range. As identified in Chapter 1, market sectors are very sensitive to general economic conditions and contractors have the ability to identify falling markets and move to different sectors as opportunities present themselves.

The size of projects that a contractor will undertake will be influenced by the projects available and whether the contractor's turnover has been achieved. Turnover made up of many small projects can cause difficulty as they usually require proportionally higher levels of overhead and the contractor is required to work harder to achieve the turnover target. A company's tender range will be influenced by its experience, its ability to take the risk and the opportunities within the market place. When a client draws up a list of tendering contractors, it will consider a range of factors such as:

- general experience, skill and reputation, assessing recent examples of building at the required rate of completion over a comparable contract period;
- the technical and management structure including the control of subcontractors for the type of contract envisaged;
- competence and resources in respect of statutory health and safety requirements;
- approach to quality assurance;
- environmental credentials;
- capacity at the relevant time.

Within a geographic region there will be representation from national, regional and local contractors. Each will have its preferred range of bidding. However, it will often need to tender for projects outside this range in order to meet turnover requirements. As indicated in Figure 7.1, smaller than usual contracts may be considered by larger contractors when times are tight and movement into territory usually occupied by smaller firms is often seen.

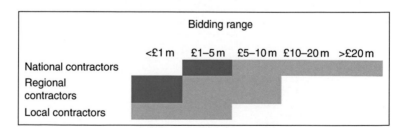

Figure 7.1 Company bidding ranges.

Competition for projects is very keen at the time of writing, it is unusual for up to ten contractors to be tendering for projects. A typical success rate of a contractor is one winning bid in eight submitted. The consequent costs of getting the tender enquiry and then securing the project are very high, and most of the time spent by the business development, estimating and planning staff ultimately ends in failure.

Recent research suggested that up to 15% of an organisation's turnover is related to securing future work. Contractors therefore have to be careful how they deploy these resources and one of the main factors informing the decision to tender or not is the approach taken by the client in procuring a price for the project.

There are obvious implications for the estimating process following the efforts made by the business development staff. Promises to return a bona fide tender are kept wherever possible to maintain the business development staff's credibility. The decision to tender is not as straightforward as it might seem and is based upon factors internal to the organisation regarding need for turnover and other factors relating to the external environment, such as maintaining relationships with clients.

7.2 Decision to tender

After the business development manager has identified a potential opportunity, the estimating manager gathers information about the project to decide upon whether the project is to be pursued. The stages the project goes through are:

- formal evaluation of opportunity phase;
- prequalification phase;
- tender.

It is important that the management team are aware of continuing projects and also has the opportunity to sign off the key stages. The risk to the organisation is often a function of the project size and, consequently, the higher the value the more senior the individual signing off the key stages. A simple table, as shown in Table 7.1, indicates the sign off of the various phases and the seniority of the director.

Table 7.1 Tender stage sign off.

Sign Off / Approval			
Sign off Stage (delete as applicable)	Formal Evaluation of Opportunity Stage	Pre-qualification Stage	Tender Stage
Prepared by:		Date:	
Business Director / E&D Director:		Date:	
Managing Director:		Date:	
Finance Director (if over £50 m):		Date:	
Chief Executive (if over £50 m):		Date:	

The amount of time available to develop the bid is critical to the company's allocation of bidding resources and potential success. Key dates of pre-qualification submissions, interviews, tender documentations due and the tender submission dates are gathered. This is used to allow the estimating manager an opportunity to review how the core resources of estimator, planner and buyer could be employed. A judgement has to be made about whether the chances of winning the project are worth deploying resources to develop the bid. An estimator also has to assess the likelihood of providing a robust bid in the time available, for example a bid that is to

Table 7.2   Bid costs and potential return.

Bid Costs and Potential Return				
Projected Turnover per year, Expected Margin	Year 1	Year 2	Year 3	Year 4
Estimated Bid Budget				
		Bid Affordability		
Return on Capital Employed				

be submitted in September may be affected by the summer shut down and the inability to receive subcontractor prices.

A wider perspective of how the potential project contributes to the company's continuing turnover and profitability targets is also required before the decision to bid is made. Typically a table, as shown in Table 7.2, will be completed indicating the turnover per annum and associated costs in bidding as well as expected margin and **return on capital employed**.

The financial position is not the only criteria used to decide whether to pursue the project. A judgement is required on the risks associated with whether the client has secured funding and is not going to default at a later stage, the procurement route and form of contract, whether any liquidated and ascertained damages are included in the contract, the extent of design and the key client drivers. Most projects are won in competition and it is important for bidding organisations to identify the number of competing organisations, who they are and whether they have a competitive advantage. This information is available from tendering subcontractors and suppliers who are common to the competing contractors.

It is important to log key information at this early stage, as communication is often key to successful bidding. The decisions made during the tender period are often taken in light of information that has been provided after the tender documents are received and it is vital that a log of where the information originated and its status is developed. At this early stage a log of key client/project team information is typically entered onto a table, as shown in Table 7.3.

Once the decision to bid has been taken a tender launch meeting is held and the key personnel are allocated responsibilities for the tender development. The key staff involved at this stage are the:

- business development director, who has a clear idea of the background to the project, the client and the competitors;
- construction director, who has to consider how the project will be built;
- the commercial manager, who can identify and assess risk and commercial opportunity;
- the bid manager, who is responsible for pulling together all the various strands of the bid in the time available;
- the managing estimator, who has an overview of the resources required for the bid development and the supply chain;
- the planning engineer, who considers sequence and timescales.

Table 7.3   Register of stakeholders.

Professional Team / Key Client Relationships			
	Company Name	Company Key Contact	Our Key Contact
Key Client Contact			
Project Manager			
PQS			
Architect			
Structural & Civil			
M&E			
Other Consultants			

A tender launch meeting is required at the early tender stage so that the tender team has clarity of what the project is, what the client's/other project stakeholders' drivers are, what the tender deliverables are and who will deliver them, the tender team roles and responsibilities and who is responsible to ensure that the bid is submitted to the client.

The agenda would focus up on the following key areas:

- project history,
- what will be contracted to build,
- risks identified so far,
- commercial review,
- project package strategy review,
- tender resources required, and
- bid strategy.

A full understanding of all these areas by the tender team will provide the company with the best opportunity to submit a winning bid, and will contribute to meeting the turnover forecast and the minimum profit target.

Usually the estimator will undertake a mid-tender review, which will focus upon progress made. It is important that tenders are tracked through the services department, as many bids are being developed concurrently.

At the mid-tender review meeting an agenda as shown below would be typical. Each section of the agenda will be reviewed and discussed by the person indicated:

1. Presentation of Project History – Business Development Manager
2. Description of Project – Construction Director
3. Review design audit, quality and clarity of information – Design Manager
4. Review consultants' scope of service, fees and responsibilities – Bid Manager
5. Review and Sign-off Winning Bid Strategy – Managing Estimator
6. Bid Resources – Construction Director
7. Bid Budget Control – Managing Estimator
8. Document Control – Bid Manager
9. Commercial review – Commercial Manager

10. Supply chain management – Managing Estimator
11. Risk and Opportunities – Bid Manager
12. Draft tender milestones and 1st weekly plan – Project Planner
13. Draft tender
14. Agree communication methods and protocol – Construction Director
15. Preparation of submission document– Bid Manager
16. Review agreed action and action owners– Bid Manager

Following the meeting the bid manager will provide a set of meeting minutes to the construction director for sign off and they will then be issued to those attending. The bid manager will use the minutes produced from the tender launch meeting and the mid-tender review to monitor actions following the meeting and take corrective action as appropriate.

Before the decision to bid is made, an initial assessment of a tender enquiry received is required, irrespective of where the enquiry originated from. The assessment will identify any unacceptable risks or obligations placed on the contractor that cannot be priced. If, after enquiries to the design team or client, the risks remain the contractor will decline to bid. There is no value in allocating limited service team resources to price projects that the company legal department will find unacceptable.

If, after this initial assessment, the decision is to bid, the service team will look at every opportunity to win the project. As price is still a determining factor in whether a bid is successful, methods of reducing price will be considered. The risks and opportunities presented by each project differ and are mostly determined by the **procurement route** chosen by the client. A design and build project places more risk onto the contractor than a traditionally procured contract with a bill of quantities and, consequently, the pre-tender team will assemble a **bid programme** which takes this into account. Each member of the tender team will have an expertise that can be used to make the bid more attractive to the client: the planner may consider alternative programmes which may reduce time-related preliminaries costs, the construction manager may look at alternative formwork approaches which maximise re-use, the commercial manager may consider the integrity of the contract and tender documentation, the estimator would look at subcontract prices and market cover risk and the buyer the supplier rates.

The commercial team identifies opportunities for making additional margin by considering what changes may occur from the 'theoretical' position expressed within the contract documents. With its knowledge of the technologies, contract procedures and programme it can carry out multiple scenario analyses and evaluate the financial impact of these prior to tendering. A theoretical amount of opportunity can be calculated which, if deducted from the net price, will potentially reduce the price submitted below its competitors. This is called **commercial opportunity** or **scope** by some organisations.

The construction project that is envisaged by the client and design team at inception is rarely the same as that which is handed over at completion. Change is an inevitable consequence of the design and construction process and can originate from any of the stakeholders to a project. It is for this reason that construction contracts have mechanisms within them to manage the consequences of change. The contract conditions refer to contract documentation which is used to specify, price and value the works. This documentation has contributions that need to be cross-referenced and coordinated from many sources: an engineer may specify structural requirements and produce structural design and details, an architect will provide outline drawings for planning purposes and tender drawings for the tender documentation, a quantity surveyor may quantify the works and produce a bill of

Table 7.4   Risks and opportunities.

Client	Design team	Documentation	Design	Suppliers	Subcontractors
Known	Reputation	Completeness	Design Stage; completeness	Volume of quantities	Specificity of subcontract
Repeat business	Ability to meet programme	Extent of provision for undesigned work	Undesigned aspects: extent of development needed	Extent of competition	Time available to re-send out
Position re negotiation	Alliances on other projects	Integrity between specification and design	Contractors design liability	Opportunities for post-contract discount	Possible future discounts

quantities. It is inevitable that errors occur during this process and they can give rise to post-tender variations. These risks and opportunities are considered in Table 7.4. The contractors commercial team will consider opportunities in the following areas:

- Design team capability: The design team is often made up of a number of different organisations which need to be able to produce working drawings in accordance with the contractors programme. It will have an established reputation, good or bad, in its ability to project manage each other and meet contractor's programmes. Often elements of design are incomplete at tender stage; the reasons for this may be budgetary or that information is not available. This is allowed for within the tender documentation issued to the contractor as defined or undefined provisional sums. If a design team cannot meet a contractor's reasonable request for information it can delay the works, which can give rise to claims for extensions of time, disruption or acceleration. If this occurs the contractor is in a position to negotiate recovery of costs and margin which will provide a margin over and above that envisaged at tender stage.
- Contract documentation: If an error is identified in the tender documentation that will require a variation to the contract once the project is underway, it will be valued using the procedures set out within the contract. If the work is of similar conditions and quantity, the valuation will use the tendered rates as a basis for agreeing a new rate; if the works differ from this a market rate would be agreed. The commercial team will re-measure the bulk quantities of the project to identify any under or overmeasure, check the descriptions against the specification and rules of measurement that apply and carry out some scenario analysis. This may identify areas of undermeasurement, which when the contract is re-measured will be based on tender rates; if the tender rates are loaded this will provide a margin contribution for this part of the works in excess of the overall rate. As identified by Rooke et al. (2004) a minor amendment at tender stage can have a large impact once the work is being undertaken. The contractor would, of course, carry the risk of not generating this additional margin if the work which had been priced with the loaded rates was omitted and that the re-measure led to quantities lower than tendered.
- Work which has been incorrectly described or measured within the tender documentation. Often the design team does not have the time to check the impact of developing design on the tender documentation, which can end up being based on superseded specifications and drawings. If an item is incorrectly measured, this

can lead to a variation, which can lead to an opportunity to recover additional margin. Seemingly innocuous errors in classification of excavation depths or not taking account of excavations below ground water or incorrect sizing of voids can all lead to considerable claims.

The supplier and subcontractors are assessed for their potential to offer discounts on the quotations received at tender stage that may be offered if the job was secured. The use of e-auction for supply chains is considered in Chapter 14, section 14.7. This assessment is informed by the specificity of the subcontractor or how 'tied in' it is to the project. The source of this specificity may be the technology offered by the subcontractor, its geographic location or the extent of competition for the works. It would be unlikely that a contractor would attempt to 'muscle' additional post-tender discount from a mechanical and electrical subcontractor that they had a back-to-back arrangement with on a design and build project. The same contractor may, in contrast, anticipate additional discounts from painting subcontractors that have less specificity to the project.

The result of these considerations are summarised by the commercial manager into a strategy for margin recovery and a total. The amount of commercial opportunity taken at tender stage is usually as a result of discussion between the tender committee members. There is a risk that an overly bullish approach can place unachievable targets on the construction team and also sour relationships with clients, design teams and suppliers alike.

7.3 Competitive tendering

Competitive bidding in construction is used by a range of procurement routes to satisfy the client's needs in ensuring value for money is achieved via a market price. Codes of tendering practice exist that prescribe a series of processes that should be followed by client advisor and tendering contractor in order to ensure compliant bids are received. Clients would ideally like to get the best bid by maximising competitive intensity. Contractors, on the other hand, would like to make the best return from their resources. It is important for clients to act responsibly in their procurement practice; it is questionable whether it is ethical to invite more than six contractors to bid for work under traditional contracts and more than four using design and build. Unfortunately, some private sector clients use the opportunity of a falling market to send out many enquiries in the hope of securing a cheap bid. Contractors are, of course, selective over which job to bid for and generally select from a range of opportunities that come their way.

To assist in achieving a competitive bid, contracting organisations have attempted to develop long-term strategic relationships with subcontracting organisations. The benefits of this approach are obvious to both parties; contractors that have a smaller number of suppliers are likely to be more responsive to requests for quotations at bidding stages, the prices received are more likely to be more trustworthy and the subcontract organisation undertaking the work is likely to do so to a higher quality and on time. The benefits to the subcontract organisation are that less time is spent bidding speculatively in the hope of future success and a better working relationship develops, which should improve payment time and understanding of the needs of the subcontractor. Supply chain management has been a term that has been used by many in manufacturing and there have been lots of calls commencing from Latham (1994) about more integration of subcontractors into the project team at an earlier stage.

Many contracting organisations make claims to undertake supply chain management. However, the realities of the competitive market in construction mean that few do so to anywhere near the level of maturity seen in other industries. The approach adopted by the leading exponents of construction supply chain management is to have competition from a smaller number of subcontractors which have been drawn from a **select list**. The list is often maintained by the buying department within the company which gathers information on the post-contract performance of the subcontractor. Information on quality of work, responsiveness to change, ability to meet programme, health and safety requirements and commercial approaches is collected and analysed to classify the subcontractor. Some observers have suggested that the benefits are skewed towards the main contractor as they are the market maker and the cost of compliance with the schemes can lie with the subcontractor.

The estimating department sends out enquiries to subcontractors which have been selected from the contractor's preferred list. The enquiries will, as far as possible, include all the relevant main contract information, such as contract conditions, specification and bills of quantity, and give a return date and also specify a period for which the subcontract tender will be open. The enquiry will be tracked by the estimator to ensure it has been received and that a bona fide price will be received. The subcontractors often raise queries relating to their work packages which relate to the design or tender documentation. These queries are generally passed to the design team or, if unanswered, may form a qualification to the main contractor's price.

It is essential that the estimator is able to assess the integrity of the quotation received from the subcontractor, as it will form a structural part of the estimate. If the subcontractor was to withdraw its quotation after a project had been secured it could have a big impact upon the potential projects profitably. In order to do this a comparison is made with competitor quotations to identify discrepancies and to ensure the contractor is not exposed in this way. It would be a risky position for a contractor to submit a bid on the basis of one quotation received from an unknown subcontractor. It is for this reason that one of the important skills of the estimator is market understanding and relationship building. Without the input from subcontractors in the form of quotations it would be impossible for the main contractor to price and secure work. Often quotations will be received at the very last minute before the tender is due and after the tender committee has decided on a bid. In this case, if the quotation has a material impact upon the price, the estimator will adjust the tender.

The two-stage bidding approach adopted by contractors with their subcontractors can make it difficult for the estimator to convince subcontractors to price work at tender stage. Many subcontracts are let to subcontractors who have submitted a lower price as a result of a post-contract send out. This results in the subcontractor used at tender stage having to stand the costs of bidding only to provide the main contractor with a price and it would be justifiably reluctant to continue doing so in future.

7.4 Tender lists

There are generally three approaches clients can use to establishing a list of contractors to tender for work:

- open competition
- frameworks and approved lists
- *ad hoc* list.

7.4.1 Open competition

There is no restriction to the number and type of organisations tendering. The approach clients generally use for open competition is to advertise in local, trade and national publications inviting contractors to request tender documents. The risk to the contractor is an unawareness of the number of competitors tendering and, consequently, the chances of winning the tender are incalculable.

As the costs of tendering are high and they are generally borne by the winning tender, the client can end up paying a premium and may also find that the more reputable contractors are uninterested in tendering. The bona fide contractor may find itself tendering against competitors who may use commercial means during the post-contract stage to recoup costs that have not been included initially. The approach does have some benefits to the client; these are that all tenders should be genuine as there is no obligation to tender, there is no restrictive list so projects are open to new contractors and there can be no sense of favouritism.

7.4.2 Frameworks and approved lists

Frameworks can be considered as the development and maintenance of approved lists of contractors which have been through a pre-selection process. The pre-selection process has been refined to include health and safety record, financial standing, experience, turnover. Often repeat build clients categorise contractors in order to match them against project criteria such as size, complexity and location. The approved lists are then used to run 'mini competitions'. These are sometimes referred to as **frameworks**.

An approach to extending the project partnering ethos espoused in Latham and Egan was developed principally by the public sector and entitled frameworks. It allowed for a single project partnering approach to be broadened to include wider portfolios of projects. The overarching benefit to clients was that it would reduce the price they paid for their projects by increasing the levels of trust between the parties, giving an indication of future workload and, consequently, reducing the **'transaction cost'**. Contractors pre-qualify for the framework, using an approach which seeks to capture information about their expertise to undertake projects; these are normally categorised by value and sector to avoid the criticism of 'one size fits all'. A framework is typically limited to four years duration. One of the difficulties contractors have faced in the process of getting onto frameworks is the variety of Pre-Qualification Questionnaires (PQQs) used by the different client bodies. This has been addressed in the latest Government Strategy for Construction (May 2011), published by the Cabinet Office, by the use of a standard form.

One of the disadvantages of a framework is that there is no binding commitment from the client to use the services of a contractor. Often a great deal of cost is incurred by a contractor in pre-qualifying only to find that there is little chance of future work, this has also been addressed in the strategy whereby future projects are to be published in advance. Also, the process of evaluation of the PQQs has been variable and some contractors have found it difficult to gain access to projects that are sponsored by clients they have worked satisfactorily for in the past.

Even with these disadvantages, contractors are major beneficiaries, as there should be a greater continuity of workload and greater certainty during the tender process. In addition, teams can be retained from one project to another which should reduce the transaction costs. It should also enable relationships that seek to develop value within projects to be developed between project stakeholders.

7.4.3 *Ad hoc* list

Clients can use a similar approach to the approved lists above with the intention of procuring a contractor for a single project. Contractors can be invited to pre-qualify through notices in the press or the ***Official Journal of the European Union*** (**OJEU**) if the project is a public sector project over £4 m.

The purpose of the European Union (EU) procurement rules was to open up the public procurement market and to ensure the free movement of supplies, services and works within the European Union. In most cases the procurement rules require that contracts are let via a competitive process. The rules reflect and reinforce the value for money (vfm) focus of the government's procurement policy. This requires that all public procurement must be based on value for money, defined as 'the optimum combination of whole-life cost and quality to meet the user's requirement', which should be achieved through competition, unless there are compelling reasons to the contrary. To make bidders aware of opportunities, contracts covered by the Regulations must be the subject of a call for competition by publishing a Contract Notice in the OJEU. Use of the Standard Forms for OJEU notices (e.g. a Contract Notice or a Contract Award Notice) is mandatory.

There are four procedures that can be used; these are open, restricted, competitive dialogue and negotiated dialogue:

The **open procedure**, under which all those interested may respond to the advertisement in the OJEU by tendering for the contract.

The **restricted procedure**, under which a selection is made of those that respond to the advertisement and only these are invited to submit a tender for the contract. This allows buyers to avoid having to deal with an overwhelmingly large number of tenders.

The **competitive dialogue procedure**. Following an OJEU Contract Notice and a selection process, the client then enters into dialogue with potential bidders, to develop one or more suitable solutions for its requirements and on which chosen bidders will be invited to tender.

The **negotiated procedure**, under which a purchaser may select one or more potential bidders with whom to negotiate the terms of the contract.

Under the restricted, competitive dialogue and competitive negotiated procedures (those where a call for competition is required by advertising in the OJEU), there must be a sufficient number of participants selected to proceed to the tender stage to ensure genuine competition. The Regulations require a minimum of five for the restricted procedure, and three for the competitive dialogue and negotiated procedures.

There are three procurement stages – specification, selection and award:

Specification stage. This determines how requirements must be specified. The Regulations make it clear that authorities may use performance specifications rather than technical specifications. They also provide clarification on the scope to reflect environmental issues in specifications. Guidance on technical specifica-tions is available on the UK Office of Government and Commerce (OGC) website.

Selection stage. There are restrictions on the use of post-tender negotiation under the open and restricted procedures. The European Commission has issued a state-ment on post-tender negotiations in which it specifically rules out any negotiation on price.

Award stage. The award of contract is either on the basis of 'lowest price' or various criteria for determining which is the 'most economically advantageous tender (MEAT)' to the purchaser. Government policy is to use the latter criterion, as this is consistent with the obligation to achieve value for money

Table 7.5 gives an indicative set of headings for the information sent by a client to a contractor in accordance with the EU rules to inform them of an *ad hoc* project. Indicative timescales are also shown.

As can be seen from the timescale shown above for a typical two stage selective tender approach, the time for appointment of a contractor is ten months.

Table 7.5 Pro forma for an *ad hoc* project.

Purchasing Authority	
General: Nature of Contract:	Appointment of Principal Contractor
Award Procedure	Restricted Procedure
1. Background to Project:	
The Proposed New Facility.	The Design and Novation Process The Project Team Division into lots

Indicative Timescales:	Return of completed PQQ submissions
	   3.08.2012 PQQ Evaluation 28.08.2012
	Short listing 11.09.2012
	Issue 1st Stage Tender Documents 12.10.2012
	Return 1st Stage Tender Documents
	   18.12.2012
	Secure planning permission 15.01.2013
	1st Stage Tender Evaluation 22.01.2013
	Select preferred Contractors 22.01.2013
	Issue 2nd Stage Tender Documents
	   25.01.2013
	Return of 2nd Stage Tenders 05.03.2013
	Appoint Contractor 23.04.2013
	Start on site 24.05.2013
	Completion 25.05.2015

Deadline and content of Pre-Qualification Questionnaire (PQQ)	The selection criteria will be addressed under the following headings.
Selection Criteria: The purpose of this PQQ is to allow the objective selection of a Tender list based upon experience, track record, technical and financial capability, and empathy with the Client's requirements	General Information and Good Standing 15% Technical Capacity & Ability 60% Financial Standing 25%

7.5 E-bidding and reverse auctions

7.5.1 Auctions

The term auction brings to mind an auctioneer with a gavel pointing to a lucky bidder and shouting sold. Auctions have been in existence for years and are used extensively to buy and sell anything from antiques to fish. There are four types of auction:

(i) Dutch
(ii) English
(iii) sealed first price
(iv) sealed second price.

The usual approach to English auctions is that the auctioneer calls for bids that are over a specified reserve and the bidders submit bids over a period of time until a maximum bid is achieved whereby no-one wants to bid more and then the bid is accepted. A lot of readers will have bought something this way on the Internet, web-based auction sites such as E-bay were set up in 1995. The price paid reflects the value ascribed by the winning bidder.

A Dutch auction works somewhat differently. The auctioneer starts with a high price which is progressively lowered until a bidder is prepared to accept the offered price. This practice has dogged the construction industry for many years.

Where bidders submit one bid which is concealed from others, this is called a sealed bid first price auction. The bids are compared and the highest bid wins. There are a number of variations to this in that the number of potential bidders is either known or unknown, the price of the winning bid may be announced or simply shared between the winner and seller. It is not usual for negotiation to occur after the auction.

A variant of this is a sealed bid second price auction which is identical to that described above except that the winning bidder pays the second highest bid rather than their own. This approach is rarely used.

7.5.2 Online and reverse auctions

Online auctions are used a great deal in other industries and are becoming more popular in the construction industry.

In **reverse auctions** competitors seek to win the opportunity to provide goods or services at a minimum price. There has been a critical debate within the industries that use this approach. The detractors suggest that it is undermines the seller and buyer relationships and increases the transaction costs on the bidders. The supporters suggest that the technology should be used to make the bidding quicker and open up opportunities to wider competition. A recent example used by a bank with four contractors gave each contractor 30 minutes to review their prices two weeks after submission, each having had feedback regarding its respective position. The lowest bidder trimmed 12% of its bid and was perceived as buying the project by other bidders.

Research in other industries has shown that shorter cycle times from bid invitation to quotation receipt can lead to improved responsiveness. It has had success in the purchase of commodities but there has been little evidence of its success for the purchase of relatively complex services such as construction. The tender documentation clearly identifies the work to be undertaken, the contractual terms, timing, location and specification of the work to be undertaken. This is exactly the same as would be the case for a normal send out; however, the opportunity for multiple prices is allowed.

7.5.3 The process

The software is generally provided by a third party which hosts the auction. The contractor invites subcontractors to take part in the bidding process. They are informed of the closing date of the bid, the details of the work to be undertaken and the number of bidders in the auction. The details of the bidders are kept confidential from each other.

Once the auction commences the bidders can submit as many bids as they wish during the auction period, these are submitted online. Each bid is compared with competitors and the bidder is informed as to its relative position at regular periods. Some systems have graphical displays which show the relative movements in bidder's prices during the auction period.

7.5.4 Advantages

It has many advantages, such as:

- it can standardise the bidding process;
- bidders can be monitored and the number of bona fide bids increased;
- it can increase the competitiveness amongst tendering contractors;
- it can widen participation to unknown subcontractors;
- it makes comparison of bids simple and reduces the costs of sending out enquiries and chasing up quotations;
- the largest benefit to the tendering subcontractor is that is allows it to amend its bid.

7.5.5 Disadvantages

The disadvantages are:

- tender period too short, giving too little time for bidders to review their prices;
- it can provide the opportunity for bidders to commence with bids which are too high;
- a fictitious contractor could be introduced, shill bidding, to artificially reduce prices and increase competition.

An example of e-bidding and reverse auctions for the procurement of subcontractors by a main contractor is given is Chapter 14, section 14.7.

References

Latham, M. (1994) Constructing the team: Joint review of the procurement and contractual arrangements in the UK Construction Industry. HMSO, London.
Rooke, J. Seymour, D. and Fellows, R. (2004) Planning for claims: an ethnography of Industry Culture. *Construction Management and Economics*, **22**, 655-662.

8

Corporate governance and management

8.1 Definitions

8.1.1 Corporate Governance

'Corporate Governance' was firstly and authoritatively defined in the Cadbury Committee Report of 1992, paragraph 2.5:

> *Corporate governance is the system by which companies are directed and controlled. Boards of directors are responsible for the governance of their companies. The shareholders' role in governance is to appoint the directors and the auditors and to satisfy themselves that an appropriate governance structure is in place. The responsibilities of the board include setting the company's strategic aims, providing the leadership to put them into effect, supervising the management of the business and reporting to shareholders on their stewardship. The board's actions are subject to laws, regulations and the shareholders in general meeting.*

Corporate governance therefore concerns:

- the activities and actions of the board of directors of a company;
- the system of values set by the directors on behalf of the company;
- the structure, strategy, leadership and supervision necessary for effective and prudent management of the business.

8.1.2 Management

'**Management**' is separate and distinct from corporate governance and concerns the day to day operational running of a company by its full-time executives or managers in pursuance of the objectives of the business.

After their corporate governance role, company directors wear a second 'hat' in performing executive management roles and are usually found in positions of executive responsibility, such as finance director, contracts director, human resources director and so on.

8.1.3 Directors

The Companies Act 2006 requires that a private company must have at least one director and a public company must have at least two.

There are basically two sorts of directors – **executive** directors and **non-executive** directors – and they each play different but nonetheless important roles. Not all companies have non-executive directors but all firms of a meaningful size do have them. Sole traders – the smallest of small firms – often have a 'sort of' non-executive director, probably without realising it! A small plumbing and heating firm, run by the husband, will often benefit from the guiding hand of the wife who has no role in the day to day affairs of the business.

Executive directors

Executive directors have a dual role:

- Stewardship:
  - ○ looking after the business on behalf of the shareholders

- compliance with statutory duties under the Companies Act 2006:
  - to promote the success of the business
  - to exercise reasonable care, skill and diligence
- ensuring that key information is submitted to Companies House
- ensuring that the company produces a set of annual accounts that present a true and fair view of the business.
- Day to day management:
  - Setting budgets
  - Managing resources
  - Ensuring legal compliance:
    - H + S legislation
    - Environmental legislation and so on
  - Product and service delivery
  - Financial management.

Non-executive directors

The non-executive director takes an overview and acts as a 'whistle blower' by ensuring adherence to good practice, respect for the interests of shareholders and other stakeholders and ensures that boardroom matters are conducted correctly. It is a supporting principle of the UK Corporate Governance Code 2010 that non-executive directors are responsible for setting appropriate levels of remuneration for the executive directors (Code: Section A4 refers).

Whilst non-executive directors have a role that is less 'hands on', they bring the benefit of objectivity and external awareness to the board. Non-executive directors are not usually involved in day to day management but, in smaller the companies, they may carry out some 'hands-on' work. Non-executive directors are bound by the same legal duties and responsibilities as the executive directors and act independently in matters of corporate governance.

8.2 The UK Corporate Governance Code

8.2.1 Application

The **UK Corporate Governance Code 2010** ('the Code') provides a set of best practice guidelines for good corporate governance and is the successor to the first Code produced by the Cadbury Committee in 1992. Strictly speaking, the Code only applies to big firms and requires public listed companies to 'comply or explain' by:

(a) disclosing how they have complied with the Code, and
(b) explaining where they have not applied the Code.

Whilst the main focus of the Code is on companies listed on the London Stock Exchange, private companies can and do conform but without the obligation to make any disclosure of conformity in the annual accounts. A simple Google search can reveal several examples of large private construction companies which obviously take corporate governance seriously and apply the principles of the Code but make no mention of the Code in their accounts.

8.2.2 Principles

The underlying principles of the Code are:

- Accountability
- Transparency
- Probity
- Long-term sustainable success.

8.2.3 Other approaches to governance

Notwithstanding the provisions of the Code, companies not required to comply normally have their own approaches to corporate governance. Matters such as the company's approach to risk and uncertainty may be explained in a corporate governance statement within the annual accounts, wherein it is also usual to find a definition of the principal risks and uncertainties facing the company.

With regard to issues relevant to this book, the corporate governance statement may refer to the company's risk management strategy concerning:

- **Procurement risk**, including the company's supply chain strategy, the development of long-term strategic partnering agreements with clients, suppliers and subcontractors and the use of key performance indicators.
- **Credit risk** and customer credit worthiness validation policy and procedures to ensure that deferred payment terms are not granted where there is a risk that a customer may default on a contract payment or fail to discharge an obligation to pay a debt resulting in financial loss to the company.
- **Liquidity risk** and company policy with regard to liquidity risk mitigation, such that the company is able to avoid encountering problems in meeting its financial obligations through the effective management of cash generation from its operations.
- **Cash flow risk** management, so that inflows and outflows of cash and cash equivalents are sufficient to finance the day to day operations of the company and its ability to meet its liabilities as and when they fall due.
- **Creditor risk**, including the company's policy regarding payment terms for suppliers, for example that suppliers will be paid 45 days after the end of the month of the date of the invoice.

The accounts may also include a statement concerning directors' responsibilities. These will include:

- Preparation of the annual report and financial statements as required by company law.
- Selection of suitable accounting standards to be adopted within the accounts and ensuring that they are applied consistently.
- Ensuring that the accounts give a true and fair view of the state of affairs of the company and of the profit or loss made by the company in the relevant period.
- Making judgements and accounting estimates that are reasonable and prudent.
- Preparing the financial statements on a 'going concern' basis assuming business continuity.

8.2.4 Corporate governance and contracts

Making the link between directors' responsibilities for corporate governance and their management functions within the company revolves around two issues alluded to in the Cadbury Report definition of 'corporate governance':

1. the system of values set by the directors on behalf of the company;
2. the effective and prudent management of the business.

The first issue concerns matters of openness and transparency in the accounts, such that they reflect a true and fair view of the business, and the second relates to the robustness of the company's reporting system, thus ensuring that the accounts are not overstated so as to obscure the true picture to stakeholders. These are matters that largely rest on the set of values adopted by directors, and inculcated in management and employees, and touch upon issues raised in the Preface to this book.

The significant parts of the annual accounts that relate directly to the reporting of the financial performance of contracts are the statements made by directors regarding:

- turnover
- profit
- debtors
- the valuation of work in progress.

8.3 Turnover

8.3.1 Definition

The turnover figure stated in the annual accounts is the total value of construction work done during the year exclusive of VAT.

The Companies Act 1985 requires that the accounts must state the turnover for the financial year and that which is attributable to substantially different classes of business. Therefore, where a contractor carries out project management work, provides construction services, engages in facilities management and is involved in infrastructure investments, such as Public Private Partnership (PPP) concessions, power stations and so on, a segmental analysis of performance by sector activity will be found in the annual accounts.

The method by which turnover is calculated will depend upon the accounting conventions of the company which should be applied consistently throughout the organisation.

8.3.2 Calculating turnover

The obvious route to take when calculating turnover is to take the total invoiced value of sales to customers as the turnover figure. In contracting this would be the total of all payment applications made on the various contracts carried out in one year's trading. This approach, however, fails to recognise that construction contracts do not conveniently finish at the point the accounts are struck and that even small contractors and subcontractors will have continuing contracts spanning more than one accounting period. It also fails to recognise that a significant part of the turnover of a contracting business is represented by work in progress. A further frailty of this approach is that payment applications and valuations are notoriously inaccurate for a variety of reasons and using them as the basis for the turnover figure would not represent a 'true and fair view'.

Barrett (1992) suggests that *the sum declared by the contractor in his accounts as the value of work done may differ considerably from the amount claimed in interim applications, the amount certified, or the cash received.* This statement recognises the extent of the adjustment necessary to interim valuations in order to arrive at 'true value'.

Consequently, the calculation of the true turnover figure involves complexities such as the valuation of work in progress and the recognition of profit on long-term contracts. Normally, turnover and profit on short-term contracts are recognised when the contracts have been completed and, if not complete, they are valued at the lower of cost and the net realisable value of the work done, exclusive of trade discounts. Long-term contracts should be reflected in the accounts by recording turnover and related costs as contract activity progresses (SSAP9, paragraph 28) and turnover should be ascertained according to the stage of completion of the contract. Profit from long-term contracts is not taken into the accounts until there is a reasonable expectation that such profits may materialise.

Determining the turnover figure for the accounts, therefore, is not straightforward and inspection of a sample of company accounts will reveal that the definition of 'turnover' varies according to the accounting policy of the company in question. Turnover and profit recognition may be expressed in the accounts as:

- a fair value for the consideration received or receivable for the services provided, or
- the net invoiced sales of goods and services (after trade discounts), or
- the turnover and profit of completed short-term contracts and the value of work done on long-term contracts, or
- some other definition.

Some companies include monies received and work invoiced and certified, but not yet received, together with work in progress (i.e. work done but not invoiced) in the turnover figure and others do not. Some accounts state turnover as the net amount of monies received on contracts but do not include monies 'in the pipeline' whether certified or not. Turnover is always stated in the accounts as net of VAT.

8.3.3 Work in progress

Accounting practice (SSAP9) recognises that turnover should be *'ascertained in a manner appropriate to the industry'* and that this can be done *'by reference to valuation of the work carried out to date'*. Accounting standards also suggest that *'companies should ascertain turnover in a manner appropriate to the stage of completion of the contracts ...'*.

There are potential complexities in determining the annual turnover in contracting companies. This is because work in progress may have been under or overvalued or there may be unresolved claims on a contract or variations which have been paid for 'on account' but not accurately valued or agreed.

Also, the rates and prices stated by the contractor in the tender bills of quantities are often manipulated to take advantage of potential claims and variations and to 'earn' profit in advance to aid positive cash flow. Therefore, the 'value' of work done on a contract may not be a true value upon which to base an assessment of turnover.

8.3.4 Cost of turnover

The cost of turnover is reported in the annual accounts and can be found in the profit and loss account. It is found directly after the 'top line' figure for annual turnover and before the figure for administrative costs. Interest payments on loans and so on are stated later on in the profit and loss account as a separate item.

Administrative costs include items such as rent for office premises, company cars, marketing costs, wages, telephone, postage, stationery and so on and are normally called 'overheads'. The cost of financing the business is also an overhead and this is shown separately in the accounts as 'interest payments'. By definition, therefore, the cost of turnover is the direct cost of production, which is made up the **cost of purchase** of raw materials (such as concrete, bricks, roofing materials etc.) plus the **cost of conversion** (e.g. labour, plant, subcontractors). Additionally, SSAP9 identifies that there are various categories of overhead that should be considered when distinguishing between the cost of turnover and administrative expenses:

1. overheads as functions of **production**, for example those site overheads which are part of the construction costs and, therefore, should be included in the cost of turnover;
2. overheads associated with **administration**, which can be directly attributed to the site, and consequently the cost of turnover, which include insurances, head office safety advice and visiting head office staff such as quantity surveyors and contracts managers;
3. overheads associated with **marketing**, **selling** and **administration**, which would be generally regarded as administrative expenses and, therefore, not part of the cost of turnover.

Effectively, the cost of sales is the direct cost of generating the stated revenues, including raw materials and goods, direct labour costs and so on, and administration costs are the indirect costs associated with producing annual turnover, including marketing and other overheads.

8.4 Profit

8.4.1 Definition

Profit may be defined as the financial benefit that is realised when the revenue from a business activity exceeds the costs, expenses, and taxes expended in completing the activity. In a construction context this might be explained as the money left over at the end of a project or at the end of the financial year after paying all costs and overheads. When a loss is made the Americans usually call this 'negative profit'!

Contractors undertake construction projects in the expectation that a profit will be made, as this is the way to grow a healthy and viable business. When a contractor tenders for work it is normal practice to add a percentage for profit based on the return required by the owner/directors in order to:

- pay an attractive **dividend** to the owner/shareholders;
- make an acceptable return on the **capital employed** in the business (compared, say, to investing the money in stocks and shares or in a building society account);
- provide money to buy new company cars, vans and wagons, plant and equipment, computer equipment and software or a new fish tank for the managing director's office!

8.4.2 Corporate profit

Unsurprisingly, profit is stated in the profit and loss account 'bit' of the annual accounts. This represents profit made in the relevant accounting period (usually covering one year's trading) ending at the accounting reference date.

The word 'profit' appears three times in the profit and loss account:

- Operating profit
  - this is profit on turnover after deducting production costs and general overheads;
  - items also deducted before declaring the operating profit include depreciation of fixed assets, hire of machinery and equipment and auditors fees and expenses;
  - it indicates how well the company has managed its resources and overheads.
- Profit before taxation
  - this is the operating profit less the cost of interest payments on loans and so on;
  - it represents 'pure' profit after accounting for payment of all costs, overheads and interest;
  - part of this profit will be paid out in tax and the remainder will belong to the shareholders.
- Profit after taxation
  - this is the remaining profit available for distribution to shareholders;
  - A proportion of this will be paid to shareholders as dividends less tax on the dividend payment and some will be retained in the business.

From an accounting point of view, it is important to be prudent regarding the expectation of profit. Profit must not be included in the annual financial report and accounts unless it has been earned or there is a reasonable expectation that it will be earned. This is because contingent liabilities or unknown future costs may erode future profits, which may turn out to be less than expected. The accounts must give a true and fair view of the business at the time they are struck.

However, on long-term contracts, failure to recognise profit until the end of the contract could distort the accounts and lead to non-compliance with the 'true and fair view' principle. Therefore, profit can be recognised if a positive balance results from the matching of costs to date with turnover. When the outcome of long-term contracts cannot be predicted with reasonable confidence, however, no profit should be taken.

8.4.3 Project profit

There is an old saying in the financial world that *a profit is not a profit until it is banked!* The same is true with construction contracts. Management reports throughout the project may suggest that the contract is 'in profit' but it is not until the final account is settled that the contractor can be sure that a profit has been made by comparing all the costs and revenues.

At tender stage the amount of margin represented by profit will be dictated by several factors, including the market conditions at the time, how badly the contractor wants the work and how he sees the risks attached to the project. However, the simple act of adding a profit percentage to the estimated cost does not guarantee that a profit will be made and loss-making contracts are not uncommon in the contracting world.

Losses can be made for all sorts of reasons, including failure to control preliminaries expenditure, failure to identify and control risks, underpayment for variations carried out, site management inefficiency, the insolvency of subcontractors, failure to control materials waste and so on. Making a 'loss' means that not only has no profit been made but also that head office overheads, site preliminaries or even the direct costs of production may not have been fully recovered.

A profitable contract is not a function of the tender profit margin but depends on many factors, including:

- lower costs of production than was anticipated at tender stage due to higher outputs or efficiency gains on site;
- quicker or more efficient construction methods compared to the estimator's allowances;
- faster completion of the contract, thereby saving on time-related preliminaries;
- better discounts or buying margins on materials and subcontractors;
- additional revenue generated by extra work ordered by the customer/client during the contract, such as variations or expenditure of provisional sums;
- better than expected settlement of contractual claims.

8.4.4 Profit distribution

Tax

Companies pay corporation tax on their profits at various tax rates depending upon the turnover of the company. The tax is levied after all allowable costs and expenses have been deducted from annual turnover but before the distribution of dividend to shareholders. Corporate tax is a complicated subject beyond the scope of this book.

Retained profits

At the annual general meeting, the directors will propose a dividend and the shareholders will vote as to how much profit should be distributed as a dividend and how much shall be retained in the business.

The portion of profits retained in the company will be used to replenish assets, including the purchase of new construction plant and equipment such as excavators, cranes, wagons and scaffolding.

Retained profits belong to shareholders as they effectively own the business.

Shareholders

Shareholders provide the equity capital in a business as opposed to loan capital which has to be repaid to the lender. The equity capital is provided in return for shares and, depending upon the class of share, voting rights in the business. Shareholders are generally interested in the long term and they invest their money with the prospect that the company will grow and pay a dividend well into the future.

When profit is earned shareholders are paid a dividend per share owned. When losses are made a dividend, or reduced dividend, may still be paid according to how the directors view the prospects for the company. Income tax is deducted by the company at source and shareholders are paid net of tax.

Equity capital is also known as 'risk capital' because shareholders' money is unsecured, unlike loan capital which may be secured against the assets of the company. Shareholders stand to lose all their investment if the company fails because they will be 'last in line' to be paid should the business be liquidated.

8.5 Long-term contracts

Special consideration needs to be given when declaring turnover and profit on long-term projects because of the length of time taken to complete such contracts. Whilst profit cannot be declared prematurely, deferring the declaration of turnover and profit in the accounts until completion of long-term contracts may result in the profit and loss account not representing a fair view of the results of the company. It may, therefore, be appropriate to take credit for ascertainable turnover and profit while contracts are in progress and guidance on the approach to this is provided in SSAP9, which explains how stocks and long-term contracts should be treated in the accounts.

The determination of profit for an accounting period involves the correct allocation of costs to the period in question and also the application of the correct revenue and margin recognition policies which, in turn, require forecasts to be made of the outcomes of long-term contracts. Consequently, assessments and judgements need to be made concerning the recovery of pre-contract costs, changes in scope of work, outcomes of contract programmes, defects liabilities and changes in costs.

These are management functions which are the ultimate responsibility of the directors, who may have to estimate the value of long-term contracts so that attributable profits are not recognised until the point at which the outcome of the contract can be assessed with reasonable certainty. On the other hand, provision should be made for losses on all long-term contracts as soon as such losses become apparent.

8.6 Management accounts

8.6.1 Control

In the same way that annual accounts are meant to give outsiders a true and fair view of the affairs of the company, so management accounts perform the same function internally.

Managers need to keep a close eye on how the business is doing financially; this is true irrespective of the size of the company. Small companies admittedly employ simplified procedures and larger companies are much more complex from a financial management point of view. The object of financial management is, nevertheless, a simple one and can be summed up in one word – **control**.

Financial information helps managers to see what is going on in the business and to take action if things are not going to plan. Forecasts (or budgets) are made for the trading year to predict turnover, profit, overheads and other types of expenditure and these budgets need to be compared to actual figures. This is the essence of management control. The control process needs to be done regularly (at least monthly) throughout the trading year so that appropriate action can be taken to redress problems; for this management needs to be supplied with accounting information.

At a basic level this information will include details of:

- revenue (income) and expenditure (costs);
- how much money the company owes (creditors);
- how much is owed to the company (debtors);
- the cash flow situation;
- costs and revenues on individual contracts.

Larger companies will also produce a forecast balance sheet and profit and loss account and publically quoted companies will make statements to the Stock Exchange when the current financial position is likely to affect forecast turnover and profit for the year end.

8.6.2 Cost value reconciliation

Contractors naturally expect to make a profit on their activities but it may be difficult to decide how much profit has been made if contracts are incomplete or not concluded (for example, where final accounts have not been settled). It is also difficult to decide on the amount of turnover when contractual claims, extras and unagreed amounts are included in the contractor's applications for payment.

For work in progress to be included in the balance sheet, with profit attributed, the expected profit has to be justified (not just hoped for) and any potential future losses need to be recognised. Therefore, to give a true and fair view of the contractor's activities, and to respect the need for prudence in the accounts, it is necessary to consider the issues which could distort the figures. This is achieved through the process of **cost value reconciliation** which enables the true value of work done to be married with the true costs of that work (attributable costs). When the figures have been properly reconciled, then the true position can be reported.

To be able to report profit for a particular accounting year, SSAP9 explains that costs must be matched with related revenues. To do this, work in progress has to be measured and valued and the costs of doing that work determined. The site quantity surveyor normally prepares the valuation and the accounts department provides information relating to the costs. The slight complication in all this is that the valuation needs to be a 'true valuation', thus overmeasured or undermeasured work has to be accounted for and any distortions in the valuation caused by the contractor's tendering strategy and pricing of the bills of quantities have to be allowed for.

Once the 'true' value is established, this can be compared to the cost records and the following conditions can then be considered:

1. value exceeds cost
2. value equals cost
3. value is less than cost.

Unless the contract in question has been completed, it cannot be said that these conditions represent a profit, a breakeven or a loss but, if the annual accounts are struck at this point in time, the work in progress must be given a value for the balance sheet, because it is an asset. The work in progress must also be given a figure for the profit and loss account as it is part of turnover.

SSAP9 says that, if future revenue is likely to be less than the cost already incurred, then the irrecoverable cost should be charged to revenue and also that stock (i.e. work in progress) must be stated at cost or net realisable value where value is less than cost as in condition 3 above.

In SSAP9, the comparison of cost and net realisable value has to be made for each item of stock (work in progress) but they can be aggregated where appropriate. In construction, this means reporting the position on the overall contract and not individual site operations.

Chapter 18 develops this important topic in much more detail.

8.7   Accounting for contracts

8.7.1   The role of directors

As explained earlier in this chapter, company directors have two roles:

1. corporate governance
2. management

The first role of directors is to ensure that the annual accounts present a true and fair view of the business at the time the accounts are struck. Directors have many duties and responsibilities but, in relation to the company's portfolio of contracts and the performance of their corporate governance role, the directors must report **turnover**, **profit**, **debtors** and **work in progress**. This needs to be done in such a way that the accounts are prepared prudently and in accordance with generally accepted accounting practices and those of the company.

In their second role, directors must ensure that the company is managed effectively so that its objectives are achieved in the interests of all stakeholders – shareholders, customers, employees and so on. With regard to the contracts that the company has, this requires a reporting system that provides directors with current and accurate information so that they may take decisions where variances with the budget are highlighted and where contract profitability may be threatened. Consequently, and preferably every month, directors need to know:

● the contract profit/loss situation;
● the value of outstanding work;
● the projected cost to complete;

- the contract cash position;
- the extent of contractual claims and their chance of success;
- the level of contingent liabilities including defective work and subcontractor claims.

It is crucial to understand that the profit reported by directors in the annual accounts is not in any way, shape or form anything to do with the monthly profit forecast on contracts. The profit/loss situation on contracts IS NOT real profit – it is simply a forecast – whereas the profit stated in the annual accounts IS real profit and it is distributed to tax, dividends and retained profits as explained earlier in this chapter.

In their corporate governance role, directors must also ensure that debtors and work in progress are reported correctly in the annual accounts. Both items appear as current assets in the balance sheet because debtors represent money certified and owed to the company and work in progress represents work done, but not invoiced, that nevertheless has a value to the company.

The reporting of profit, debtors and work in progress is undeniably complex and largely the province of those qualified and competent to deal with such matters. We trust, therefore, that accountants will forgive any oversimplification we may be guilty of in trying to make the subject clearer to non-accountants.

8.7.2 Work in progress

One difficulty that contractors have in reporting their financial position is that they will probably have several contracts in operation at any one time and there may well be completed contracts and further work in the pipeline not yet started. This means that, in any one accounting reference period, there may be:

- contracts started and completed;
- contracts started which will not be completed until a later accounting reference period;
- contracts completed but started in a previous accounting reference period;
- completed contracts for which the final account has yet to be concluded.

This situation was illustrated in Figure 2.1.

As a consequence of such complexities, contractors will have continuing work at the end of the accounting reference period which has to be valued for the annual accounting process. This **work in progress** appears in the company balance sheet as 'stocks' or, under European accounting standards, as 'inventories'.

Barrett (1992) advises that a company should make a clear statement in its accounts defining how work in progress has been valued. Therefore, as part of their corporate governance responsibilities, directors need to decide what to say about this issue.

There are various approaches that may be adopted for short-term contracts, one example being that:

- work in progress is valued at cost less any provision for foreseeable losses;
- 'cost' comprises direct expenditure with an appropriate proportion of production overheads;

- progress payments certified and receivable by the year end are deducted from work in progress balances; and,
- if progress payments exceed work in progress balances, the net amount is included in current liabilities in the annual accounts as payments on account.

Another approach referred to by Barrett (1992) and explained in SSAP9, is that stocks (work in progress) are valued at the lower of cost and net realisable value. Net realisable value represents the estimated amount at which stock could be sold after allowing for the cost of completion and future losses. In a contracting context, net realisable value would be the contractor's net allowances in the bills of quantities less future losses and contingent liabilities such as defects in the work.

8.7.3 Short-term contracts

SSAP9 makes no reference to short-term contracts, preferring the phrase 'contracts other than long-term contracts'. Practically speaking there is no difference between the two.

Contracts that commence and reach completion in a particular accounting period are normally considered to be short-term contracts. Considering the lengthy defects correction periods in construction contracts – normally six months minimum and commonly 12 months – short-term contracts will have fairly short contract periods and are likely to be relatively small projects, fast-track high-intensity projects or subcontracts.

Short-term contracts are normally accounted for at cost or net realisable value, unless the contract has been concluded whereupon turnover and profit may be entered in the accounts. It must be said that different companies take different approaches to this issue and that SSAP9 is neither obligatory nor universally applied.

If a contract with a duration of less than one year is not concluded in the accounting period, SSAP9 requires that the accounts state the position as if it were a long-term contract, provided that the size or importance of the contract is such that the accounts would be distorted by not reporting turnover and attributable profit.

8.7.4 Long-term contracts

Long-term contracts are defined in SSAP9 as those where the 'contract activity' falls into different accounting periods. Due to the time taken to complete long-term contracts, SSAP9 says that deferring the reporting of turnover and profit until such projects are complete may distort the accounts. Consequently, where turnover can be ascertained it should be reported in the accounts and where profit can be assessed with reasonable certainty, attributable profit should be reported 'on a prudent basis'. Where the situation is less certain, SSAP9 states that no profit should be taken to the profit and loss account and reported turnover achieved may be apportioned to the overall contract value 'using a zero estimate of profit'.

Looking at some published accounts, directors seem to have taken the view that long-term contracts are included in the balance sheet at the value of turnover less the value of progress payments certified and receivable. Should the turnover figure exceed the total of progress payments, the net balance is included in debtors as amounts recoverable on contracts and where progress payments exceed turnover the net balance is included in current liabilities as payments on account.

Long-term contract work in progress is given by costs not yet taken to the profit and loss account less related foreseeable losses and payments on account.

8.7.5 Worked examples

Example 1

Figure 8.1 illustrates the valuation, certification, payment and work in progress situation on a contract of four-months duration with a defects correction period of six months. The contract starts at the end of month 1 of the accounting year and finishes on time. The defects period finishes at the end of month 11 of the accounting year. The contract details require:

- monthly interim certificates;
- interim valuations prior to each interim certificate when required by the architect;
- final certificate two months following certificate of making good of defects;
- final payment 28 days from final certificate.

Figure 8.1 shows that:

- work in progress for the first three weeks or so of the contract is valued in valuation 1 and then certified at the end of the month - payment is made 14 days later;
- work in progress at valuation 2 is the work carried out between valuations 1 and 2;

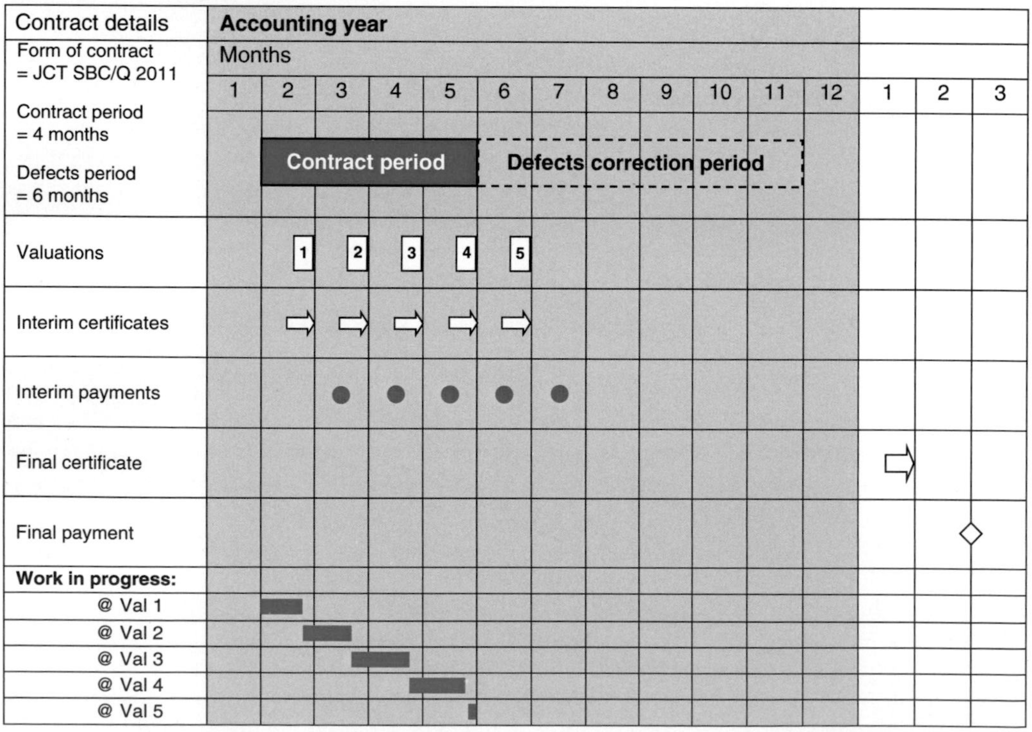

Contract details	Accounting year														
Form of contract = JCT SBC/Q 2011	Months														
	1	2	3	4	5	6	7	8	9	10	11	12	1	2	3
Contract period = 4 months Defects period = 6 months		Contract period				Defects correction period									
Valuations		1	2	3	4	5									
Interim certificates		⇨	⇨	⇨	⇨	⇨									
Interim payments			●	●	●	●	●								
Final certificate													⇨		
Final payment														◇	
Work in progress:															
@ Val 1		▬													
@ Val 2			▬												
@ Val 3				▬											
@ Val 4					▬										
@ Val 5						▪									

Figure 8.1   Work in progress – Example 1.

- work in progress at valuations 3 and 4 is the work carried out between valuations 2 and 3 and 3 and 4, respectively;
- work in progress at valuation 5 is the work carried out between valuation 4 and practical completion of the contract;
- the final certificate is issued two months after the end of the defects period and (importantly) one month into a new accounting period;
- final payment is made 28 days later, that is two months into the new accounting period.

The fact that final certification takes place in a new accounting period effectively makes this a long-term contract, which has implications for:

- the value of work in progress stated in the annual accounts;
- the recognition of profit made on the contract in the annual accounts.

Had the final payment been made before the end of the accounting year then this would have been treated as a completed short-term contract with no question that any profit/loss would not be taken into the annual accounts in full.
Consider the situation at the end of the accounting year:

- there is no work in progress because all the work has been completed;
- there are no debtors because all interim certificates have been honoured and paid to the contractor;
- strictly speaking, there is no profit because profit should not be taken on long-term contracts without reasonable certainty that it will be earned;
- however, there may be residual monies to come when the final account is agreed and the directors have to decide whether a profit should be reported on the contract (if there is one, of course!) at this point;
- because the final certificate is due only one month after the end of the accounting period, SSAP9 would probably say that the outcome of the contract may be assessed with reasonable certainty and, therefore, profit (if any) may be taken into the accounts – the provisos would be (a) that relations with the architect/ professional quantity surveyor (PQS) were good and (b) there were no significant disputed items in the final account application;
- turnover cannot be reported as the final account figure because this has yet to be agreed and, therefore, a reasonable estimate will have to be made for the profit and loss account.

Example 2

Figure 8.2 shows the same contract but with a different set of circumstances:

- the contract has not reached practical completion by the end of the accounting period
- the defects correction period has not started.

Consider the situation at the end of the accounting year:

- there is some work in progress, that is the work done in the last week of the accounting year, which will be later included in valuation 4;

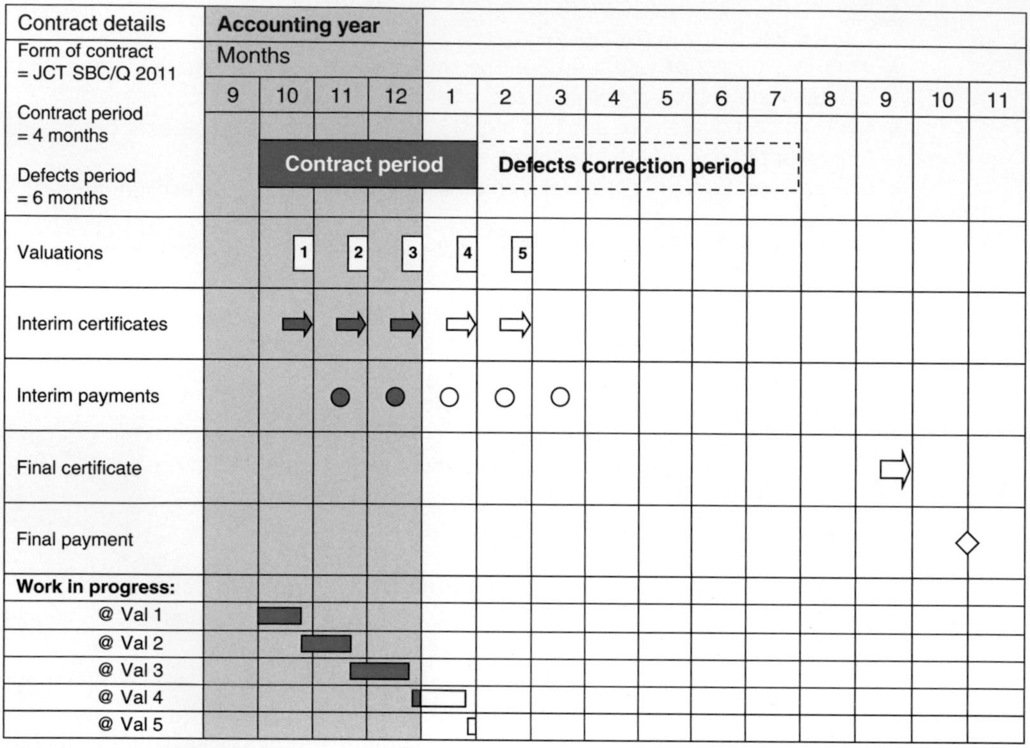

Contract details	Accounting year														
Form of contract = JCT SBC/Q 2011	Months														
Contract period = 4 months	9	10	11	12	1	2	3	4	5	6	7	8	9	10	11
Defects period = 6 months		Contract period				Defects correction period									
Valuations		1	2	3	4	5									
Interim certificates		⇨	⇨	⇨	⇨	⇨									
Interim payments		●	●	○	○	○									
Final certificate													⇨		
Final payment													◇		
Work in progress:															
@ Val 1		▬													
@ Val 2			▬												
@ Val 3				▬											
@ Val 4					▭										
@ Val 5															

Figure 8.2 Work in progress – Example 2.

- there are debtors because the work in progress at valuation 3 has been valued and certified but not yet paid;
- there are no further debtors because interim certificates 1 and 2 have been honoured and paid to the contractor;
- as there is still outstanding work on the contract and the final account is a long way in the future, there is no profit as far as the annual accounts are concerned because profit should not be taken on long-term contracts without reasonable certainty that it will be earned;
- a reasonable estimate will have to be made for turnover in the profit and loss account, which should be reported on the basis of the net value of work done on the contract but taking into account any under or overvalue, monies in advance, contractual claims and contingent liabilities that there may be.

Reference

Barrett, F.R. (1992) *Cost value reconciliation*, 2nd edn. The Chartered Institute of Building, Ascot, UK.

9

Company structure

9.1 Management functions

Construction projects are technically difficult and require the expertise of professionals who are skilled in the technological, logistical and contractual aspects of everything that makes up a project. On first reflection, it may seem that the staff numbers in construction companies are high. However, the reader should reflect upon the complexity and bespoke nature of construction projects. Although the employment churn rate within the industry is high when economic times are buoyant, most of the movement is from one company to another where personnel seek promotion within their discipline. Construction companies rely greatly upon the experience of their staff and it is unusual for senior staff to be drawn from other industries. It is also rare for staff to move from the commercial department of a company to a construction department. Each department has an individual culture and way of doing things which is adopted by new members of staff, and whilst the success of a project requires a high level of interdependency there is also a healthy rivalry.

Financial Management in Construction Contracting, First Edition. Andrew Ross and Peter Williams.
© 2013 Andrew Ross and Peter Williams. Published 2013 by John Wiley & Sons, Ltd.

A company may be owned by shareholders who select directors to look after their interests; the directors will ensure the key management functions are provided by capable staff supported by the relevant information system. The company governance management function is fulfilled by the directors, they are generally lead by a managing director who will be supported by a financial director. Most construction companies will have commercial and construction director roles which provide the link between the strategic and operational management of the company.

The key management functions are:

- Getting work – this requires a business development department whose aim is to get the company onto tender lists and obtain tender opportunities.
- Winning work – The pricing of the tenders requires estimators and planners. Their role is to win work, usually in competition with other companies, and to provide the financial budget and programme by which a margin will be made.
- Construction of projects requires construction managers, engineers and a buying department. Their role is to ensure the project is completed to the satisfaction of the client and to procure the resources within the budgets set by the estimator.
- Getting paid – ensuring that the contractor gets paid and pays its creditors is undertaken by quantity surveyors (QS). The quantity surveyor's role is to ensure the contractor gets the entitlement due by valuing the works in progress, and by financially managing the supply chain within the budget the quantity surveyor will monitor and report the margin set by the estimator.
- Reporting costs, paying creditors and chasing debtor – this function is carried out by the accounts department.

9.1.1 Principles

Each management function is undertaken by specialists who are organised into departments and supported by specialised information systems. Normally, the department will have a director or head who is responsible for the performance of the department and reports upwards to a strategic management board. The head of the department will be involved in the recruitment, appraisal and allocation of staff within the functional area. In areas such as accounts, the workload is relatively constant when compared to areas such as estimating, where workload can fluctuate enormously.

Staff are recruited on the basis of specialism and a simple search on the internet for estimating, planning and quantity surveying jobs will indicate how the roles are differentiated. This differentiation is generally consistent between construction organisations; an estimator's role in one firm is likely to be similar in another. The only significant difference may be the information system that is used to support this function.

Staff often belong to different professional bodies; construction managers are encouraged to become members of the **Chartered Institute of Building** (CIOB) and quantity surveyors aspire to become corporate members of the **Royal Institution of Chartered Surveyors** (RICS). Technical staff are generally educated to degree level in construction management, civil engineering and quantity surveying. No undergraduate degrees exist which are solely for planners or estimators. The extent of commonality between these degree programmes is relatively small and that which exists has been driven by efficiency drives through modularisation rather than by educational ideology. The degrees are accredited by professional bodies such as the RICS, ICE (**The Institution of Civil Engineers**) and CIOB, which monitor both the programmes and higher education institutions for content and quality. These professional bodies require members to demonstrate their competence against a range of headings which reflect the respective skills that are required.

9.1.2 Estimating and tendering

The structure of a company is influenced by the project nature of its business. Separate departments exist to 'service' the company's requirements; the estimating, planning and buying departments are often referred to as core service departments. They provide a pan company service which is cross- functional and is required before a project is awarded. The staff who work in these departments are head office based, form part of a general overhead and are not usually allocated to projects.

The estimating department, planning and buying departments report to a services director, who is responsible for ensuring that bids for future work are responded to and that site requirements for planning and materials supply are met.

Before the decision to bid is made, a commercial report will have been undertaken. This will identify the major project risks and the directors will use their judgement whether the likelihood of winning the job is high. Typically the 'strike rate' is one winning bid to six submissions. As the cost of pricing the losing bids has to be absorbed by the company and the estimating department has a limited capacity in how many bids it can undertake, it is important for the contractor to assess all the risks and opportunities before making the bid decision. Recent research suggests that for small companies a strike rate is 1 in 8 and that it takes 100 man hours to submit a bid. The cost of winning work can be as much as 4%, which is significant in an industry where tender margins are only 1-2%. Once the decision has been made the estimator will develop a programme for the estimate. This will identify the deadlines for the contributions by others involved in the estimate:

- A buyer will identify the major suppliers of materials and provide the estimator with a schedule of purchase prices. Usually the company will have a strategic alliance with suppliers for popularly used material. If the specification requires materials to be sourced from a supplier outside of the company's normal pool of suppliers, the buyer will send out for quotations.
- The contracts manager may provide a method statement outlining how the project is to be completed, the major activities that are to be undertaken and the construction risks. Companies use this early involvement of the construction staff in tenders to ensure that there is a continuity of responsibility from tender to project if it is won.
- A commercial manager will provide a tender report indicating the risks and opportunities presented by the design, specification and documentation for the project. This will inform the discussion on the margin to be added at tender stage. As with the case above, the project, if won, would be managed by one of the commercial manager's project surveyors and the responsibility for the commercial decisions at tender stage would be retained.
- A planner will compile a contract programme which will reflect commencement dates and completion dates and will be used to develop a price for the preliminaries or general expense for the project.

- Project quantity surveyors may be called upon to take off quantities for elements of the work which require pricing. This is often the case for design and build or drawings and specification projects. These tend to be builders quantities and do not strictly comply with the rules of measurement. They are compiled with specifications into work-package tender enquiries and sent out to subcontractors (who may not return a price if it requires too much time).
- An estimating assistant will support the estimator and is often allocated to a number of estimates. A great deal of time is spent sending out enquiries, chasing subcontractors for promised quotations, completing subcontract quotation comparison sheets and undertaking general administration such as scanning documentation.

A project estimating process is supported by software specifically designed for the purpose of producing an estimate, tender and allowance bill. Most estimating software systems are based on a number of relational databases which allow for project specific libraries of items to be developed. The software supports the development of work packages of measured items and corresponding specifications to be sent by email to selected trade subcontract organisations. It supports a reminder function that automatically reminds the subcontractors to return their price by the required deadline. Recent developments allow for integration with contractors categorised supply chain subcontractors and allow for automatic updates about forthcoming tenders to be sent to prospective subcontract organisations.

The estimating system also allows for margin to be distributed around the tender prior to submission in order to tactically manage the prospective projects cash flow.

9.1.3 Purchasing

The purchasing or buying function makes an important contribution to the profitability of a company. A buyer is usually based in a head office and the proportion of buyers to other service staff, such as estimators, tends to be low. A buyer has a detailed knowledge about the local materials suppliers prices and the scarcity or otherwise of typical resources. The department is managed by a head buyer who will represent the department at tender committee meeting. Most organisations will have arrangements with large suppliers for stock prices for common items, the buyer will have historical records of how much has been used in the past and negotiate new prices as the market rises or falls. The stock prices will be used by the estimator when compiling the estimate and also by the buyer when placing the contract order.

The buyer may undertake bulk material quantities checks against the drawings to get an idea of the amount needed and the potential discount supplier may give in the future. Materials can form up to 40% of a contractor's cost and mistakes at tender stage can easily wipe out margins once orders are placed.

The buyer will schedule all the major materials quotations used by the estimator in production of the estimate. This schedule, combined with knowledge of the intensity of the competition for supply of materials, will inform the tender committee about potential future discounts that may be available if the job is won. These discounts may make the difference between winning or losing the contract.

One the job has been won, the site will complete a materials requisition schedule indicating the quantity, location and date for the key materials. The buyer will have a copy of the project specification and will be responsible for placing an order for the specified materials to arrive at the site at the correct time and for a price within the allowance agreed at the tender stage. The buyer has a good knowledge of latest innovations in materials which are developed by close contact with the market. This knowledge is passed to the estimator or site team who may discuss alternative specifications to those originally considered with the design team and client. This knowledge can also provide a strategic advantage to a contractor working in a competitive market.

In some companies, the buyer is responsible for production of material cost accruals. This is a logical connection of functions with the accounts department, as the buyer will have a record of the materials requisitions, the materials received sheets and the order rates. It is an onerous task, however, and can tie up expensive knowledgeable resources in undertaking accounting administration work.

An important function undertaken by the buying department is the production of buying reports. As discussed in Chapter 18, it is essential that the site team disentangle buying losses and gains from those connected with the production process. If they are not disaggregated, problems in production may be masked by financial gains made via the procurement process.

9.1.4 Production

There is a hierarchy to a company structure; the construction department is headed by a construction director who reports to the managing director. The construction director takes overall responsibility for the ensuring that projects are resourced and completed to the satisfaction of the company's clients. The construction director usually has a background in site management, which is invaluable in assessing the technical aspects of a project, considering whether a projects progress is satisfactory and also the performance of the construction staff.

Contracts managers also have a background in site management and are responsible for the construction aspect of a number of projects. They support the site managers in their role, often attend in project site meetings to provide advice or make decisions that are beyond the remit of the site manager.

A site manager is project based, usually has a qualification in construction management or civil engineering. The site manager's experience is wide ranging, from technical problem solving, which encompasses everything from setting out buildings to managing design teams input, to quality assuring subcontract work.

9.1.5 Quantity surveying

The commercial department is staffed by quantity surveyors; they have usually completed a degree in quantity surveying or commercial management and once they have decided upon a career as a contractor's surveyor rarely transfer employment to a client quantity surveying organisation. It is also rare for a quantity surveyor who works for a client to transfer to work for a contractor although a lot of the skills are similar. A commercial director would head the commercial department; this post requires someone who has had experience in identifying and managing contractual and commercial risks. These risks can arise from both up the supply chain from the client or from down the supply chain from the sub-contractors and suppliers. In a similar fashion to that of the contracts director, the role involves managing subordinate surveyors who are allocated to projects. One of the main roles of the commercial director is to assess and report on the aggregate financial position of the company's projects and report this information to the managing director.

A managing surveyor will be responsible for a number of projects and fulfils a similar role to a contracts manager. The role is generally one of scanning the project environment for future risks that may put the margin in jeopardy. If a risk is spotted that requires senior level intervention the managing surveyor will discuss it with the project surveyor and intervene where required. The interventions may be attendance at a meeting to resolve a problem, writing a contractual notice or allocating more resources. The managing surveyor is responsible for the recruitment, appraisal and allocation of project surveyors.

The project surveyor's role can be considered to have three facets. One facet is client orientated and involves ensuring that the contractor has been paid appropriately for the works completed; another facet is procuring and paying the labour and subcontractors who are undertaking the works; the last facet is internally orientated and involves completion of company systems that are needed to report the financial progress of the projects.

9.1.6 Supply chain management

Supply chain management is a term used a great deal in other industries to describe continuing relationships with subcontractors and suppliers. To have a truly effective supply chain, buying companies need to be in a position to offer selling companies continuity of business. This continuity of business allows the selling company to save on its transactions with the buying business and allows it to enter into long-term relationships with its own suppliers which should bring about cost saving benefits. Construction companies have great difficulty in offering continuity of business to their subcontractors and suppliers, as their clients rarely have a steady supply of work. This is not the case in all sectors, public housing repairs and maintenance, a sector which is significant in terms of value and whose clients are committed to long-term contracts, has the pre-conditions that support supply chain management. Organisations such as Fusion, which acts for clients and contractors, have achieved high levels of savings by offering aggregated demand from a number of clients to single suppliers of, say, kitchens or bathrooms.

Some of the larger contractors would state that they undertake supply chain management. However, usually they use a form of category management and use competition between pre-qualified subcontractors to achieve the best price. Some subcontractors complain, with some justification, that the contractor achieves more benefit than they do by their membership of the category management system, in that the subcontractor is guaranteed the opportunity to price all the enquiries that have a requirement for a particular trade but, however, they do so in competition and only win the work if they can beat other subcontractors within their category. As it is in the interest of the main contractor to only keep market competitive subcontractors in their categories they are often in competition with the most competitive!

Constructionline is an example of how some of the costs of assessing subcontractors have been reduced by centralising the pre-qualification process. The Constructionline system uses eight criteria for registration and, ultimately, assessment. It is becoming a major part of supply chain management quality assurance, particularly for public sector works.

A company's centralised supply chain management function is usually managed by the buying department. A list of preferred suppliers which will be called upon to undertake the work as required is kept.

9.1.7 Accounting

The accounting function is central to the financial management of the organisations and is often much maligned by construction project staff for some of the errors it makes and for the time it takes to make promised payments to subcontractors and suppliers. The accounts department is usually within the management remit of a financial director who will have an accounting rather than a construction background. It will have well documented systems for recording and categorising costs, payment of creditors, the chasing of debtors for payments and producing reports for audit purposes. It will produce regular strategic and operational management reports that can be drilled into for deeper analysis when required. Most medium sized organisations will employ a financial director who will have an accounting rather than a construction background. It is the responsibility of the financial director to ensure that the company has systems in place for the recording and reporting of costs.

The accounting information system will have pre-defined inputs and is central to how the company request information from site and from its suppliers. Most companies use an 'off-the-shelf' system, such as Jobmaster or Summit, and adapt their forms to provide information to enter into the system. The investment in the system is significant, as it not only includes the hardware and software required to run the system but also the training for the accounts staff as well as the site staff who are required to understand how the system records and reports costs.

Most systems provide **sales and purchase ledgers** systems which include contract and subcontract and payroll ledgers. They allow for the reporting at company level, contract level or at the level of individual subcontractors or suppliers. The heart of the system is a number of relational databases supported by a coding system that has the ability to be extended for future contracts, suppliers and subcontracts.

The site manager and quantity surveyor input information that supports the costing of the project, such as **subcontract liabilities**, materials received sheets, plant returns and labour returns. The buying department supplier inputs order details. The senior management provides details regarding overhead allocation. The accounts department function is to process all this information and provide monthly project cost reports. These are used by the project staff to reconcile the project's value and it is important that they can be checked. Often an unexpected discrepancy between the cost and value will require the site team to drill down into the accounting system to get to the root of a problem. The cost system will allow multiple slices of information to be retrieved and analysed to identify where a problem may lie.

9.2 Organisation structures

The structure of a company in influenced by the type of work it undertakes, its history, the risks within its business environment(s) and its size. It can be organised by function, by regional area, by specialised divisions or use a combination of all three. A company such as Carillion is structured around the sectors it works within – rail, civil engineering, building, defence and so on are different divisions within the company. Other structural arrangements are influenced by the geographic concentration of a company's activities, for example Shepherd Construction has three regional divisions – east, west and south.

In the case of large firms, a combination of geographic region and sector may be used to subdivide the business into semi-autonomous management units. Often the sectorial divisions are linked to future work prospects; it is advantageous for both marketing purposes as well as internal efficiency to gather together the expertise in a company into a specialised sector. Kier Construction, for example, has a geographic division as well as a large projects division. Size, as measured by turnover, is the most influential factor that influences how a company is structured. As described in Chapter 1, the majority of the businesses that make up the construction industry directly employ fewer than ten staff and would have more simplified structures and systems than are described in this book. A detailed examination of how the external business environment influences how a company structures its operations is beyond the scope of this text. However, to provide the reader with the organisational context for financial management, an example of how a typical medium size construction company with a turnover of approximately £40 m is organised is given below.

9.2.1 Structure

The group consists of four separate companies that are semi-autonomous; it uses group-wide systems and resources for efficiency and standardisation. Its radius of operation is about 100 miles, The companies that make up the group, with the last annual turnover indicated in brackets, are:

- construction (£40 m)
- small works (£8 m)
- property services (£7.5 m)
- homes (£8 m)

The cross-group services, which are shared by all the companies, are accounts, human resources, health and safety , administration and IT. Figure 9.1 indicates the group structure.

Each company reports separately and is headed by a managing director who takes overall responsibility for the company's operations and represents it at the monthly group board meetings.

The company has 41 employees; the numbers of staff in each department are given in brackets in Figure 9.1. The company is structured in a typical fashion to others of its size. The construction and commercial function are managed separately by directors who are responsible for the recruitment, appraisal and allocation of staff to the company's projects. Although these two departments are highly interdependent, the culture and working practices within each one is significantly different.

The company's managing director oversees all the company's activities and ultimately takes responsibility for everything that happens within the company. The position is an onerous one and requires a broad knowledge of all aspects of the construction business, everything from contract conditions to employment problems are all likely to require decisions. The managing director reports upwards at group monthly board meetings and also has monthly meetings with the divisional directors of the company.

The managing director has a strategic role in planning for the year ahead and reporting on the year just gone. The company above has a year end which falls at the end of September. The planning for the annual report and future commences in July. At this stage the managing director will compile a future workload forecast which is categorised into continuing and future. The turnover budgets for continuing projects are relatively straightforward to calculate; each project will have a turnover projection

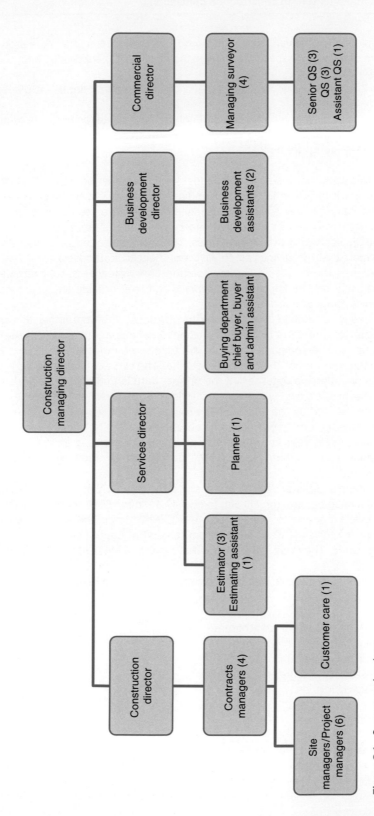

Figure 9.1 Company structure.

to completion which is updated each month. Future work is a little more problematic, as some of the work is speculative and in the pipeline. It is subdivided into:

- definite (projects secured and agreed commencement date) – 100% taken to future budget;
- probable (projects being tendered/negotiated) – 70% taken to future budget;
- possible (projects identified as strong leads by the business development staff) – 20% taken to future budget.

It is at this point that the overhead budget would be reviewed. At the time of writing the industry is in economic recession and overhead budgets are being trimmed. The managing director has some difficult decisions to make regarding where to trim. Experienced staff can be expensive and can be seen as an easy saving to make. However, junior staff can make mistakes. The cost consequences of these mistakes can be more that the savings made. Also, experienced staff have a great deal of tacit knowledge of the company systems, relationships with clients, subcontractors and suppliers that can be impossible to replace and may provide an advantage to the company's competitors.

The budget is refined over the months of August and September and will form the basis for the targets for the business development director. The annual production of the company accounts can take three months. For a company with a financial year end of the end of September, the cycle will begin in July with the process of gathering together all the project data is undertaken, a close scrutiny of all the project cost value reconciliations is undertaken to ensure that the picture presented by the site team is an accurate one. As will be discussed later, many surveyors tend to protect some autonomy by 'hiding' money in accruals and it is important at year end that these practices are exposed as the accounts the cost value reconciliations support will be subject to audit by an independent external body.

The financial director reports the costs to the company at the end of September and the managing director will aim to give an audited set of accounts ready for the group board meeting by the second week in November. The auditors will be with the company for a six week period from the end of September until the accounts are completed. They visit the company full time for a block of two weeks at the end of October.

The managing director will also attend monthly group board meetings, a typical agenda would be:

- opportunities and bids in progress;
- management accounts;
- performance and bonds;
- safety quality and the environment;
- strategic priorities;
- staff;
- any other business.

The managing director assigns two days of every month to overview the progress of the company's projects. A meeting is arranged to consider each project. The managing director chairs the meeting and in attendance are the commercial director, contracts director, the site manager, the managing quantity surveyor and the project surveyor.

Typically, it takes approximately forty minutes to assess the project; the construction manager commences by reviewing the progress and programme to completion, the quantity surveyor will explain the value and cost provisions and the construction and commercial directors highlight areas for management action. Often ratios of a projects' progress to value and estimated final value, costs and margin are gathered at this meeting. These are summarised by the managing director for onward reporting at the monthly group meeting.

9.2.2 SMEs

The earlier description about company structures applies to medium sized firms with a turnover of £10–50 m per annum. Most organisations in construction fall into the category of small or micro sized enterprises. The definition of an SME was expanded in 2005 to include medium <250 employees, small <50 employees and micro <10 employees. Organisations of this size do not have the turnover to warrant structures of this nature. Often a micro business will rely on one or two individuals to undertake everything from business development, pricing the work, procuring and paying for the resources, invoicing and chasing for payment. In some cases they even undertake the work themselves as well as managing the business.

In this case a lot of use if made of external service providers, such as accountants or freelance quantity surveyors, to fulfil the essential functions of getting paid and completion of the statutory annual returns to Companies House.

9.2.3 Large firms

The regional nature of the demand for construction services means that large firms are often structured with regional divisions. Regional clients prefer to place their business with organisations that have an established reputation within their region and are 'known'. Equally, a lot of subcontract and supply organisations are of an insufficient size to operate at anything other than a regional level. As a lot of work is subcontracted, it makes good business sense to employ staff with a good knowledge of the region. Consequently, large firms will have regional offices for marketing and business operations purposes. A region may have a regional director, contracts and commercial director; however, this would depend upon the size of the region. They will report up the line to area and board levels as required.

The cross-company services, such as accounts and HR, are generally centralised and provide support for the different regional offices. This allows the regional offices to minimise their overhead requirement whilst maintaining a presence within a region. Most of the regional overheads are chargeable to projects and will be project-related staff.

9.2.4 Very large firms

The very large firms are relatively few; the top five in the United Kingdom are Balfour Beatty, Carillion, Laing O'Rourke and Morgan Sindall. They all have specialised subsidiaries which are organised to support the business or to win work within a competitive market. For example, Balfour Beatty Building Services, the expanded piling company, Heyrod Construction, Lovell partnerships. They are often structured along sectorial lines with regional offices.

Most subsidiaries are operated as virtually autonomous business units and make independent decisions regarding their structure, organisation and information systems. Although the model discussed above of the separation of the commercial function from that of construction appears to be replicated in subsidiaries which deal with building services, such as piling. The group company will dictate dates and formats for reporting turnover, profit and cash that the subsidiary company has to comply with.

10

Service departments

Financial Management in Construction Contracting, First Edition. Andrew Ross and Peter Williams.
© 2013 Andrew Ross and Peter Williams. Published 2013 by John Wiley & Sons, Ltd.

10.1 Estimating and tendering

Construction projects are designed and constructed by temporary multi-organisational teams (Cherns and Bryant, 1984) that are usually geographically distant and functionally disparate (Eccles, 1981; Murray *et al.*, 1999). The formation and inter-organisational dynamics of these teams is a function of their formation and structure, and early decisions made can have a significant influence over the performance of such a team.

One of the most critical tasks of the contractor pre-construction professionals, such as the estimator, buyer and planner, is to design project organisational structures that are suitable for the project environment (Moore, 2002). These include internal resources as well as subcontract and supplier resources. The organisational[1] structure has to take account of the technology of the project, the procurement arrangements of the clients, the internal environment of the contractor and the supply organisations involved in the project (Hughes, 1989; Shirazi *et al.*, 1996; Murray *et al.*, 1999). The type of tender documentation used to gather price data, the judgement of the estimator, the content and format of the price information received from the supply chain and the contractor's general management costs are all synthesised into an informational model that is used for estimating, tendering and the economic organisation of project during the post contract phases.

The development of the temporary multi-disciplinary organisation that will build the project is unique to each bidding contractor and goes through a series of stages, as shown in Figure 10.1. Each bidder has established preferred subcontractors it uses for quotations and each project has unique features, which means that some subcontractors are more likely to eventually undertake the work package than others. This can be referred to as specificity and can be influenced by project related factors (e.g. ease of replacement of a particular technology), subcontract factors (e.g. the number of subcontractors available) and main contract factors (such as the extent of interdependency required by the project).

The importance of organisation of the estimating process is stressed in the industry guides (CIB, 1987, 1989a, 1989b, 1993, 1997) and this textbook does not aim to replicate this advice. It is important for the reader to understand how the estimator and planner's contributions influence the post-contract control of a project.

The process of estimating has changed over the years. Whereas it was once considered as primarily unit rate build-ups in response to bill of quantities items, it has now moved to one of coordination and synthesis of numerous subcontract quotations and coordination of design contributions(Abdel-Razek and McCaffer, 1987). The management of the subcontractor input into the estimate and construction programme is central to the company's chances of securing the work. The responsiveness of the subcontractor to tender enquiries is also a key factor and will be related to the perceived specificity to both the contractor and the project.

Consequently, the role that price information plays within a project structure is an essential one and is used for formation of contracts, coordination of parties inputs and outputs to processes and also the economic control of subcontract organisations should misalignment occur. Thus, the estimating process can be considered as

---

[1] The term organisational has two different meanings in this context; it relates to the process of organising (the way the work is arranged and allocated amongst the group so that the goals of the organisation are achieved) and the type of organisation structure, which relates to how responsibility is divided.

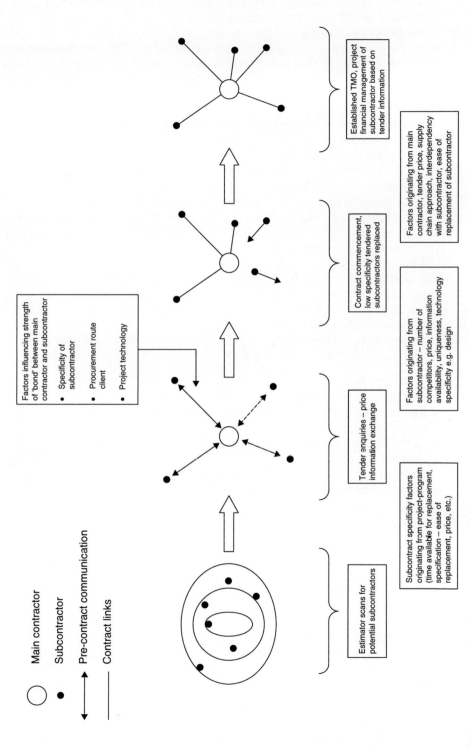

Figure 10.1 Establishment of a temporary multi-organisation.

a system of structured information flows which can be considered to influence an organisational design (Murray *et al.*, 2000).

The development of the project estimate is a critical process for any construction company, as it is used as the basis for the tender, establishes the budget allowances for the various resources that are to be used and provides the essential information required to predict the turnover that the potential project will produce if won. It is interesting to reflect that whilst being a critical process which takes deterministic, fixed data such as quantities, time periods and resource levels, the role of the judgement of the estimator is central to developing a credible estimate. Statistical techniques of regression analysis to identify factors that influence production have been applied in manufacturing very successfully; techniques such as parametric estimating, which purport to be able to clearly identify the parameters of output rates given certain conditions, have not been popularly applied to the construction estimating process.

This chapter refers to the Chartered Institute of Building (CIOB) code of estimating practice as the basic framework for the development of the estimate. However, it discusses the reality of estimating as a part of a commercial process which involves the judgement of a range of disciplines.

An organisation has to maintain turnover year on year. This is assessed by using a **turnover budget** which takes account of current projects that are making a contribution to turnover in the financial year, those projects which have been secured and are to commence in the financial year and an assessment of works that are likely to be allocated to the organisation.

The approach virtually all organisations take to the development of an estimate is a combination of a set of basic functions:

● decision to tender;
● development of a tender timetable;
● development of enquiry documentation to suppliers and subcontractors;
● analysis of quotations received;
● identification of any temporary works needed;
● development of production rates;
● pricing the preliminaries;
● conversion of the estimate into a tender.

10.1.1 Preparing the estimate

As can be seen above, the context of a tender varies enormously and it is the skill of the contractor's estimator, planner and commercial team that allows for a bona fide estimate to be prepared. The first decision to make is whether to return a tender or not. The factors that influence this are numerous (recent research identified 75!); the most important are:

● current and future workload;
● capacity of the estimating and planning department to price the work and return the bid by the due date;
● the extent of design, construction or finance risk that is to be borne by the contractor;
● the likelihood of winning the project;
● the availability of suppliers and subcontractors.

The production of the estimate does vary with different procurement approaches and the reader is recommended to reflect on how project information is communicated using the different approaches. The estimate can be considered to be separated into five separate components:

1. supply chain prices
2. preliminaries
3. contractors measured items
4. attendances
5. profit.

10.1.2 Tender enquires

The majority of procurement systems are linear in nature and require a contracting organisation to submit a price for carrying out construction works that is based upon design information. The form of communication of such a price is usually pre-scribed by the design team or client organisation and comprises deterministic and non-deterministic information. The final cost is usually a single point deterministic figure that is based on a number of differing cost models that are used by the supply chain; these cost models and their precepts are obscured by information communi-cative practice within the industry. The reader is referred to Chapter 7, where the client's tendering approach is considered in more detail.

Although companies have categories of subcontractors within their supply chain systems, estimators have a list of preferred subcontractors, preferred in that they can be relied upon to provide a price within the timescale of the tender.

The estimator has to ensure that all aspects of the estimate are **'covered'**, this means that they have a supporting price which has been provided by a reliable sub-contractor or supplier, preferably more than one. The planning of obtaining prices will start at the earliest stage of the estimate, a list of work packages and a schedule of items to be included will be drawn up. The estimator's understanding of the tech-nologies used is essential to ensure that the work packages are as complete as pos-sible and any queries regarding 'interfaces' between them has to be noted, as they can be significant at later stages of the estimate. Figure 10.2 indicates the market

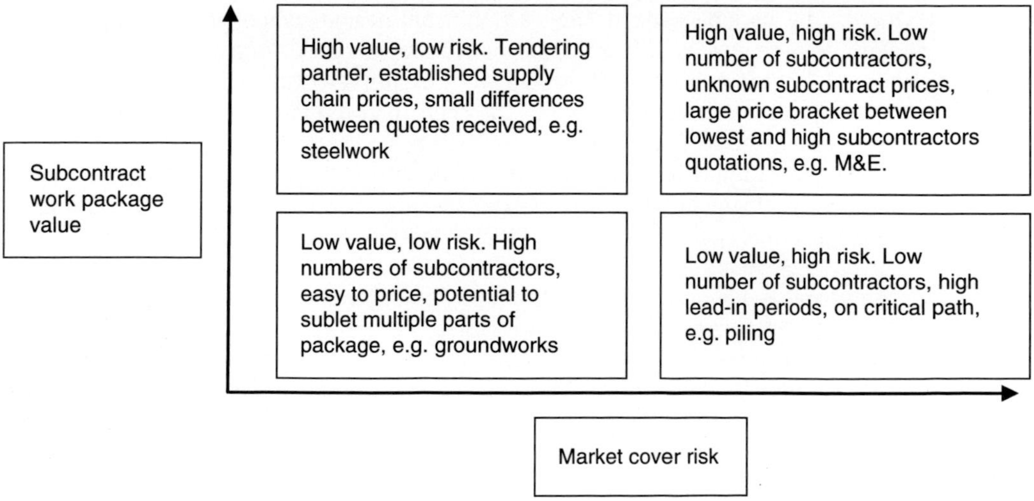

Figure 10.2   Market cover risk.

cover risk that the estimator aims to manage through the process of scanning the environment, sending out subcontract enquiries and selecting an appropriate quotation to include within the estimate. The estimator has a good knowledge of which subcontractors are likely to price enquiries and a working knowledge of the technologies they use; it is unlikely that he will have a detailed understanding of the subcontractors' estimating processes. As a consequence, the analysis of quotations received can be difficult; some subcontractors will bracket items together and provide a lump sum, others may only allow for materials and plant, some may qualify regarding attendances and offer alternative specifications.

The **tender committee** has to decide upon how much of this market risk can be accepted and will carefully analyse the estimator's assumptions about the robustness or otherwise of the subcontractor's prices. Often a quotation will be received very late from a subcontractor which may have priced for a competitor or may be unknown and the price may be significantly below those the estimator has received. The dilemma is whether to include the quotation in the estimate and improve the competitiveness of the bid but risk the subcontractor withdrawing its price after securing the project or whether to stick with the subcontract quotations already received.

Contractors will use pre-qualification criteria which may be developed at a company level to assess and award subcontractors '**preferred**' status. Pre-qualification questionnaires can be used and subcontractors can be classified according to the type or category of work they can undertake, financial thresholds which can include professional indemnity insurance for design, geographical location or a mixture of these. Third parties are also used to maintain the lists and monitor subcontractor's performance after completion of the contract (www.constructionline.co.uk). This places the onus on the subcontractor to keep the necessary statutory information stored, such as insurance certificates and so on, up to date. This compliance with the numerous pre-qualification approaches used by different contractors can be off putting, as there is usually a cost involved to be registered and there is a cost incurred in collecting and uploading the information needed.

The tendering subcontractors will be contacted to determine their interest in returning a bona fide bid, explanation of the tendering timescale and the bid requirements. If a work package includes a significant design component, a design coordinator may be involved. It would be normal for a 'one to one' arrangement to be agreed and the subcontractor would work exclusively with the main contractor in developing the bid. The recovery of the costs incurred in undertaking the design is usually deferred to if and when the contract has been won.

The next stage is to send the tender documentation to the list of work package bidders; the documentation can be extensive and is usually sent as a CD. If the work package needs quantifying, the estimator (or a quantity surveyor) may produce a schedule of quantities; this reduces the cost of bidding by the subcontractors and also provides a uniform basis for the returned quotations to be assessed against.

Subcontractors have a similar win rate to contractors, (1 in 8) and often do not have the turnover to support extensive estimating and planning staff. Consequently, they are very selective in what they price and who they price for. The estimator will ring around the subcontractors checking that they have the documentation and that a price will be submitted by the required date. The estimator is often caught in a bind. A subcontractor will want to be assured that if it returns a price that it will be guaranteed the job if the contract is secured. However, these assurances cannot be given as the award of the work package is often within the remit of the post-contract construction team. Vague assurances are usually given as both parties are aware of the difficulty.

When the quotations are received, they are compared with the information sent out for completeness and compliance with a particular focus on any qualifications that may have been included. Contractors do not have tender protocols, such as those used by the employers, and although a qualified bid can be considered as a counter-offer, the tenders are not often rejected. The qualifications often reflect the subcontractors specialist knowledge of a technology used and can provide the basis for queries to the design team. Other qualifications may highlight that have not been included and that require pricing by the contractor.

A schedule for each work package tender will be developed. This will identify the enquiries sent out, quotations received and the detail of the price, programme and any qualifications. The estimator will select the most appropriate quotation and include it within the estimate. The 'contexts' or stories behind each of the trade work package quotations will be given to the tender committee. The estimator will communicate a view as to the subcontract market, the merits (or demerits) of the selected organisation, the completeness and competiveness of their quotation and whether the quote is likely to be withdrawn or not post-contract.

10.1.3 Preliminaries

There is no formal definition of preliminaries, they can be considered as site overheads and are priced separately. The Royal Institution of Chartered Surveyors (RICS) new rules of measurement uses 16 categories to describe the general contract conditions and preliminaries. The first categories (drawings, site information, existing buildings, description of the works and description of the contract documents including phased or sectional completion) are used to record the project information and are generally not priced. The remaining categories are priced as either fixed or time-related charges. They are in two main categories: employers requirements (any specific requirements regarding quality, security working conditions, protection of the works, site accommodation, site records and completion and post-completion records) and main contractors cost items, which are more extensive:

- management and staff
- site establishment
- temporary services
- security
- safety and environmental protection
- control and protection
- mechanical plant
- temporary works
- site records
- completion and post-completion requirements
- cleaning
- fees and charges
- site services
- insurance, bonds, guarantees and warranties.

Under the traditional procurement route it is usually the client's quantity surveyor who defines the preliminaries and includes them as clauses to be priced in the bill of quantities. Often standard clauses are used from previous projects. However, this can cause some difficulty, as often the extent of contractor's design varies, as does the extent of the consultant's duties.

In design and build and Private Finance Initiative (PFI) projects the preliminaries position can be complicated, as often specialist warranties, insurance and subcontract conditions are included that need to comply with performance specification or drawn information.

It is important for the contractor's estimator and commercial team to ensure that the preliminaries wording does not contain any implicit risk. Disputes often occur as a result of an ambiguous word being interpreted by one party to mean almost the opposite by another. Often large sums of money are involved in the definition of seemingly simple words, such as approved.

As identified above, the contractor sends out the main tender preliminaries documents to subcontractors to ensure a back to back approach to risk. If there are any costs associated with the sequencing of the different work packages, the contractor needs to include these in the appropriate section within the preliminaries to ensure that it is included within the appropriate work package. Protection of the works is a good example of a cost that can be disputed.

The priced preliminaries form an important part of the contractual agreement between client and contractor and are used for payment of interim valuations, assessment of variation costs and also payment for prolongation and acceleration. As a consequence of this, contractors are very careful in how they present this information.

The basic approach to assessment of the costs of each of the subheadings is given in the following. For a more detailed approach to the estimation of the costs of these items the reader is referred to the CIOB codes of practice and the textbook of Brooks (2008).

10.1.4 Employer's requirements

Employers specify what accommodation and services are required; site cabins are generally hired and charged on a weekly basis. There would be a delivery and removal lump sum cost that will need to be included. A telecoms company will give an estimate for installation of a line to the cabin; an allowance for line rental and usage is usually included. Services such as electricity and water are usually nominal costs when considered to the overall project costs and often based on previous records. If the costs were considered to be significant, subcontract quotations would be sought.

10.1.5 Contractor's requirements

Management and staff

The staff resource needed to construct a building will vary from commencement to completion. For example, a project with extensive groundworks may need setting out by engineers who would not be required at the later stages. Additionally, a contracts manager may look after a number of projects whereas a site manager may be allocated to one project at any single point in time.

A contractor's estimator will allow for an ideal scenario of staff on a project; this may be varied at tender stage and is discussed later in this chapter. Staff cost allowances are based on the typical monthly cost of a member of staff, which would include any additional costs associated with national insurance, pension contribution, holidays and car. If the person was to be employed fully on the project, a 100% allowance would be made. If, as in the example of the contracts manager above, the project was only part of their job a percentage allowance may be made.

The service staff costs associated with the development of the estimate is usually part of a head office overhead that is added to the total project cost. In projects where additional programming may be required, such as phased completions or complex construction sequences, an allowance based upon a percentage of a planner's time will be included.

Site establishment

The budget allowances for this section are generally made up of the accommodation required for the site management; this includes offices and welfare facilities such as canteen and toilets. The estimate is based on the programmed duration, a weekly hire cost for the cabins and an allowance for the services required (telephone, electricity, water and waste).

Other costs associated with this section may be the establishment of a temporary road, car park, providing a hard-standing for the site accommodation or provision of fencing. These items can simply be measured and priced in accordance with the measured work approach given below.

Services and facilities

These include power, lighting, fuels, water, telecommunications, IT facilities and administration. Power and water to the site are provided by the utility companies, which will charge for establishment of the service and also for the amount of service used. The contractor is responsible for ensuring that the services are sufficient to test the building before hand-over. The estimator includes an allowance, usually supported by quotations, for provision of the service and an estimate of how long the service will be needed and extent of usage, which would be based on previous records.

Security

Allowances for costs associated with security vary enormously given the location and type of project. Typically, a subcontractor will be asked to provide security staff that would vet entry onto the site and also provide a presence overnight. This is a time-related cost. There may be some other costs, such as fencing the site, provision of lighting, CCTV and alarms. The allowances for these costs would normally be supported by quotations from specialist suppliers; it would include dismantling and removal of any equipment used once the project was complete.

Safety and environmental protection

The Construction (Design and Management) Regulations 2007 (CDM2007) came into force from 6 April 2007. The Approved Code of Practice (ACOP) Managing health and safety (2007) provides practical guidance on complying with the duties set out in the Regulations. It replaces HSG224, the ACOP to the Construction (Design and Management) Regulations 1994. The contractor will require a detailed health and safety plan, which will have both fixed (establishment) costs and time related costs in monitoring compliance.

Control and protection

Protection costs can be extensive and can be associated with work that is to be protected from the weather, for example re-roofing, protection of brickwork from overnight frost and so no. Protection of completed works to ensure following trades do not cause damage can include temporary floor and wall finishes. The allowances are usually calculated using the programme and quantity of works as a basis.

Mechanical plant

Mechanical plant, including cranes, hoists, personnel transport, earthmoving plant, concrete plant, piling plant, and associated allowances that are not included in the rates are often allowed in this section. They are usually priced based on a time requirement and a weekly hire/ownership cost.

Temporary works

The cost of temporary works can be significant and can make the difference between winning and losing a contract. They are priced by the estimator based on information provided by the planning engineer, who has a significant role in these costs. The planning engineer will consider sequencing and associated costs, will develop formwork re-use systems that minimise in the cost of propping, will consider access systems and commission engineers to design propping systems where required. The contractor needs to allow for the inclusion of costs for temporary roads, walkways, access scaffolding, support scaffolding, propping, hoardings, fans, fencing, tempo-rary roofs, hard-standings and working platforms and traffic control. Where the work is to be undertaken by subcontractors, enquiries and quotations will be sought as above. The method statement will provide details of the approach to be taken and will be used post-contract for the valuation of the time related fixed costs of these works.

Attendances

The contractor must allow for special attendance for nominated subcontractors and suppliers. When a nominated subcontractor is invited to provide a quotation to the design team for its work, it will be informed of the general facilities available and will be asked to provide details of any special or specific attendances that it requires. The communication of these requirements will be within this section of the tender documentation and includes access roads, hard-standings, craneage, temperature control and so on. The costs of these may be fixed or time related.

Provisional sums can be defined or undefined; the implications of this distinction are considered elsewhere in this textbook. The estimator would note them as they give an indication of the extent of completeness of the design and the uncertainty that the design team has regarding some of the works. The rules for the valuation of works instructed against provisional sums are discussed in Chapter 17. A provisional sum for daywork (labour, materials and plant) would be included, the rules for valuation are included within the contract and the RICS's Definition of Prime Cost of

Daywork carried out Under a Building Contract (September 2007 edition or current at the time of contract) is used to establish a rate. The contractor has an opportunity to add a percentage addition to cover overheads and profit.

10.1.6 Measured items

The responsibility for the accuracy and completeness of description of measured items depends upon the procurement approach. If the traditional approach is used the quantity surveyor produces a bill of quantities, the measured items represent the quantity and quality of the works shown on the tender drawings. It should be measured in accordance with a **method of measurement** – SMM7, NRM, CESMM4 – which categories items into work sections and also identifies what is included or excluded. If the works are procured through a design and build arrangement, the contractor will often produce quantities from the drawings. These are often not in accordance with a standard method and are sometime referred to as builder's quantities.

The measured items are enumerated (Nr), measured by volume (m³), by weight (tonnes), by area (m²) and also by length (m). These are referred to as the units of measurement.

Each item is then priced using a unit rate which is derived from a production function. The production function can be referred to as the relationship between the output of a process and the resources required to complete such a process. For example, the output of a process may be a brick wall; this will have a quantity and specification, which will be factors that influence the resources required. Some of the resources can be derived through simple mathematical relationships. For example, a half brick thick wall requires 59 bricks per m²; other resources, such as the time taken to get the materials to the place of installation, the mortar required, any temporary works needed to allow for access to the works, wastage of materials, labour and plant, require a combination of professional judgement and analysis of previous projects records.

As discussed earlier, most contractors subcontract over 80% of their work and rely upon the specialists to provide them with lump sum quotations based on the measured bill of quantities. The specialists sometimes rely upon their subcontractors and suppliers and somewhere along this chain will be an estimator who uses skill and judgement to arrive at a price for the work. The total cost is the summation of the products of the quantities multiplied by the corresponding unit costs. The **unit cost method** is straightforward in principle but quite laborious in application. The initial step is to break down or disaggregate a process into a number of tasks. Collectively, these tasks must be completed for the construction of a facility. Once these tasks are defined and quantities representing these tasks are assessed, a unit cost is assigned to each and then the total cost is determined by summing the costs incurred in each task. The level of detail in decomposing into tasks will vary considerably from one estimate to another.

The tight timescales often mean that it is very difficult to get every measured item priced and estimators will sometimes use **'plug' rates**. These are rates that have been provided by subcontractors on previous jobs for similar items.

10.1.7 Attendances and profit associated with domestic subcontract works

The allowance for an attendance can be a fixed cost, time related or method related. A subcontract quotation may expressly exclude some aspect of work associated with the trade; for example cladding subcontractors often require a 1m working area around the perimeter of the building that is 'stoned up', a piling subcontractor may exclude the removal of spoil from the estimate. Sometimes the attendances are specified and included within the measured works section of the bill or, more often than not, the attendances are simply entitled attendance.

10.2 Tender submission

The form, content and date and time for submission of a tender are specified along with the location to where is it to be received. Recent research undertaken the BCIS (RICS, 2009) identified that greater use of electronic document transfers was evidenced. The size of the files required physical media to be used, such as a CD or memory device. The five key benefits identified were

1. lower administration costs/effort (printing, copying and distribution);
2. better contractor access to information for subcontractors;
3. reduced effort in issuing clarifications;
4. reduced timescale of tendering;
5. reduced effort in analysing tenders.

Usually the form of tender will be in a simple form and will be a statement of a sum that is required to undertake the works specified in the tender documentation; it will be open for a period prior to acceptance by the client. By the time the tender is submitted all of the competing contractors will know each other's identities, often they use common suppliers and subcontractors, particularly for specialist items.

10.2.1 Tender margin

The tender margin is a percentage that represents the profit element of the project – the site overheads have been calculated as preliminaries, the general overheads have been included as a percentage that relates to the turnover of the company. Some companies have a policy that the tender margin percentage cannot decrease below a certain level, which aims to avoid departments 'buying' work. The tender margin can be considered in the same way as other allowances within the estimate, it is a separate heading to be reported upon and any variances from that within the tender need explanation post-contract.

The margin is often added to the net estimate, which has been subject to **commercial adjustment**. This aspect of tendering is rarely mentioned in textbooks or codes of estimating practice. However, the consideration of commercial opportunity is an integral part of the tendering process. **Commercial opportunity** can be defined as ... 'the opportunity that arises through the incompleteness of the construction contract to deal with every eventuality that occurs post-contract and an amount that represents the opportunity within a project is deducted from the net allowances prior to calculation of the tender margin'. The commercial team generally produces a report which identifies the risks and opportunities within a project; some examples of these are given in the following sections.

10.2.2 Design risks

This is the likelihood of the design organisations to meet their contractual obligations for provision of construction information on time; if they fail in this important duty the contractor is entitled to an extension of time and recovery of associated costs. If the documentation has errors, these can be undermeasures or overmeasures within the bills of quantity; the contractor can tactically price these in order to over-recover profit once the items are re-measured when the work in completed.

Clashes between drawing, specifications and bills of quantities often lead to variations which allow the contractor to negotiate .

During the course of tendering contractors become aware of cheaper alternatives to that specified in the documentation and will offer alternative prices based on the revised specification. This can lead to a qualified bid, which may be problematic (Section 10.2.6).

10.2.3 Construction risks/opportunities

Often market conditions change and suppliers are prepared to look again at their estimates once they know that the contractor they provided the bid for has won the project. Unfortunately for the suppliers and subcontractors which provided a quotation at tender stage, the winning contractor is often contacted by other subcontractors/suppliers interested in bidding for the work. This can lead to a practice where subcontract packages are resent out during the post-contract phase to a new set of subcontractors to seek additional discounts.

10.2.4 Tender committee

The estimator presents the estimate to the tender committee for discussion. There is usually representation from the buying department which provides comparisons of bulk material prices, the planning department which will have developed a construction programme for the works that will show sequence, critical path and durations of key activities, a commercial manager who will present the commercial position and a contracts manager who represents the team which is to build the project. The estimator is 'challenged' to justify the allowances included, assumptions are tested and the alternative subcontract/supplier prices considered. A reconciliation between resources within the programme and those included in the estimate is sometimes undertaken for those projects that require a lot of in-house resource.

The tender committee usually meets within a few hours of submission of the tender to ensure it has the most up to date market information.

If the contractor is very keen to win work and keep within its self-imposed rules of tender margin limits, monies will often be deducted from the bid as either commercial opportunity or, as often happens, the staff allowance is reduced. When considering how staff costs are allocated to projects, it will be seen how this can be an expedient way of reducing the net cost without reducing the more critical allowances for labour, plant and materials.

Computer-based estimating information systems

The technological support for the estimating process relies upon basic look-up software systems and there is no evidence of the use of expert systems or knowledge-based decision support systems. **Computer aided estimating** (CAE) systems use a database/ library approach of rates and outputs that are adjusted through the use of judgement to develop project-related data. The technological support provided by the CAE system structures the organisation's response to a request for a tender and ensures a logical set of processes are used to develop the estimate. Recently developed systems, such as Conquest and Causeway, use Internet technologies to send enquires, and to record and analyse quotations. The CAE systems can integrate with the post-contract control and financial management systems used by the contractor, which can reduce the estimator's time in the production of allowance bills.

Once the decision to bid has been made the contractor will put in place a tender strategy. This was discussed in Chapter 7. The pre-tender team's role is to offer the best price, programme and technical proposal to the client. This is unique for every project and relies on the coordination of contributions from outside and within the organisation. Mistakes at tender stage can ruin companies and there are a number of checks and balances to ensure the company is not at risk.

The commercial team will have a strategy to recover any opportunity identified at tender stage. This may require the movement of monies within the estimate, ranging from global loading of sections to enhance cash flow to individual rate manipulation to enhance valuation of change. These manipulations will not take place until the contractor has received notice that it was the lowest bidder and that a detailed estimate was required.

10.2.5 Final adjustments

The process of estimating and tendering can take up to six weeks and involve hundreds of organisations pricing documentation which can be ambiguous in parts. Therefore, it is no surprise that, even after the tender committee has met, there are some last minute adjustments; it may be a last minute quotation or clarification from a supplier about a rate which may make the difference between being the lowest or coming second.

10.2.6 Qualification

The code of practice of single-stage selective tendering expressly disallows qualification and states that qualified bids shall not be considered. This can be considered as ideal practice. However, the reality of projects often means that there are risks within a project that need limiting and the protection of a commercial position requires the contractor to qualify.

Sometimes a qualification is tactically included to artificially lower the tender to get to a position where an invitation to negotiate emerges. This is obviously sharp practice and the risk is that the bid may be discounted as it is non-compliant. The client's pre-contract advisors are well aware of the tactics of qualification and will require their removal before a tender is considered.

10.2.7 Production of allowance bill

Once a contract has been secured the estimator produces an allowance bill of quantities which represents the budgets for the different work items. The allowances are expressed in net terms; the difference between the net value and the 'external' value represent the on-cost. The client's representative is presented with the external rates and the internal rates are used to monitor the project financial progress. The allowances for all sections of the estimate are subdivided into labour, materials, plant and subcontractor. The subcontractor allowances are occasionally subdivided into domestic, which includes labour only, and nominated. The use of allowances to cost control the project is considered in more detail in Chapter 18.

10.3 Planning

A medium sized organisation with a turnover of about £50 million will probably only have sufficient works to employ one planning engineer. Producing a programme which embodies the logic of the construction process requires knowledge of the sequence of construction, the construction technology used by the project, typical outputs from resources and an ability to design temporary works. Cooke and Williams (2009) provide an excellent explanation of the planning function and techniques used.

10.4 Buying

Construction organisations have taken the opportunity offered by integrated financial management systems to centralise the buying function. A medium sized contractor with a turnover of approximately £50m will employ a chief buyer, a buyer and a buying assistant. The buyer has a number of roles within a company:

- Strategic roles:
  - the management and maintenance of relationships with suppliers of materials and plant,
  - development and management of the category management systems used for procurement of subcontractors;
- Project roles:
  - pre-tender roles in identifying alternative specifications and suppliers which may provide a competitive advantage and knowledge of the market,
  - post-contract processing the requisitions from site and arranging for materials and plant to be delivered in accordance with the construction programme, completion of cost accruals for items scheduled on the materials received sheets,
  - completion of monthly buying gain/losses reports used within the cost value reconciliation (CVR) process.

The service teams are usually represented at senior level by a services director who manages their workload and assesses their ability to respond to tender enquiries when received.

10.5 Accounting, costs and information

The processes of cost data collection, allocation, reporting and analysis are costly and rely upon well established procedures that have been developed to meet a company's financial and management accounting needs. For the directors to have confidence in reporting profits, the cost recording and reporting system must be applied uniformly across all the projects that make up the turnover of the company. When the range of suppliers, subcontractors, plant hire companies, labour, staff and clients, is considered, this is a significant undertaking. Before considering the attributes of the IT system that make this possible, it is important to have an understanding of the distinction between value, cost and profit as previously discussed.

10.5.1 Definitions of costs

Cost has been defined in Chapter 8 as 'that expenditure which has been incurred in the normal course of business in bringing the product or service to its present location and condition' (SAPP9, paragraph 17). This simple definition belies the processes that are required to report upon a projects cost. There will always be an element of judgement when arriving at the approximation of the costs of a project, and in order to ensure that the costs are as objective as possible and to remove, as far as possible, any subjectivity, a robust information system is required. The system of capturing and reporting costs is at the heart of the company's decision making.

The recording of costs relies upon a robust system of information capture from a range of internal sources, such as site and head office, and external sources, such as subcontractors, suppliers and clients. **Cost reporting** requires the establishment of a method of classifying costs and aggregating (or disaggregating) into headings that can be used for future analysis. The large number of sources, as well as the range of individuals involved in input into the system, can make cost recording and reporting problematic. However, it is essential that a company can demonstrate to its stake-holders and auditors that it has a robust and reliable system.

The cost report is generally produced by an accounting department and takes account of both **direct costs** (costs that have been paid for) and **accrued costs** (costs that have been incurred for which a judgement is made about their scale). The cost information system allows for cost reports to be produced in three common formats: year to date, cumulative and accounting period. The database design of the systems also allows multiple levels of report to be generated, some examples being:

- project level (usually monthly to support the CVR);
- supplier level (to support audit of individual suppliers accounts);
- resource level (to aid the *ad hoc* reconciliation of costs and budget during an accounting period).

The term to '*its present location and condition*' is key to the identification of costs and value to be applied to the project. A simple interpretation of this may be all the labour, plant, materials, subcontract, site overhead and head office overhead attributable to the contract. However, as discussed throughout this book, how do we measure the costs and when do we report them? Construction work is always **in progress**. Some questions come to mind such as:

- What cost should be reported for materials which have been delivered to site but not incorporated into the works?
- What cost should be included for work carried out by a subcontractor for which an application has been made to the client but has yet to pay for?
- What consideration should be given to costs incurred after the valuation date?

It is essential that the contractor has a set of guiding principles to ensure consistency in the reporting of project costs. These principles need to be able to be applied to the variety of costs incurred throughout the project and include the recording and reporting of direct and indirect cost and accruals, which are considered in Section 10.7.2.

10.5.2 Timing of cost information flows and reporting

Cost information originates from internal sources, for example staff costs, and from external costs, for example materials supply cost. Irrespective of the source of information it needs to be collected, categorised and reported at company level, contract level, resource level and even activity level. The information is generated at a time which is often unconnected with the company's reporting timetable; for example, concrete deliveries are programmed to meet project requirements, plant is hired to undertake an activity with commencement and completion dates which are independent of the accounting reporting periods.

To make some sense of these seemingly random timing of flows of cost information and to provide a framework within which to report costs, a company will establish a series of **cost cut-off dates**. These will apply to all projects and provide a structure to the cost reporting, payments and reconciliation processes that are undertaken. However, due to the extensive use of credit within the industry and the consequent delay in timing of payments, not all costs 'fit' with these cost cut-off dates and the cost reporting system will have to have the flexibility to not only report those costs which have been incurred and paid but also those costs which have been incurred and yet to be paid. This is essential to comply with accounting regulations and the central principle of matching expenditure with income. It is important to note that expenditure does not just mean what has been paid for and an approach to cost accruals is required (Section 10.7.2).

10.6 Company management accounting systems

10.6.1 Company information systems

All companies will have some form of a financial management information system. The range of systems available is vast, the most sophisticated are supported by software companies that developed and maintain them and, in some cases, aim to integrate all the company's functions from looking for work to chasing retentions after final certificate. The most basic may be a simple look-up system that can is used to produce company monthly statements of cost and cash flow. Most systems use an approach of relational or object-oriented databases.

The coding and design of such systems is beyond the scope of this book but it is useful for the reader to understand a little about database design and operation. Most enterprise-wide systems use a number of pre-designed **relational databases** for storage, analysis and reporting of information. A database is used for storage of data in a series of fields and records. The definition of a field is a 'data store with defined characteristics'. A key field contains an item of data that is unique to the record. The data stored within the fields can have values such as text based, date based, numerical or can be calculated from other fields. A record is a series of fields of data that is unique to the item being stored.

In Table 10.1 the text-based field stores the name of the subcontractor, the accounting period end date is a date field and the rest of the fields are numerical. The data within the numeric fields are either input directly or calculated from other fields. For

Table 10.1 Subcontract ledger for payment.

Subcontract Ledger						
Subcontractor:		XY201	Williams Excavations Ltd		Accounts period 2	
Job No:		AB1234A	LJMU Byrom Street			
Order Number:		1	Domestic Groundworker			
Gross Cert Val	20000		Order Val	0.00	R/tn Paid	0.00
Discount	475		R/tn Limit	0.00	R/tn Held	1000.00
Retention	1000		R/tn %	5000	Taxable	0.00
Net Invoice	18525		Disc %	2500	Variable	18525
VAT	3211.88					
Tax	0.00					
CITB	0.00					
Contract	0.00					
Payable	21766.88					
Paid	0.00					
Unpaid	21766.88					
Unrealised	0.00					

this record the input fields would be order value, retention percentage, discount percentage and gross cert value. All the other fields that are wholly numerical are calculated from other fields.

The advantage of having a relational approach to data storage is that the data only has to be entered once and is used many times within the system. This has advantages for efficiency and consistency and avoids duplication of records. In order for the system to be relational, a primary key is required; this is unique to a series of records. In the case of this example, the **primary key** would be job number. All the related databases that linked to this job would contain this field and the data within the databases could be combined to create reports.

The data are usually input into the databases using a series of predefined forms; the forms are designed to ensure that the data are of the correct type, that is date or number, and that they fall within expected ranges, for example a retention percentage outside the normal range of 2–10 % would be flagged as unusual. The strength of this **'forms'-based approach** is that the forms can be designed to allow only certain users in different departments to enter data that are relevant to their job function. In the example in Table 10.1, the buying department may have a form for entry of order value, retention and discount percentages. If the forms are password protected and the data fields cannot be overwritten by other users this provides for a security of input. This is essential to maintain the integrity of the data. Two separate functions may enter data into one database; in the example above the quantity surveyor on site may input the gross certificate value field and the accounts department may complete the paid field.

A database management system can be programmed with a series of queries using structured query language (SQL). SQL queries can be used to sort or search within the records of the databases. SQL queries can also be used to combine records from different databases and generate reports. These reports can be generated monthly within predetermined designs or can be produced on an *ad hoc* basis with user defined fields.

Table 10.2 Company ledgers.

Company-wide ledgers	Nominal	Sales	Purchase	Payroll
	Profit and loss Balance sheet	Cash book Contract Payments	Materials Plant Subcontractors	
Contract Ledgers		External valuations	Requistions Stock control	Overhead costs
Typical reports, period and cumulative to date		Valuation, certificates and retentions	Contract cost reports at resource level Stock levels	Labour costs Staff costs

Most users of the system are unaware of the database structures that lie behind the predefined data input forms and the system defined reports. Most users also do not possess the skills to develop a query in SQL to interrogate the databases and, if they did, the company IT staff would not allow them to practice on the business critical systems! The systems used have been designed over many years and tend to be bug free. Most software companies will develop the systems to create company specific logos or particular reports that are relevant to the business.

How is this relevant to a text on contract financial management? The financial management reports and management account reports are all generated from the data held within the databases. Senior management has access to reports that can aggregate all the contracts together and analyse trends and ratios, which their older colleagues could have only dreamed about twenty years ago. **Auditors** can 'drill down' to site records to check accruals against order value and ascertain whether contracts are correctly costed.

The 'system' becomes a major tool in the senior management's control of the company and the site staff's control of the project. The site quantity surveyor has to understand the system's design and be able to track back from a report to find the point where the data were originally entered. It is only through this understanding that it will be possible to identify where errors in costing may have been made. The systems, therefore, have an ability to share data, provide access to information which meets the varied needs of the finance and commercial teams and, above all, must be simple and easy to use.

The system may have a number of relational databases or **'ledgers'** that support the financial and commercial management of the organisation (Table 10.2). Typically, procurement databases will support the maintenance of preferred lists of suppliers and subcontractors, send tender documents to them for pricing and then support a comparison of quotations between the subcontractors against the allowance rates.

Although some systems will integrate the pre-contract part of the business by tracking business leads through to tender submission, it is rare for this to happen effectively in practice, as the tender requirements for every project are different and the functionality required by a system for pricing work is quite different from that of financially managing an organisation and commercially managing a project.

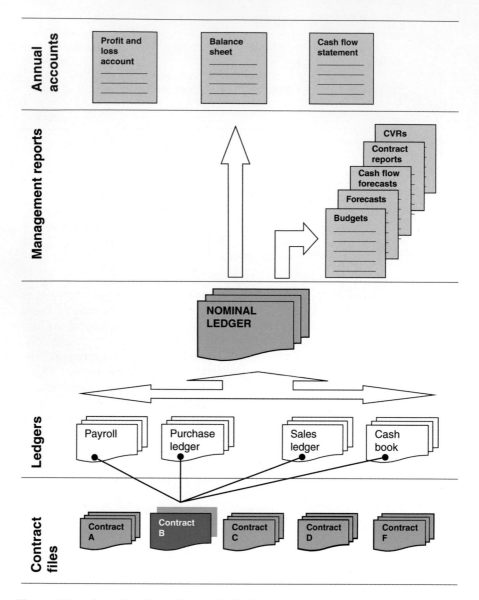

Figure 10.3 Information flow – the nominal ledger.

10.6.2 Contract operational ledger

A contract operational ledger gathers all the information relating to a contract for enquiry and reporting. It allows for quick access to cost, value, budget and committed cost information and most systems will allow the user to define the outputs. The database systems allow for multilevel cost and budget reporting, which allow for reports to be generated at a summary or transaction level. If information has been transferred from the estimate, the system would allow for budget/cost comparisons from a range of different perspectives. Figure 10.3 illustrates the relationship of the contract ledgers to the other ledgers held and also the published accounts.

Table 10.3   Contract summary ledger.

Contract Ledger					
Job No:	AB1234A		Name:	Byrom Street	
Sale Acc:					
				Labour	21408.00
Cash Goods:	17100.00			LO S/C	0
VAT	2002.50			Materials	12250.04
Total:	20092.50			S/C	28525.00
				Plant	1739.25
Retention:	900.00	WIP:	63922.29	Sundry	0
Ext Value:	18000.00	Profit:	-45922.29	Prelims	0.00
					0.00
				Adjustment	0.00
Sundry Inv	0.00	O/S	0	Other	0.00
Certified	18000.00	O/S	0	Total	63922.29
Applied for	22000.00	O/S	0	Subsidy	0.00
				G. Total	63922.29

The benefits of a flexible database approach allow for the data to be interrogated at resource level, for example materials or material type, labour or trade, and at activity level, for example completion of phase 1. This can allow for quick reports which support reconciliation to be produced.

The **contract operational ledger** can have links to sales, procurement, requisitions and subcontract ledgers. It is through these links that summary reports of contract listings, cost libraries, sales and cost transactions, retentions, applications, valuations and cost summaries can be produced. An example of a contract operational ledger is given in Table 10.3, which illustrates how the costs can be summarised to give an indication of contract profitability.

The **subcontract ledger** will maintain and report on all relevant information relating to the suppliers and subcontractors. The databases can usually be adapted to store details of subcontract registration, insurance expiry date details and also rating systems which link to contractor's category management system (Table 10.4).

In the example shown in Table 10.5, a subcontract valuation certificate amount can be input by the site quantity surveyor; the system will automatically deduct discount and retention before adding VAT. A payment date can be generated by the accounting system to maximise cash flow and the payment would be made via BACS to the subcontractor. The ledger approach will allow a search to be made at subcontract or project level or both.

The **sales ledger** is used to record the sales of the company. Contracting requires a flexible approach to be used in the definition of sales, for example an application for payment rarely has the same value as the agreed valuation and consequently would not be liable for VAT (Figure 10.4). The person who undertakes the credit control within the company also requires a reporting function which indicates which payments are overdue and a tracking function to identify who to contact with the relevant details (application, certificate, retention position etc) available should there be a query that may cause delay.

Table 10.4 Supplier database.

Supplier Database						
Code	Name	Address	Post Code	Phone No.	Acc. Type	Bal This Month
CGL001	Cedar Green Ltd	164 Body Shape Lane	BL12 3QA	0112 332145	B	0.00
PWS005	Pat's Window services	74 Brownhill	SE6 9 PH	0201567 111	L	0.00
MOS009	Moonlight Suppliers	81–85 Bypass Road	NH 67 0 WE	1721237778	B	0.00
ASE002	A&S Electricals	17 High Street	L2 3FY	0844 124333	L	0.00
BOB010	Bob & Co. Ltd	1A Business Park	OL90 1KM	0115 567890	L	1200.00
STG003	Steve Group	34 Save Journey Lane	WV5 6JK	204748867	L	900.00

Table 10.5 Subcontract payments.

Subcontractor Transactions									
Subcontractor:			XY201	Muckaway Excavators					
Job No:			AB1234A	LJMU Byrom Street					
Order Number:			1	Domestic Groundworker					
Trans Type	Cert	Date	Gross	Retention	Discount	Tax	VAT	Contract	Pay date
Cert	201	17:07:12	20.000.00	1000.00	475.00	0.00	3241.88		30:07:12

The **purchase ledger** can be used to manage the payment of the suppliers to the company (Table 10.6). It usually integrates with the procurement and stock control ledgers, which store the agreed order rates and record resources delivered to the site. This is the key ledger for the reconciliation of accruals with direct costs and, as such, allows users to 'drill down' to the cost records for contracts over a period. Most of the original invoices and delivery tickets will be stored as scanned images and can be retrieved should a dispute arise.

The **procurement** ledger is used to manage the production and management of orders placed with suppliers to the company. Some systems can have the facility to store scanned images of delivery tickets and invoices and match them to materials received sheets. These types of ledgers allow for summaries of committed costs to be calculated, which allow the commercial team a quick view of the percentage of orders to be places and an idea of the future scope for commercial opportunities. The database approach allows for different searches to be undertaken, for example by contract, order, supplier and even buyer and date. This approach makes the tracking of order to delivery to payment very straightforward, as illustrated in Table 10.7.

SALES LEDGER UPDATE

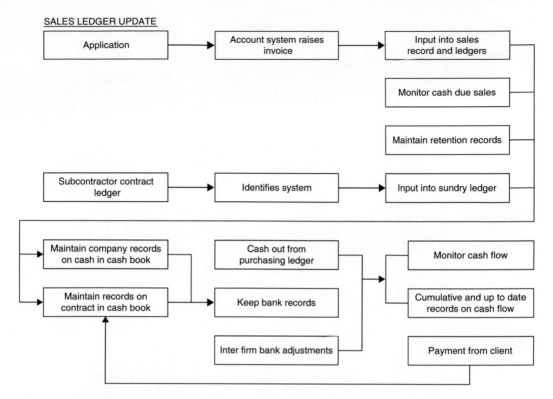

Figure 10.4 **Sales ledger inputs.**

These systems allow for the centralisation of the buying function. However, this can also constrain the site from procuring materials quickly. Most systems will have some form of requisition system that allows the site-based staff to generate goods requirements, which have a specification and date required attributes. The requisition is accessed by the buyer, who may have agreed rates with a strategic supplier and who will be able to place the order and confirm to site. The site can then track the order and access the suppliers contact details should there be a query.

A **payroll ledger** manages the contract-costed and overhead-costed employees and usually has the flexibility to cope with the different payment terms used with staff (weekly, fortnightly or monthly). Systems have been created for the entry of time sheets and calculated fields designed for PAYE and NI deductions. The systems integrate with the company's bank to allow for automatic payment by BACS into employees bank accounts (Table 10.8).

All of these ledgers support the **nominal ledger**, which aggregates all the information into financial management reports. Most systems will allow for categorisation into divisional or sectional subsummaries to allow for subsidiaries to use the same system. Profit and loss and balance sheet reporting is usually a standard feature. This can be reported monthly and the actual compared with budget generated year-to-date, with a given period and with comparisons to previous years (Table 10.9).

The strength of the integrated database system is that the outputs from the various ledgers can be used to monitor cash flow. Figure 10.5 indicates how an agreed valuation would be used to raise an invoice to a client and how the various ledgers accept the inputs from this process.

Table 10.6 Purchase ledger: invoices.

Purchase Ledger: Invoices

Contract Name	Contract	Order No.	Supplier Name	Order Date	Del Date	Exc. VAT	Status
Byrom Street Project	BS10044AA	1	Make Sure Suppliers	17:07:12	17:07:12	12100.00	Confirmed
Byrom Street Project	BS10044AA	2	P&S Groundswork	23:08:12	23:08:12	9000.00	Partly Invoiced
Byrom Street Project	BS10044AA	3	Posh Ceiling Ltd	19:08:12	19:08:12	3400.00	Fully Invoiced
Byrom Street Project	BS10044AA	4	Lion Metalworks	25:07:12	25:07:12	1250.00	Fully Invoiced
Peter Jost Project	PJ10022CC	5	Industrial Cleaning Ltd	30:07:12	30:07:12	500.00	Fully Invoiced

Table 10.7 Supplier account report.

Supplier Transaction

Supplier Account:	RMCO35	Ready Mixed Concrete						
Period	Batch	Date	Doc Type	Ref1	Ref2	Inv Val	VAT Disc	Goods/Cash	Due Date
12:11:12	100234	01:07:12	Invoice	1002341	1CCC	1250.00	250.00	1000.00	30:01:13
15:12:12	100432	08:08:12	Invoice	1004326	2DDC	700.00	140.00	560.00	15:02:13
16:12:12	100567	15:08:12	Invoice	1005671	7SSS	24100.00	4820.00	19280.00	18:05:13
16:12:12	100231	24:08:12	Invoice	1003219	3QQS	90.00	18.00	72.00	20:02:13
18:12:12	100789	29:08:12	Invoice	1007890	1GGH	1500.00	300.00	1200.00	24:02:13

Table 10.8 Payroll ledger report.

Code	Type Desc	Calc Description	Code Desc	Hourly Desc	Amount1	Amount2	YTD
1001	Pre Tax Pay	Emp standard rate	Basic Hrs	8.20	0.00	0.00	2,648.60
1002	Pre Tax Pay	Emp standard rate	O/T @ 1.50	8.20	0.00	0.00	1,020.90
1009	Pre Tax Pay	Emp standard rate	O/T @ 2.00	8.20	0.00	0.00	377.20
1014	Pre Tax Pay	Company Amount	Tool Money	0.00	0.00	2.25	20.25
1017	Pre Tax Pay	Employee Rate	Trip Time	4.00	0.00	0.00	130.00
1025	Pre Tax Pay	Company Rate	Attendance	2.00	0.00	0.00	90.00
1028	Pre Tax Pay	Variable Amount	Subsistence	0.00	0.00	0.00	410.00
1029	Pre Tax Pay	Variable Amount	Bonus	0.00	0.00	0.00	170.00
1032	Post-Tax Pay	Variable Amount	Expenses	0.00	0.00	0.00	0.00
1033	Post-Tax Ded	Employee Rate	Loan Repay't	0.00	0.00	15.00	135.00
1039	Post-Tax Ded	Reducing Balance	CSA	0.00	20.00	0.00	180.00
1045	Cost Only	% of Gross	Hol Cr Reserve	0.00	0.00	0.00	486.74

Table 10.9 Summary of costs for a group of projects.

				YTD	Prev	Q1	Q1
Code	**Name**		**Cur. Per**	**Actual**	**Yr Act**	**Actual**	**Budget**
		Nominal Ledger Summary					
BSP/1/90	Byrom St. Project Sales	Sales	−11700.00	−11700.00	0.00	−11700.00	0.00
BSP/1/100	Byrom St. Project Labour		80080.70	80080.70	0.00	81056.95	0.00
BSP/1/120	Byrom St. Project Material		1150.00	1150.00	0.00	1150.00	0.00
BSP/1/130	Byrom St. Project Subcon		9000.00	9000.00	0.00	9000.00	0.00
		Total	78530.70	78530.70	0.00	79506.95	0.00

10.7 Contract cost reports

10.7.1 Principle of cost cut off

To provide an aggregate position of all of the costs incurred by the company at a particular time, a series of dates whereby all the project and overhead costs are recorded is required. These dates are referred to as the cost cut-off dates (Table 10.10). They are usually at the end of each month and relate to an accounting period, for example 30 April (period 1), 31 May (period 2) and so on. These dates do not mean that all the creditors will be paid by this date but that a liability will be recorded for payments due.

10.7.2 Direct costs and accruals

To allow for the lag between costs incurred and payments made, accounting systems use a convention called cost accruals. A **cost accrual** is the 'provision for expenditure that has been incurred but yet to be paid'. The use of accruals to produce cost reports is not exclusive to the construction industry and it is important to understand how they are calculated. The scale of payments made in an accounting period can be a large percentage of a contract value and an inaccurate accruals can cause great difficulty in the cost value reconciliation process that follows the cost reporting process (Table 10.11).

The calculation of accruals is based upon site records which record the liability of the site for a cost; this could be for materials as shown in Table 10.11 or could be for plant or subcontractors. The table shows a weekly delivery total of m³ of concrete; at the end of the first four weeks 184 m³ has been delivered to site and recorded on the materials received sheets. These will be passed to the accounts (or buying) department for calculation of accruals. The supplier will send an invoice for the first month's materials supply which will be based upon the signed delivery tickets. This invoice is often received after the cost report has been produced. The accounts department has to produce a cost report for period 1 for the materials received by the project and will base its calculation upon the materials received sheets (MRS) and the supplier's order rates, which will be stored on the company system.

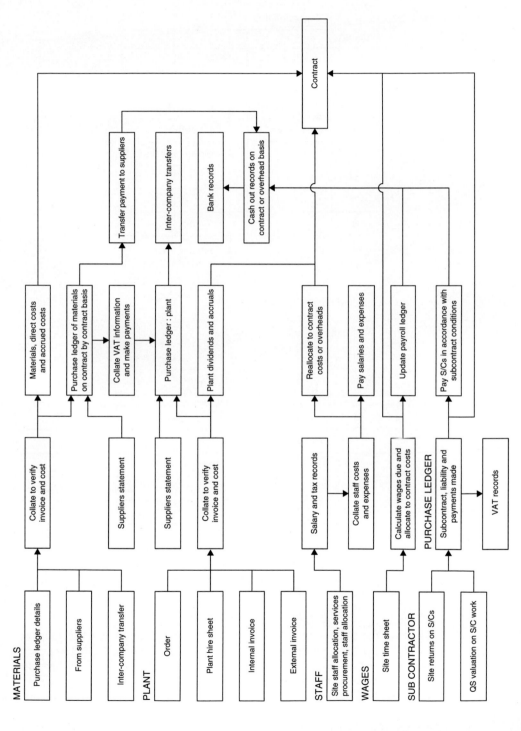

Figure 10.5 Cost information flows.

Table 10.10   Company 'cut-off' dates.

		Reporting dates		
Week				
1				
2				
3				
4		Cost report ⎤		Cost cut off
5		⎬		Work in progress
6		Valuation ⎦		
7				
8				
9				Cash cut off

Table 10.11   Accruals to account for costs incurred.

Contract XYZ						
	RMC Concrete accrual				Cost report	Total
	Week	MRS records (m³)	Supplier invoice	Direct costs	Accrued costs	
Period 1	1	36				
	2	48				
	3	50				
	4	50	22 080	0	22 080	22 080
Period 2	5	62				
	6	50				
	7	36				
	8	6	1680	22 080	1680	23 760

In period 2, the cycle repeats. The only difference is that by now the supplier's period 1 invoice will be entered into the system, will be reconciled with the period 1 accruals and will be paid. The period 1 accruals are, consequently, annulled and replaced by a direct cost (equivalent to the supplier invoice), the period 2 accruals are based upon the site MRS for deliveries received between weeks 5 and 8.

There are frequent errors in the calculation of accruals. Some relate to the company's systems for capturing data from site, others relate to the accounting process itself:

- materials received sheets incorrectly completed;
- lost delivery tickets;
- order values missing and accounts clerks assuming rates;
- supplier records incorrect.

The consequence of **incorrect accruals** is inaccurate costs. If the costs are taken to the project CVR it can lead to overly optimistic (or pessimistic) reports about the project's profitability and undermine the integrity of the system. Consequently, most organisations place the responsibility for ensuring that the costs are correct with the site quantity surveyor. This means that the recording and accrual system should be **internally auditable** and provide a basis for *ad hoc* error checking.

The site produces information that relates to the amount of resource used in 'bringing the product to its present condition'. The resources can include those that are internal to the organisation, such as labour or supervisory staff, and also those resources supplied by other organisations, such as material, plant or subcontractors. The information provided by the site allows for the computation of accrual information, which in the case of materials suppliers is based upon the physical deliveries of materials to site, supported by documentation, order values and a form of recording typically a materials received report. The use of standardised forms collecting information on supplier details, initial order values and deliveries or plant hired that are regularly compiled and passed to a central department is common across the industry. The accounting department uses the information received from the site and the procurement department to calculate accrual values and process payments. This process is relatively robust and transparent; the main difficulties arise with either missing data (material delivery tickets or plant hire reports missing) or incorrect calculation of accrual values (often caused by missing order values). The site will also produce a schedule of materials on site, which will be identified on the cost statement (often as a negative accrual) and be excluded from the total costs, as these material should not be consider as being incorporated into the works and, consequently, do not meet the accounting policies requirements of 'bringing the product to its present condition'.

The process for producing information to support the reporting of costs of subcontractors' work is similar to that of materials and plant. However, it is subject to more judgement and, consequently, is less transparent and can provide, in some companies, the site team with an opportunity to protect some of its decision making autonomy. Subcontractors are procured under a wide range of contractual arrangements. However, they are usually reported under three categories: domestic (this includes labour only, labour and plant, and supply and fix), nominated or named. The cost heads used for these are often separate, as the profit/losses reported are often related to the nature of the relationship that exists between the reporting organisation and its subcontractors. The process for the completion of a cost accrual for a subcontractor would typically refer to the subcontractor's application for payment for work completed, materials delivered to site and variations and claims. The site team would assess the valuation and compile a statement of accrual based upon its judgement as to the entitlement of the subcontractor. The summary accrual figure would then be entered onto a standardised form, which would usually include headings for order value, discount, **contra charges** and last payment details. This would then be passed to the accounting department for aggregation into summaries of costs for each of the cost heads.

The head office's role in the production of cost information tends to relate to the allocation of costs of company and project overheads. At tender stage the organisation would have predicted the project overheads required; these normally would include supervisory staff such as site agents, engineers and commercial managers. These **allowances** would generally be time related and based on the expected contribution that the staff were to make on the project. During the project execution, the staff costs allocated to the project tend to be for those who are directly employed as well as a percentage contribution to staff who are involved with a number of projects.

The interim valuation of construction projects and the uncertainty that surrounds the evaluation of work is a central issue to the variance of approaches taken by construction companies and their teams working on projects. Projects often incur costs that may be paid for by another party at a future date, the uncertainty about whether recovery will take place is identified by the accounting policy which states that 'if it is not likely that any reasonable expectation of future revenue to cover total cost is forthcoming, the total cost should be charged against the accounting period in which the cost is incurred'. Consequently, a judgement has to be made regarding the likelihood of future revenue when deciding whether to include a project cost provision within an accounting period. Chapter 18 considers project costing and the reconciliation in more detail.

10.7.3 Cumulative and period reporting

As discussed in Chapter 5, the convention within the industry is to value projects using gross cummulative figures. This is useful for clients to see how much of the budget remains as the contract progresses although they pay interim payments in monthly increments. A contractor needs to consider the project on a 'to date' or cummulative basis as well as on a period or monthly basis. A comparison of a project's cummulative net value with cummulative costs will provide an assessment of whether the project is progressing to the financial plan developed by the estimator at tender stage. However, the cummulative picture reports on everything that has happened from commencment to date. It is at too coarse a level to provide a basis for anything other than a holistic analysis. A report is therefore produced which gives both cummlative and period figures that can then be reconciled with the respective resource allowances to provide a level of detail which can support management action.

An example of a cummulative and period cost report indicating the direct and accrual costs for the various resources is given in Table 10.12.

The value of the cummulative approach to reporting at a project level is useful as it gives a macro picture of how a project is performing. Most companies will undertake monthly calculations of net value and report upon both the cummulative and the accounting period positions. If the **net sales value** (NSV) is below the cost, the project may be in trouble and the site team will need to know where the problem originates from.

10.7.4 Cost provisions

Cost provisions can be defined as a 'cost included within the accounts for a future liability for which there is little information regarding its amount or when it may become liable for payment'. In Table 10.12, a cost provision of £25 000 has been included within the subcontract area of the accounts. This may be included to

Table 10.12 An exemplar cost report.

Desc	Direct costs	Site Material Accrual	Site Plant Accrual	Sub Contract Accrual	Labour Accrual	Staff Accrual	Site cost prov	Total Accrual	Cumm Costs	Prev Period	Period
Labour	388 829				4291			11 455	404 575	345 600	58 975
Materials	9667	41 500						41 500	92 667	75 645	17 022
Plant	11 588		18 678					18 678	48 944	11 588	37 356
S/contract	25 447			65 789			25 000	65 789	182 025	25 447	156 578
Site overheads	14 555					12 500		12 500	39 555	14 555	25 000
O/heads	22 504							22 504	45 008	22 504	22 504
Totals	472 590	41 500	18 678	65 789	4291	12 500	25 000	172 426	812 774	488 629	317 435

highlight a possible claim from a subcontractor that the site team feel may have a justification. An argument could be made that it should be included within the relevant subcontract accrual; however, provisions are often kept separate to avoid confusion with claimed liabilities.

10.7.5 Individuals involved

Table 10.13 indicates the various staff within the company who contribute to the cost report. Senior management allocates overhead staff to the contract and should attempt to match the allowances made at tender stage. This seldom happens as often the workload of the company will dictate which contracts have the potential to carry additional 'unscheduled' cost. The cost information hub is the accounts department. Here the various ledgers are maintained and the accounts staff process the information received from site and buying, arrange for payment of the suppliers, subcontractors and staff and produce the monthly cost reports. Chapter 18 goes into more details about these information flows and how they support the cost value reconciliation reporting.

Table 10.13   Contributors to the cost report.

Department	Staff	Contribution
Senior management	Contacts manager	Overhead allocations
Accounts	Accounts clerk	Supplier invoicesResource records
Buying	Buyer	Supplier order details
Site:	Site Manager	Materials requisitions; Materials received sheet
		Labour returns
		Plant returns
	Site QS	Subcontract liabilities

The site produce the records of what resources have been received, materials received sheets, plant return and the cost liabilities of the subcontractors are entered by site staff and used by the accounts staff to produce the cost reports.

Responsibility for accuracy

The monthly contract cost reports often contain errors caused by inaccurate data entry from site, buying or accounts. Most companies will place the responsibility for accuracy on the site team, usually the site QS. This means that a good knowledge of the company system, who inputs the data, likely sources of error and a gut feel for the costs is required in order to be able to track back to the source of the error. Unusual monthly movements of cost trigger an enquiry, as most site quantity surveyors have a good idea of the costs before the accounts staff issue the reports.

10.8   Project audits and site processes

There is always a sharpened interest in making sure the project cost report and attendant CVR are correct when the site team knows that it is to be visited by the auditors. An auditor would drill down into the project to look for the justification for the costs included within the accounts, this can go as far as considering subcontract valuations and accruals and requesting a build up for any discrepancies found. The audit usually extends to the value stated within the CVR, which is discussed in Chapter 18. However, it would be an unusual month if the QS was not require to audit back to site records to check upon a cost within the monthly report, irrespective of whether or not the project was subject to a formal audit.

The nature of the construction process is that often a delivery ticket may have been lost into a concrete pour, which thus would not be recorded on the MRS and would not be allowed for when the material accrual for concrete was being calculated. When the invoice from the concrete supplier is received it would reflect the true cost of supply, which would show a higher than normal cost movement in the concrete cost in the next period. This would need explaining and it can only be done so by tracking through the system to identify the under accrual. The safeguard to avoid these types of problems is to have formal systems in place for recording and reconciliation at the end of every month. However, even the best systems can be frustrated by human error.

References

Abdel-Razek, R. and McCaffer, R. (1987) A change in the UK construction industry structure: implications for estimating. *Construction Management and Economics*, **5**, 227–242.

Brooks, M. (2008) *Estimating and Tendering for Construction Work*. Butterworth Heinmann.

CIB (1987) *Code of estimating practice: supplement no. 1 – Refurbishment and modernisation*. Chartered Institute of Building (CIB), Ascot.

CIB (1989a) *Code of estimating practice: Supplement no. 2 – design and build*. Chartered Institute of Building (CIB), Ascot.

CIB (1989b) *Code of estimating practice: Supplement no. 3 – management contracting*. Chartered Institute of Building (CIB), Ascot.

CIB (1993) *Code of estimating practice: Supplement no. 3 – post tender use of estimating information*. The Chartered Institute of Building (CIB), Ascot.

CIB (1997) *Code of Estimating Practice*. The Chartered Institute of Building (CIB), Ascot.

Cherns, A.B. and Bryant, D.T. (1984) Studying the Clients role in construction management, *Construction Management and Economics*, **2**, 177–184.

Cooke, B and Williams, P.N. (2009) *Construction Planning and Control*. John Wiley & Sons Ltd, Chichester.

Eccles, R.G. (1981) Bureaucratic versus craft administration: the relationship of market structure to the construction firm. *Administrative Science Quarterly*, **26**, 449–469.

Hughes, W. (1989) Identifying the environment of construction projects. *Construction Management and Economics*, **7**, 29–40.

Moore, D. (2002) *Project Management: Designing effective organisational structures in construction*. Blackwell Publishing, Oxford.

Murray, M., Langford, D., Hardcastle, C. and Tookey, J. (1999) Organisational design. In: S. Rowlinson and P. McDermott (eds), *Procurement systems; A guide to best practice in construct*. E & FN Spon, London.

Murray, M., Langford, D., Tookey, J. and Hardcastle, C. (2000) Management of the Construction Supply Chain in Pre-Construction Procurement Process. In: Information and Communication in Construction Procurement, Proceedings of the CIB W92 Procurements System Symposium, 24–27 April, Santiago, Chile. Pontificia Universidad Católica de Chile and The International Council for Research and Innovation in Building and Construction.

RICS (2009) E-tendering survey report. Building Cost Information Service (BCIS) of the Royal Institution of Chartered Surveyors (RICS), London.

Shirazi, B., Langford, D.A. and Rowlinson, S.M. (1996) Organisational structures in the construction industry. *Construction Management and Economics*, **14**, 199–212.

11

Financial management

11.1 Budgetary control

Budgeting and its use in systems of monitoring and control influence a great deal of organisational life. The discussion of organisational cybernetics is beyond the scope of this text; However, anyone who has worked within organisations which use budgets will recognise the adage 'what gets measured gets attention'. A lot of analogies are used to consider control, the most common mechanistic one being

Financial Management in Construction Contracting, First Edition. Andrew Ross and Peter Williams.
© 2013 Andrew Ross and Peter Williams. Published 2013 by John Wiley & Sons, Ltd.

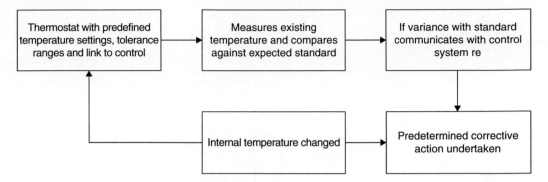

Figure 11.1   Control, measurement and remedial action.

a thermostat (Figure 11.1): a standard established through norms, a system of measurement of actual against predicted, a means of communication of this to some mechanism that can take corrective action and, finally, an action taken which is likely to have an effect upon the factors that caused the variation from the standard.

This analogy is helpful in understanding a **simple control system** and can it easily be extended to consider budgetary control within construction organisations, but it has many shortcomings as it does not reflect the complexity and interconnectivity of organisational systems, human behaviour, the role of measurement and feedback.

Construction firms have to budget and manage their budgets in dynamic economic environments; the measures they use have to be adapted to a wide range of clients, resource suppliers and project procurement approaches. The link between the corrective actions undertaken and their impact is often uncertain and the individuals who use the systems tend to be from different functional groups, which all have their own organisational subcultures and differing reward and incentive systems.

To gain some element of control of this environmental variety, firms continuously adapt and seek to improve their systems. Inevitably, the systems are adapted both formally and informally at a departmental, project or individual level where they are interpreted by their respective functional needs. The consequence of this is that different views of the outcomes of evaluation can lead to different approaches to measurement, which, in turn, affects the system's effectiveness. The levels of **autonomy** of the different functional groups in a construction firm vary; an accounts clerk producing accruals for materials would have limited autonomy whereas a project manager's role requires extensive autonomy and responsibility. The systems for budget monitoring and the range of corrective actions that can be undertaken should a variance be encountered consequently need to be flexible enough to remain effective for the differing functional groups.

A budgeting system is part of the planning within an organisation; the budgets are based upon forecasts of the outcome of future events. The budget is then operationalised into a series of measures that are then used to assess the outcome of actual events with those that were originally forecast and also, if a flexible budget is in place, to update the budget as more information is collected. There is both a feedback and **feedforward** mechanism to most budgetary monitoring systems, as illustrated in Figure 11.2.

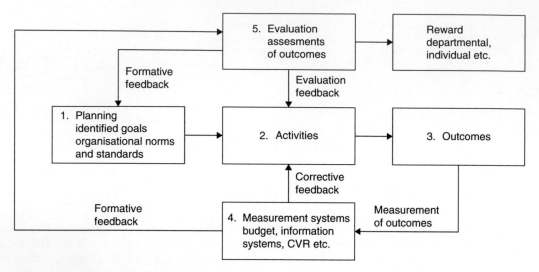

Figure 11.2 Feedback and feedforward.

11.2 Definitions

11.2.1 Cost

As has been discussed throughout this text it is important to have a clear definition of cost and also to consider the properties of cost information. The definition of cost can be considered to be the total money, time and resources that are associated with a purchase or an activity. Cost can be recorded as direct or accrued and have fixed variable or mixed properties. Direct costs have been incurred and paid. Accrued costs are liabilities for costs which have been incurred but for which payment has yet to be made. A fixed cost is one that does not vary with activity for a control period; examples could be rent, insurance and property rates. A variable cost is one that has a relationship to activity; the costs are usually considered to have a linear relationship, that is as an activity increases the cost will increase by a similar proportion. This is an overly simplistic assumption. However, the unit rate approach of budgeting is normally based on this. Most activities have a combination of fixed and variable costs and these are referred to as semi-variable costs. The total costs for an activity can be expressed as the sum of fixed plus variable and semi-variable costs.

11.2.2 Value

In accounting terms, the definition of value is the monetary worth of an asset, goods sold, service rendered or liability or obligation acquired. The value of a project to a company is more than just the monetary worth or amount paid for an activity; the value includes the tangible aspects, such as cash to reduce borrowing, turnover to demonstrate to shareholders that the company is maintaining its market share and also the amount of overhead contribution the activity makes.

The value ascribed to a construction project is difficult to measure due to the lag between the work being undertaken and the timing of monies received. A further complexity could also relate to the means of valuation, which uses approximate quantities and unit rates that approximate the unit costs of an activity.

11.3   Cash flow

The management of cash flow is imperative to the success of any organisation. However, it is more so in the construction industry given the value of projects and the low profit margins. Cash is often seen as the most fundamental and influential of resources on a construction project; as the negative consequences of inadequate cash management more than outweigh the inadequate management of other resources. Russell and Jaselskis (1992) stated that 'an excess of 60% of construction contractor failures are due to economic factors'.

11.3.1   Movement of funds

Companies prepare a record of where their cash has come from and also how it has been used. These can be considered as a summary of the whole year and will link to the other statutory reports, such as the profit and loss statement and balance sheet. These were discussed in Chapter 2. Companies report their cash flow statements in a standard fashion, as shown in the example from the Shepherds Building Group illustrated in Table 11.1, and although always reported at the end of a financial year, most will produce quarterly or even monthly statements during the year.

Table 11.1   Shepherd Building Group cash flow statement.

	Year 4	Year 3	Year 2	Year 1
Net Cash In(Out)flow Operating Activity	70800	44800	14200	-8900
Net Cash In(Out)flow Return on Investment	100	-2100	800	1500
Taxation	-5000	-1000	-600	-8800
Net Cash Out(In)flow Investing Activity				
Capital Expenditure & Finance Investment	-14800	-6400	-28000	-37700
Acquisition & Disposal	2600	500	2800	-200
Equity Dividends Paid	-6600	-6300	-6300	-6300
Management of Liquid Resources				100
Net Cash Out(In)flow from Financing	-36500	-12100	16800	15000
Increase(Decrease) Cash & Equivalent	10600	17400	-300	-45300

Net cash in

The net cash figure shown in row one of the table is the inflows less the outflows, which correspond to the operating profit reported in the profit and loss accounts. It is worth noting that the company has improved its cash flow position over the last four years. As can be seen from the trend, the company has made a great effort to improve its effectiveness in recovering its cash from operating activities. The next row indicates the returns on investments or, if negative, the payments made in the form of dividends or interest paid.

Net cash out

The next five rows indicate where the cash out has been reported; capital expenditure, finance investment and acquisition are indicated. The equity dividends paid are shown as is the net cash flow from financing other projects. The increase or decrease in cash and equivalent is then shown. Further detail in the form of notes on the cash flow statement would provide the components of these summary figures.

11.3.2 Cash flow forecasting

As well as the formal record included with the profit and loss statement and balance sheet, a company should provide a cash flow budget or forecast. This can be defined as an estimate of the timing and amounts of cash inflows and outflows over a specific period (usually one year). A cash flow forecast shows if a firm needs to borrow, how much, when, and how it will repay the loan. It provides vital information on the efficiency of the firm and also allows the firm time to plan if it requires to source finance.

Cash inflows are the receipt of cash into a business. They would include payment for work complete, interim payments for continuing work, payment for materials on site and payment from other organisations for services offered.

The cash outflows are the transfer of cash from the business to creditor organisations. This would include payment for materials, subcontract payments, staff salaries, repayment of loans as well as purchase of capital equipment.

The net monthly cash flow is simply the balance of a month's total cash inflows in relation to the months total cash outflows. Businesses need to forecast the cash flow on a monthly basis to allow them to predict their finance requirements and to estimate whether they have the capacity to undertake additional work in the future.

11.3.3 Client and contractor

A construction project client does not use a cash flow projection in the same way as a contractor, as generally the funding has been secured prior to the project commencing and the client 'draws down' from the fund to pay interim valuations. The client therefore only requires an estimate of the monthly amounts that will need to be paid. These amounts will often include costs for site acquisition, planning and design team fees, which will be incurred well before the construction project commences. A client's quantity surveyor (QS) can use fairly unsophisticated techniques to model the cash requirements of the client.

There are a number of techniques or models that are used to forecast future movements of cash, some are fairly simplistic and are used by client organisations to give them a series of monthly figures that can be used for payment purposes. The contractor's situation is a great deal more complex, as the contract payment is only one side of cash flow forecasting; payment to suppliers, subcontractors and other creditors require a more sophisticated approach.

Forecasting is slightly different to budgeting in that budgeting is more of an active approach and plans for the future whereas forecasting is more passive and focuses on past experience and estimating from this past data. It is therefore imperative that the forecasting model used is as accurate as possible. A contractor's cash flow forecast will try to predict the cash coming into and going out of the organisation, based on previous experience, often in order to produce a cash flow budget that sets a plan to generate income exceeding expenditure.

A cash flow forecast enables organisations to look into when payments are likely to take place in the future and the dates at which these will occur. It is essential to find the lead-in time between incurring expenses and paying out for these, and the time lag between receiving payments and paying the supply chain.

For the contractor, there can be a significant lapse in time from the point at which the contractor wins the work, incurs labour and material costs and is actually paid for the completed work. Consequently, forecasting of the project cash flow is essential. Cash flow projections can be revealing to contractors. A contractor's most profitable period can be the period of greatest cash flow needs from outside sources. Thus, the old adage of profit is sanity, turnover is vanity but cash flow is king often holds true.

11.3.4   Cash flow forecast limitations

Cash flow forecasts are more often than not carried out at the estimate stage and pre-tendering stage, so they are not a precise forecast of reality, they do not include for the unforeseen circumstances that can occur on a project, and the uncertainties and risks that will only come to light once the project is on site. Cash flow forecasts also fall short in that they do not account for time delays for costs and earned values, and they do not take account of potential variations and claims.

Although forms of contracts specify the extent to which clients may delay payments to contractors, the contractor will need to be able to forecast its cash requirements in relation to what will happen in practice, and this can be affected by individual client behaviour. No construction project is likely to ever be in completed in accordance with the initial planning, and as such any cash flow forecasted at the pre-contract stage is unlikely to be completely accurate. There are more than likely going to be discrepancies between forecasted interim measurements/payments and actual progress. But cash flow is a reality and it is essential to find a cash flow model that can be adapted to the specific requirements of the project and that includes for the uncertainties relating to the construction business and specific jobsite procedures.

Cash flow forecasts are needed at both the on-site stage of current contracts and contracts still to be tendered for. Forecasts are needed at the tendering stage to estimate the finance needed for the project, but contractors often do not have the time, resources or funds to plan detailed schedules before contracts are awarded.

The variables that influence the cash flow are not always accounted for in simple cost models. Contract conditions and items involved in interim valuations, such as tendering strategies, estimating errors, specific dates of interim payments, subcontractors retention and so on have been incorporated in more detailed cash flow models, and have been shown to influence the cash flow prediction. But a cash flow model based on a large scale data analysis from bills of quantities and construction programme details has yet to be developed and, as a result, there is still the potential for improvement in the accuracy of the cash flow model. During the construction process there are also many influential factors on the cash flow, namely, cost over-runs, time delays, variations and technical changes. The aim for cash flow forecasting is to provide as accurate, flexible and comprehensible a forecast as is possible. It seems that in order for this to happen, the cash flow model needs to be developed to account for these influential variables.

Factors that would be expected to influence cash flow are linked to those that affect the timing of expenses occurring for construction works. It would be expected that design variables affecting the intensity of work at commencement would influence valuations at the early part of the programme. For instance, the type of construction will influence the time needed in the construction programme for labour and materials to be deployed onto the project and accounted for in valuations. It has been reported that steel frame construction often involves greater expense at the commencement of the works compared to other forms of construction due to the high costs of steel. The value forecast for a steel frame building would therefore be different than a concrete frame.

Another consideration that influences the cash flow would be the variable of the project procurement route. Different procurement routes influence how construction teams are organised – under the traditional procurement route it would not be normal for the constructor to pay for a design team whereas a design and build procurement approach may require the contractor to appoint a design team at tender stage and pay for a proportion of the design before the site has been possessed. The consequence of this is that it is impossible to find one model that suits all projects; most organisations use a number of approaches and often combine them in a pragmatic fashion to find the one that is most likely to represent their needs.

11.3.5 Simple forecasting models

A simple technique used is the ¼:⅓ model, This suggests that the first 25% of expenditure is within the first 33% of the programme time, the next 50% of expenditure within the next 33% and the remaining 25% of expenditure in the last ⅓ of the programme time. As an example, for a construction project of £10 million over a twelve-month period, an expenditure profile is illustrated in Figure 11.3.

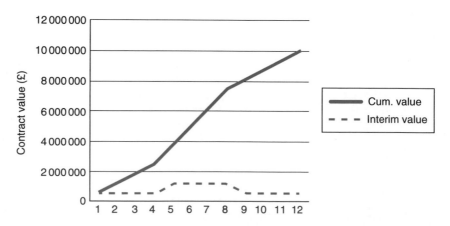

Figure 11.3   Cash flow forecast using the ¼:⅓ rule.

Table 11.2   Schedule of payment receipts based on the ¼:⅓ model.

Month	Value	Cumulative
1	625 000	625 000
2	625 000	1 250 000
3	625 000	1 875 000
4	625 000	2 500 000
5	1 250 000	3 750 000
6	1 250 000	5 000 000
7	1 250 000	6 250 000
8	1 250 000	7 500 000
9	625 000	8 125 000
10	625 000	8 750 000
11	625 000	9 375 000
12	625 000	10 000 000

As can be seen the intensity of the project, as measured by the curves' gradient, commences at a slower rate than the middle ⅓ of the project before returning to the same rate as the first ⅓ of the programme. Table 11.2 indicates the monthly gross values.

The data can be expressed in a table as shown in Table 11.3 that can be easily understood by all parties and is easy to use if a quick and rough projection is required. Little information is required to be input in order to get an idea of future project payments.

A slightly more sophisticated model was developed in the late 1970s. It also required little data input but was purported to give a more accurate output. K.W. Hudson, in 1978, was one of the first authors to develop a cash flow model for the forecasting of cash flows. He developed the two parameter **'DHSS expenditure model'** that is the model most widely used for expenditure forecasting. Hudson developed the model after the analysis of data produced from a large sample of hospital projects, providing values of parameters over a range of project sizes. The model is so well acknowledged that the Building Cost Information Service (BCIS) uses an 'S-curve' formula based on work done by the Hudson for the DHSS model to calculate cash flow forecasts.

The 'S-curve' formula is:

$$y = S(x + Cx^2 - Cx - (6x^3 - 9x^2 + 3x) / K)$$

where y is the cumulative monthly value of work executed before deduction of retention monies or addition of fluctuations, x is the proportion of contract completed expressed as the month in which expenditure y occurs divided by contract period and S is the contract sum. C and K are parameters which determine the shape of the curve.

Changing the C and K parameter values can alter the shape of the S curve; they can be values chosen to optimize the cash flow forecast and to best suit the individual construction project. According to Hudson, different parameter values should be applied to different types of construction. Altering parameter C changes the build-up time and run-down period while altering parameter K affects the rate of expenditure over the central part of the graph.

Table 11.3 Table indicating time, payments and cost.

Time (weeks)	1	2	3	4	5	6	7	8	9	10
Time (Months)				1				2		
Cum. value	48420	134738	255681	407977	588356	793546	1020275	1265271	1525263	1796980
App for payment						793546				1796980
Certificate issued							793546			
Payment received										769739
Margin	4842	13474	25568	40798	58836	79355	102027	126527	152526	179698
Cost	43578	121264	230113	367180	529521	714191	918247	1138744	1372737	1617282

Table 11.4   C and K parameters for the DHSS S-curve cash flow model (RICS, 2003).

Contract value (2000 prices) (£)	Standard parameters	
From	C	K
26 000–130 000	–0.439	5.464
130 000–260 000	–0.370	4.880
260 000–520 000	–0.295	4.360
520 000–1.3 million	–0.220	3.941
1.3–5.2 million	–0.145	3.595
5.2–9 million	0.010	4.000
9–14 million	0.110	3.980
14–19 million	0.159	3.780
19–23 million	0.056	3.323
>23 million	–0.028	3.090

(Reproduced by permission of BCIS).

Hudson focused his model on the 'contract value' variable and thus altered the C and K parameters to reflect the contract value, as can be seen from Table 11.4, taken from the BCIS.

Hudson acknowledged that his model had flaws and observed that '...Difficulties are to be expected when trying to apply a simple mathematical equation to a real life situation, particularly one as complex as the erection of a building'. Other research has suggested different approaches, Kenley and Wilson (1986) produced a cash flow model that took account of the variables, contract sum (V) and the percentage of time complete (d). The Kenley-Wilson formula is:

$$V = 100F / (1 + F)$$

where $F = e^a[d/(100-d)]^b$ (a and b are constants), V is the contract value and d is the percentage of time complete.

The slight flaw apparent with this model is that when the project is complete, hence the percentage of time complete is 100%, the value of (100-d) will equal zero, causing an error in the calculation; for example $F = e^a[100/0]^b$. Therefore, it is necessary to omit the final monthly valuation from the study, where the project is fully complete.

Once more information is available to the project team, a more accurate prediction of project values can be made. A cost plan when used with a construction programme can give a more accurate projection of project values over time.

Table 11.5 shows a group elemental cost plan with programme period. This can be simply developed to take account of retention and also introduce the important element of the lag period from completing the works to when the contractor receives the payments. Table 11.6 indicates the payment delay and month by month payments.

The forgoing models have only really dealt with one aspect of cash flow, that of project value. The nature of the constructors business is that both income and expenditure on a project require prediction and control.

As discussed in Chapter 7 the delay to a contractor receiving payment is a result of the project contract conditions. A contractor needs a cash flow forecast to identify the timing and significance of when a closing balance at the end of a month is forecast to be negative. This is to allow for the identification of possible corrective actions, which may relate to expediting payment from client or delaying payment to

Table 11.5 Group elemental cost plan with programme period.

Element	Elemental total (£)	Programme period	Cost of element per month (£)
Substructure	440684	April	440684
Superstructure	2857686	May–October	476281
Internal finish	712294	July–January	101756
Fittings	91261	June–July	45631
Services	2822904	June–December	403272
External Works	63031	December–January	31516
Preliminaries	485787		
Total	7473647		

Table 11.6 Payment delay and month by month payments.

Project month	Cost (£)	Retention 5% (£)	Net (£)	Cumulative total (£)	Cum. total inc. delay (£)
April	489263	22034	467229	467229	
May	524860	23814	501046	968274	467229
June	973763	46259	927504	1895778	968274
July	1075519	51347	1024172	2919950	1895778
August	1029888	49065	980823	3900773	2919950
September	1029888	49065	980823	4881595	3900773
October	1029888	49065	980823	5862418	4881595
November	553607	25251	528356	6390774	5862418
December	585123	26827	558296	6949069	6390774
January	181851	6664	175187	7124256	6949069
February					7124256

suppliers or subcontractors or transferring funds from one part of the business to another or discussing overdraft facilities with a bank.

11.3.6 Credit terms

A contracting business organisation has a range of creditors which offer their materials or services on varying terms of credit. The shortest period tends to be from labour. Gross hourly labour costs for employers include national insurance contributions, holiday and sick pay allowances, clothing allowances as well as discretionary bonus and non-productive payments. Consequently, the hourly rate the site operative takes home is considerably less that that paid by the employer.

Over the last three decades there has been a significant reduction in the direct employment of labour by large main contractors. The main reason for this is the cost of administering a workforce that may spend a great deal of time non-productively. The labour-only subcontractor has filled this gap. Labour costs on a project can amount to 30%. However, this could change significantly if differing SME organisations are considered. For instance, a painting subcontract may have a significantly higher labour cost component per £1000 turnover than, say, a steelwork contractor.

The extent of resources that require relatively quick payment can have a significant impact on the cash flow and financing requirements of an organisation. The payments to labour tend to be weekly, one week in arrears. At the end of each week it is not unusual for the QS to undertake bonus or payment measures with the various direct labour-only subcontract trades and agree an amount to be paid the following week.

Materials suppliers extend credit terms to a contractor; this is usually 30 working days from the contractors' receipt of invoice. Invoices are generally produced monthly from agreed materials received records held by both contractor and supplier. Materials can contribute 30–50% of projects costs and require careful management. Suppliers will deliver materials and charge a gross cost, costs associated with storage; transport and wastage are born by the contractor.

Plant can range from small tools to large items such as cranes. It can be owned by the organisation or hired, as is usually the case, from specialist plant hire companies. These organisations have sophisticated systems for recording the type of plant, the hire period and generally charge for transport to and from the site. The credit terms are similar to materials, in that they will invoice at the end of the month, listing plant on hire and the charges, and will require payment 30 days from contractor's receipt of invoice.

The costs associated with subcontractors can be as much as 80% of a project's value and it is no surprise that their payment is one of the most contentious aspects of the construction industry. Their payment terms are normally within the form of contract between contractor and subcontractor. These terms are included within the tender send out and all parties should be aware of them before the contract is commenced. Normally, subcontractors will submit an application for payment and then be paid some 21 days after the application has been agreed. This may mean that subcontractors may have a period from undertaking work until first payment of up to 40 days.

Overheads are charged to projects in two ways. The costs of overheads directly employed by the site are allocated monthly. These can range from the employment of site staff, such as the construction manager, site agent, site engineer and quantity surveyor, to the cost of accommodation, which may be hired from a plant firm. Most companies code the different overhead costs separately from costs associated with the works to provide an accurate feedback to the management team regarding the adequacy of their forecasts at tender stage. The other costs are associated with head office contribution; a fixed percentage may be applied to the project to pay for head office services such as estimating, planning, accounting and senior management. Sometimes divisional staff have percentages of their time allocated. It is not uncommon for these allocations to be rather arbitrary once the project has commenced.

The DHSS S-curve approach can be used to illustrate the how a delay in receiving payment can influence the contractor's cash requirements, as illustrated in the next section.

11.3.7 Minimum and maximum cash requirements

The contract cash flow is shown in Table 11.7.

For explanatory purposes, costs are assumed to be value less margin and are incurred from the commencement of the contract.

If the S-curve formula is applied with C and K values of 0.11 and 3.98, respectively, an S-curve which forecasts interim payments as shown in Figure 11.4 is obtained.

Table 11.7   Contract cash flow example.

Contract sum, £	10 000 000
Contract period, months	6
Retention, %	3
No. contract payments	6
Margin, %	10
First payment details	
Application	Week 6
Certification	Week 7
Payment received	Week 10

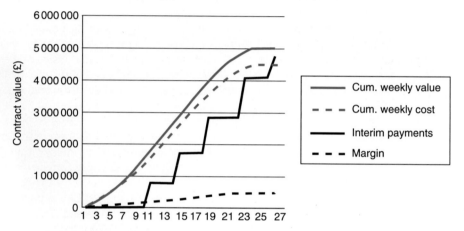

Figure 11.4   Interim payments.

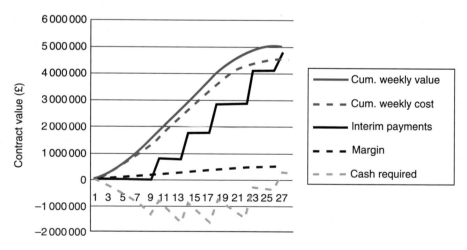

Figure 11.5   Cash requirement.

If the cash requirement is added to the contract – that is the shortfall between income received and costs paid – the contractor's cash requirement will be as shown in Figure 11.5.

As can be seen from Figure 11.5, the cash required varies throughout the contract. It peaks at week 17 at £1634151. The cash situation only becomes positive towards the end of the contract. The spreadsheet for this is on this book's companion website if the reader wishes to change some of the parameters and see the results.

11.3.8 Capital lock up

As can be seen from the example, there are periods where the contract requires financing and some periods when it generates finance. This has been termed 'capital lock up', that is capital is locked up in the project at the negative stages waiting to be released as the project is undertaken. There are a number of means of reducing the extent of lock up that are used by contractors which effectively work in one of two ways: enhancing the value of monies in and delaying the timing of monies out.

To bring forward the money into the firm to a period before it would normally be received, project activities that commence and complete at the start of the con-struction programme may be loaded with additional margin or on-cost. A corre-sponding adjustment would be made for later items, otherwise the margin would be increased to an uncompetitive level. This effectively increases the project margin when the contract would normally require the most funding.

Another method often deployed by contractors is to overclaim works undertaken at the start of the project, these can vary from preliminaries (claiming for set up and enabling costs prior to payment for these items), overmeasuring works completed by own resources or subcontractors or claiming for payment on account for varia-tions yet to be agreed. Obviously, ensuring early payment for work undertaken will reduce capital lock up as well.

The delaying of payment to debtors is, unfortunately, the one used by most organ-isations. There is a balance to be struck in this area. Many subcontractors refuse to work for certain contractors who have poor payment records. A main contractor who takes this route may find that a premium will be added to subcontract quota-tions to take account of the costs of financing a subcontract work package that are a result of tardy payment.

The above example is used to illustrate how changing some of the parameters can have an impact upon the capital lock up, cash required and cost of this cash.

11.3.9 Expediting receipts

As shown in Figure 11.6, the first payment is now received in week 7 and the design team has issued the certificate immediately. The payment period from certificate issue remains at 14 days as before. The cash required position on the contract has now altered, the maximum cash needed now falls at week 14 and now has a value of −£1367770. This is a reduction in cash required from the previous position of −£266381.

11.3.10 Delaying payment to suppliers

As discussed earlier, the costs do not get incurred at the commencement of the con-tract. Credit is offered by suppliers and subcontractors, which has the impact of delaying the movement of costs from the project. If the costs outflows lag by four weeks, this has the following impact: by delaying payment to creditors the cash situ-ation is improved, as can be seen in Figure 11.7, where the lagged cumulative cost curve moves closer to the interim payment line, thus reducing the cash requirement.

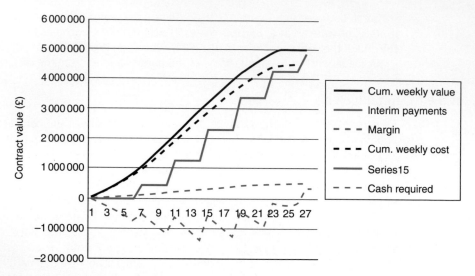

Figure 11.6 Impact of expediting receipt of payment.

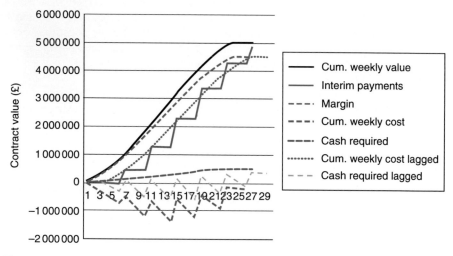

Figure 11.7 Impact of delay in paying creditors.

By delaying the payments to suppliers by four weeks, the maximum cash required is now in week 16 and is approximately £84000. This is a significant improvement on the previous situation where there was no delay. The contract can also be seen to be generating cash surpluses as receipts are received before payment is made.

The situation above is idealised, it would be a lucky contractor that could include a 10% margin in today's economic climate and the cost flows are not quite as simplistic as the above would suggest. The example in Figure 11.8 gives a better idea of the complexity that can arise when modelling cash flows using S-curve data; it illustrates a more complex example. The S-curve predicts the value at certain points during the contract; it can be used weekly or monthly. In this example it is used monthly. The processes for a contractor getting paid are discussed in Chapter 12; the application for payment can be made up to seven days before the due date for payment, it can

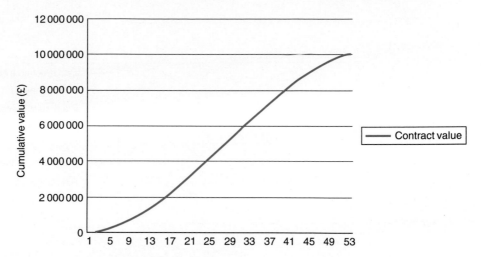

Figure 11.8 A more complex payment example.

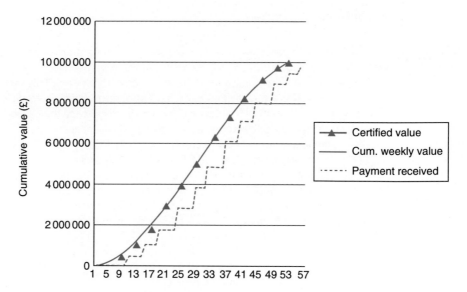

Figure 11.9 Value curve and payment delay.

then take a further five days for the valuation certificate to be issued and then a further 14 days before payment is received.

Consequently a delay of four working weeks from valuation to payment is not uncommon. The due date for payment can often be as much as six weeks after possession of the site. The consequences of this are that the contractor has to find funds to pay for the resources used before the employer pays the interim valuation. This is illustrated in Figure 11.9.

As discussed above, most contractors have credit terms offered by their suppliers. The credit terms for plant and materials suppliers are that they usually require payment within 30 days of submission of invoice. The invoices are usually submitted monthly.

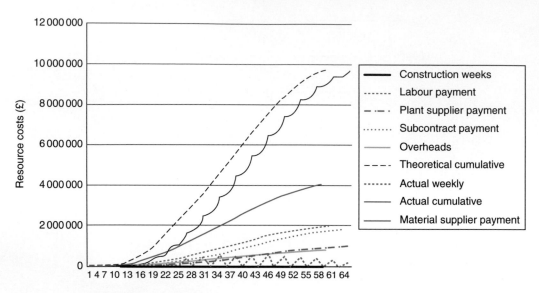

Figure 11.10 Indicative project cost flows.

Subcontract payments have been the source of contention since subcontracting began. Normally subcontractors will get paid 21 days from their application for payment but it is normal for their application to take five days to process. This means that the subcontractor is effectively giving the main contractor five weeks credit on the work undertaken. It is not unusual for the main contractor to increase the proportion of work undertaken by the subcontractor when it comes to their application to the employer.

If the example is extended now to consider the payments made as well as received, the picture becomes more complicated. Figure 11.10 indicates costs flows on the project.

As identified above, the contractor must sometimes fund the project, as there will be periods when the outflow of cash is in excess of that received. Using a spreadsheet model the cumulative outflow of monies paid for labour, materials, subcontractors, plant and overheads can now be compared with income received.

11.3.11 Project cash flow

As can be seen from Figure 11.10, the project requires funding to differing amounts as it progresses. At some periods the value is in advance of the costs whilst at others the delay can be quite significant. The implications of this are that a project may have a capital requirement and the timing and size of this can be predicted. Figure 11.11 summarises the project cash flow position over the 58-week payment period.

The cash requirement on the contract ranges from £2800 in week one to a maximum of £560 000 in week 42. If this was to be funded by a short-term overdraft at a rate of 12%, the overall cost of finance would be approximately £17 000, excluding the cost of arrangement. This might not appear a large amount. However, the project margin is only 3% and this finance cost represents nearly 6% of the margin. This should be balanced with the fact that the contract has periods of maximum liquidity also during the contract.

If there were a number of contracts where the maximum capital requirement peaked at the same time, it could take the business over its pre-arranged overdraft limit and make credit very difficult to obtain in the future.

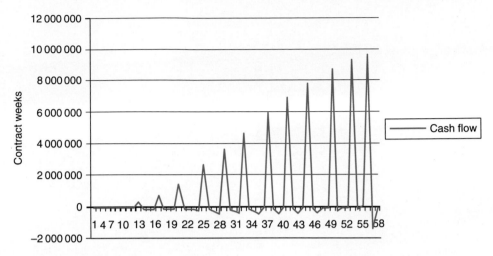

Figure 11.11 Project cash flow.

A project cash flow budget is required every month; this takes account of any projects which are likely to have a cash movement. The short-term forecast is usually based on current figures and project projections for three months in the future. The frequent monitoring of the forecast is essential to inform management action if required. Unlike the example above which uses an S-curve formula to predict value and costs, a contractor would have much more detailed design information to hand.

The following information would be available to support a detailed cash flow forecast for projects that had been won:

- a construction programme;
- a cost plan/bill of quantities;
- a priced preliminary schedule;
- a buying programme indicating major materials purchase;
- a subcontract procurement programme indicating:
  - trades
  - values
  - commencement dates.

This information would support a detailed project cash flow budget that could be updated should delays or variations occur. Table 11.8 shows a project cash flow exemplar.

Table 11.8 Project cash flow exemplar.

Contract	Liverpool	123	Surveyor	ADR	Date of projection		
MONTH	Contract valuation dates	Date cash expected from client	Payment Advice to accounts	Payment notice to subcontractor	Subcontract cheque release date	Monthly cash receipts	Cumulative cash received
1							
2							
3							

Table 11.9 Cash flow projection.

North Region – Manchester
Cash flow projection
16 months ending August

	Name	At 30/04	May	Jun	Jul	Aug	Sep	Oct	Nov	Dec
1011	Building 108 Barracks	2 068 243	–	–	–	–	–	–	–	–
1020	RMBI Poynton	1 555 871	–	–	–	–	–	–	–	–
1021	RBS, Preston	301 200	–	–	–	–	–	–	–	–
	Total									
	A = anticipated									
	B = new works									
	C = final accounts									

The project cash flow budgets are then aggregated to produce an area/divisional/company summary which would be updated monthly. Usually, the project cash flow forecasts are included with the project cost value reconciliation and discussed at the monthly progress meeting. It is important to categorise the cash flow budgets to reflect the certainty of the forecast; usually categories such as (i) new work, (ii) work won but yet to be started and (iii) final accounts are used. Table 11.9 indicates an aggregated cash flow projection pro forma used by a medium sized contracting company.

11.3.12 Organisational cash flow

The importance of keeping close control of cash flow cannot be emphasised enough if a project is delayed, a settlement postponed or a payment made early. These can all have an effect upon the aggregated position and may cause the contractor difficulty in finding the necessary funds. Consider the case where a number of projects require funding at different periods and the organisation has an overdraft limit. If a payment on a contract is delayed or a creditor requires payment before the scheduled period it could have a dramatic impact upon the ability of the contractor to find the funds required, as shown in Figure 11.12.

How does a contractor pragmatically manage all this complexity of money flowing in and out of the organisation? By the use of 'cut-off dates'.

Using cut-off dates, the organisation will seek to establish a series of cash received dates. The cash received dates indicate when all project payments are to be received. Project quantity surveyors will use these dates and attempt to agree a series of dates with the client's representatives for the application for payment, payment notice or certificate and then the date for receipt of payment. Often these dates are inconvenient for the project contractors QS who has to undertake a lot of adjustments for work in progress when undertaking the cost value reconciliation.

Simple seemingly inconsequential issues, such as delays in receipt of notices, failure to agree in a timely fashion outstanding items and incomplete applications for payment, can have a dramatic knock-on effect to the businesses cash flow. It is

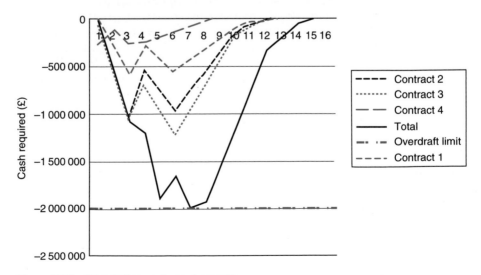

Figure 11.12 Cumulative cash requirements.

Table 11.10   Aggregate cash budget.

	Jan	Feb	Mar	Apr	Total
Receipts	550	750	1200	250	
Total	550	750	1200	250	2750
Payments					
Creditors	250	300	350	400	1300
Interest	50				50
Taxation				40	1350
Dividends				30	30
Wages	35	35	35	35	140
Overheads					
Fixed		15			170
Variable	12	16	20	15	63
Fixed assets					
Total	347	366	405	520	1638
	203	384	795	-270	1112
Opening balance	-300	-97	287	1082	972
Closing balance	-97	287	1082	812	

essential that the project QS understands the payment mechanisms and ensures all parties are aware of their obligations in making sure the processes work effectively.

To complete the picture about the aggregate position, the company accountant will produce a cash budget for the year; this will summarise the monthly income (receipts) and outgoings (payments). A format similar to that in Table 11.10 will be used.

11.4   Working capital

Working capital is a measure of a company's efficiency and its short-term financial health. A simple calculation of **working capital = Current assets - current liabilities**. If the ratio is positive it means that the company is able to pay off its short-term liabilities. If it is negative then the company is unable to meet its short-term liabilities with its current assets (these include cash and accounts receivable). An assessment of the working capital of the company is usually taken from the company's balance sheet, which provides a snapshot of the company's health and can be done annually, quarterly or even monthly.

Considering a summary of the Shepherd Group balance sheet for the last four years (Table 11.11) shows the extent of fixed and current assets.

11.4.1   Current assets

Current assets are a little theoretical when considered from a construction organisations perspective. There are two types of assets, fixed (these include land, buildings, plant and equipment) and current. Construction companies tend to have fewer fixed assets relative to their company turnover and a useful way of considering a company's health is to consider its current assets. The main kinds of current assets are cash, work in progress, stock and debtors.

Table 11.11   A typical balance sheet report indicating four years of reporting assets.

Shepherd Group Balance Sheet				
	Year 4	Year 3	Year 2	Year 1
Fixed Assets				
Tangible Assets	120000	120100	132200	123000
Land & Buildings	29300	29900	31300	30100
Freehold Land	29300	29900		
Leasehold Land				
Fixtures & Fittings	0	0	0	0
Plant & Vehicles	90700	90200	100900	92900
Plant	90700	90200		
Vehicles				
Other Fixed Assets	0	0	0	0
Intangible Assets	3700	7100	7600	8500
Investments	100	3800	4600	7900
Fixed Assets	123800	131000	144400	139400
Current Assets				
Stock & W.I.P.	150800	164400	181300	174800
Stock	150800	146200		
W.I.P.	0	18200		
Finished Goods				
Trade Debtors	74800	83500	38900	43400
Bank & Deposits	46600	18100	2000	1300
Other Current Assets	31000	26000	81200	99800
Group Loans (asset)	0	0	0	0
Directors Loans (asset)	0	0	0	0
Other Debtors	31000	25800	81200	99800
Prepayments				
Deferred Taxation		200		
Investments	2000	1400	1500	1500
Current Assets	305200	293400	304900	320800

Cash

The forgoing section has illustrated how a cash flow forecast and a cash flow budget can identify the organisations cash position. Current, future cash generated and future negative cash requirements can be calculated. The figure in the cash book at the end of the period being considered should be taken.

Work in progress

This is the price of the project work undertaken since the last valuation on each project and up to the end of the period being considered. It is often very difficult to assess work in progress as there are few external measures by which to validate its veracity.

Stock

Construction materials that have yet to be incorporated into the works have a value. They have yet to be sold and can be considered to be a current asset. Of course, the organisation may have yet to pay for them, in which case the cost of such materials will also be considered as a liability.

Debtors

Debtors is the value of money owed to the organisation by its clients at the end of the accounting period for which the current assets are being calculated. Often outstanding claims are included in this aspect of the calculation and care should be taken when considering whether or not the full value of these should be included.

Table 11.12   A typical balance sheet report indicating four years of reporting liabilities.

	Year 4	Year 3	Year 2	Year 1
Current Liabilities				
Trade Creditors	-92500	-89100	-100900	-119400
Short Term Loans & Overdrafts	-800		-1300	-400
Bank Overdrafts	-800		-1300	-400
Group Loans (short t.)	0		0	0
Director Loans (short t.)	0		0	0
Hire Purch. & Leas. (short t.)	0		0	0
Hire Purchase (short t.)				
Leasing (short t.)				
Other Short Term Loans	0		0	0
Total Other Current Liabilities	-88800	-77700	-109000	-90100
Corporation Tax	-3400	-2300	0	-700
Dividends	0	0	0	0
Accruals & Def. Inc. (sh. t.)	-49700	-44500	-62500	-53300
Social Securities & V.A.T.	-7900	-9800	-7700	-10200
Other Current Liabilities	-27800	-21100	-38800	-25900
Current Liabilities	-182100	-166800	-211200	-209900
Net Current Assets (Liab.)	123100	126600	93700	110900
Net Tangible Assets (Liab.)	243200	250500	230500	241800
Working Capital	123100	126600	93700	110900
Total Assets	429000	424400	449300	460200
Total Assets less Cur. Liab.	246900	257600	238100	250300
Long Term Liabilities				
Long Term Debt		-20000		-3200
Group Loans (long t.)		0		0
Director Loans (long t.)		0		0
Hire Purch. & Leas. (long t.)		0		0
Hire Purchase (long t.)				
Leasing (long t.)				
Preference Shares				
Other Long Term Loans		-20000		-3200
Total Other Long Term Liab.	-5000	-4200	-5800	-5400
Accruals & Def. Inc. (l. t.)	0	0	0	0
Other Long Term Liab.	-5000	-4200	-5800	-5400
Provisions for Other Liab.	-5500	-5600	-16000	-14900
Deferred Tax	-5500	-5600	-6000	-6200
Other Provisions			-10000	-8700
Pension Liabilities	1300	-5000		
Balance sheet Minorities				
Long Term Liabilities	-9200	-34800	-21800	-23500
Total Assets less Liabilities	237700	222800	216300	226800

Table 11.13   Typical financial summary showing four years performance.

Shepherd Group Financial summary				
	Year 4	Year 3	Year 2	Year 1
Turnover	251274	185212	326890	329202
Profit (Loss) before Taxation	4562	−1943	−15215	−12995
Net Tangible Assets (Liab.)	23508	19619	21228	11055
Shareholders Funds	19356	16074	16013	6562
Profit Margin	1.82	−1.05	−4.65	−3.95
Return on Shareholders Funds	23.57	−12.09	−95.02	−198.03
Return on Capital Employed	19.41	−9.90	−71.67	−117.55
Liquidity Ratio (x)	1.21	1.19	1.13	1.03
Gearing (%)	68.19	46.45	62.04	113.08
Number of Employees	525	710	908	961

11.4.2 Current liabilities

These can be categorised as loans, bank overdraft, creditors and tax liability. The loans will be based on the amount outstanding and the bank overdraft can be taken from the cash flow statement. Creditors are organisations that the business owes money to. The value of the creditors liability can be based on invoices received for services or works undertaken and accruals (value of services or works undertaken but for which payment has yet to be made). Table 11.12 indicates a four year summary on liabilities.

11.4.3 Profitability ratio

The return on net assets = net profit (before tax)/assets − current liabilities gives an indication of the profitability of the organisation, as shown in Table 11.13. This is sometime referred to as return on capital employed. At the time of writing the construction industry is in deep recession and the net profits of organisations is very low which means the return on capital employed is low and, consequently, companies are struggling to show a profit or return for their shareholders. The Shepherds Group, however, has improved its position.

References

Hudson, K.W. (1978) DHSS expenditure forecasting method. *Chartered Surveyor: Building and Quantity Surveying Quarterly*, **5** (3), 42–45.

Kenley, R. and Wilson, O.D. (1986) A construction project cash flow model – an idiographicapproach. *Construction Management and Economics*, **4**, 213-232.

RICS (2003) BCIS support FAQs (6 January 2003). *Chartered Surveyor B and QS Quarterly*, **5** (3), 42–45 (http://service.bcis.co.uk/support.html).

Russell, J. and Jaselskis, E. (1992) Predicting Construction Contractor Failure Prior to Contract Award. *Journal of Construction Engineering and Management*, **118** (4), 791-811.

12

Project governance

Financial Management in Construction Contracting, First Edition. Andrew Ross and Peter Williams.
© 2013 Andrew Ross and Peter Williams. Published 2013 by John Wiley & Sons, Ltd.

12.1 Introduction

As explained in Chapter 8, the UK Corporate Governance Code 2010 ('the Code') provides a set of best practice guidelines for good corporate governance (FRC, 2012). In this context, the board of directors is responsible for determining the nature and extent of the significant risks that it is willing to undertake in striving to achieve the strategic objectives of the business. Section C.2 of the Code requires the board to maintain sound risk management and internal control systems and points to the guidance suggested in the Turnbull Report (ICAEW, 1999) as a means of applying this part of the Code.

Accountability, transparency, probity and **long-term sustainable success** are the supporting principles of the Code and the effective stewardship of the business is influenced by the extent to which these goals are achievable in the context of the portfolio of contracts undertaken by the business.

In common with the corporate governance of a business, project governance may be defined as the system by which projects are directed and controlled. The project governance régime sets the tone and focus for the financial management of the contract which starts with the contractor's tender, develops through the external valuation and culminates in the monthly cost value reconciliation (Chapters 17 and 18).

Four principal factors determine the governance framework for a project:

1. the **procurement route** chosen by the client (employer);
2. the **conditions of contract** that will apply to the project;
3. the **method of measurement** applicable;
4. bills of quantities (or other pricing document).

Most people would recognise that the procurement route is the factor that determines the relationships between the participants in a construction project and the allocation of risk between the parties. They would also recognise that the conditions of contract formalise these risks and relationships by creating a legally binding agreement that sets out the rights and obligations of the parties and which is then subject to the statutory and civil law of the land. Many would be surprised to learn, however, that the method by which the works will be measured and valued also plays a prominent role in the governance of a construction contract, but it is nevertheless the case. This is because the method of measurement – whether a formal standard method of measurement, or a hybrid document or an informal 'in-house' system of measurement – read in conjunction with the conditions of contract, determines, to a significant extent, the balance of risk between the parties. This is then expressed in a formal document – the bills of quantities, for example – which is then incorporated into the contract (or subcontract).

Whilst the RICS (2011) reports that bills of quantities are on the decline, it is, nevertheless, the case that, firstly, this refers to 'formal' bills of quantities prepared as part of contract documentation on traditionally procured projects and, secondly, that the increase in popularity of design and build and drawings and specification arrangements means that the use of 'informal' bills of quantities, or similar pricing documents, must be on the increase. This is an inevitable conclusion, based on evidence of a decline in traditional procurement methods, because, despite the procurement route chosen, the price for a construction project must be based on some sort of recognition of the quantity of work involved in order that an estimate can be prepared. Complex construction projects cannot be priced on 'the back of an envelope'!

This chapter concerns project governance issues relating to the procurement methods commonly used in the industry, to some of the popular standard conditions of contract and subcontract and to the measurement conventions established, industry wide, irrespective of whether formal or informal pricing documents are employed. In this regard, most contractors will adopt measurement conventions at least loosely based around standard methods of measurement when preparing their so-called 'builders quantities' irrespective of the type of project concerned.

Construction is a credit-based industry that relies upon a system of payments whereby all members of the supply chain receive their just entitlement and are paid on time. However, the culture and complexities of construction work are such that this 'ideal' is rarely achieved and very few construction contracts are devoid of disagreements over money issues. Standard and non-standard conditions of contract, and a wide variety of procurement arrangements, influence the choice of payment system for specific projects but the propensity for people working in the industry to argue over money or to engage in 'sharp practice' is a key factor that often militates against the system working as it should. Life is full of people who want to get 'one jump ahead' and construction is no exception!

12.2 Procurement methods

The procurement of construction work is a well documented topic and standard texts worth reading include Morledge *et al.* (2006), Turner (1997) and Masterman (2002). Cooke and Williams (2009) provide a succinct coverage of the topic with some useful practical insights. Consequently, the focus here will be on the financial implications of procurement methods to the extent that they impact on the governance and financial management of projects.

There is no question that all authorities on the subject of procurement are of one mind with regard to traditional procurement methods and it is clear, in terms of the Corporate Governance Code, that directors' duties regarding **accountability**, **transparency**, **probity** and **long-term sustainable success** are undermined by this method of procurement. It is open to debate whether the alternatives available offer any tangible improvement.

12.2.1 Traditional

The traditional method of procurement adopted in construction is characterised by the separation of design from production and this factor alone brings into question issues of transparency and probity (moral correctness) because the two camps – the client's camp and the contractor's camp – have diametrically opposed objectives in terms of the

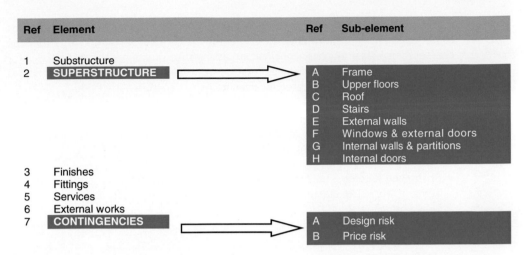

Ref	Element	Ref	Sub-element
1	Substructure		
2	**SUPERSTRUCTURE**	A	Frame
		B	Upper floors
		C	Roof
		D	Stairs
		E	External walls
		F	Windows & external doors
		G	Internal walls & partitions
		H	Internal doors
3	Finishes		
4	Fittings		
5	Services		
6	External works		
7	**CONTINGENCIES**	A	Design risk
		B	Price risk

Figure 12.1 Developing the cost plan.

outcomes from a construction project. Consequently, the measures of accountability for project outcomes are different for each camp. Neither is traditional procurement a sustainable method, as it is wasteful of the industry's resources and requires temporary organisations to be set up and dismantled for every project undertaken.

Traditional procurement effectively excludes the contractor from making any legitimate input into the design process and, more particularly, contributing ideas for more 'buildable' and cost-efficient construction methods. Once the contract has been awarded, any such suggestions can be no more than peripheral in their impact on the design or superficial in financial terms.

In building work, the financial management of a project during the design stages commences with the client's **budget** and is developed via the professional quantity surveyor's **cost plan**. The cost plan is an elemental model of the project which develops the budget, firstly in outline and then incrementally in detail, up to the point where scheme drawings are available; from this point, outline quantities can be taken, thereby enabling cost checks to be carried out on the cost plan. The process of cost checking facilitates modifications to the design so that the overall project stays within the ambit of the client's budget. The final stage of the cost planning process is to reconcile the cost plan with tenders received from the contractors which are vying for the contract.

The usual elements of an outline cost plan are illustrated in Figure 12.1 which also shows an example of how the elements are developed in more detail, by creating sub-elements, as the design develops. The contractors tendering for the contract are always interested in the cost plan but they are not party to it nor do they ever see it. If they were able to see it, they would be interested in the overall budget and in the contingency allowance made by the professional quantity surveyor (PQS) which is for **price risk** (inflation and market fluctuations) and **design risk** (allowances for potential design changes and as yet unknown expenditure). The client's budget would give a useful indication as to where to pitch their tenders and knowledge of the contingency allowance would assist in making informed judgements as to possible variations and the expenditure of provisional sums during the contract.

Some of this information can be picked up by contractors from the bills of quantities, where approximate quantities, provisional sums, daywork and general contingencies are stated and where mistakes in the quantities could indicate possible variations to the contract. This information then informs the commercial judgements that the contractor makes prior to tender finalisation and these judgements set the scene for the contractor's pricing strategy as regards front-loading the rates, moving monies into early items and loading preliminaries items to generate early cash.

Harris and McCaffer (2006) explore contractors' pricing strategies and explain some of the common terms used in the industry. It should be emphasised, however, that the contractor's probity is tested by traditional procurement and this leads to practices aimed at maximising profit and minimising negative cash flow:

- **Front-end loading** – maximising positive cash flow by moving margin from later items of work (e.g. plastering, painting) to earlier items (such as excavation and filling, drainage etc.).
- **Back-end unloading** – moving margin and net cost from later items of work to earlier items.
- **Loading rates** – transposing monies from one item to another so as to inflate unit rates for items where variations or increased quantities are expected.
- **Loading preliminaries** – moving monies from unit rates to early stage preliminaries items such as mobilisation of the site, establishing the site set-up, fencing and security, constructing access roads and so on.

Some of the disadvantages of traditional procurement may be overcome by adopting an accelerated traditional procurement strategy, which Morledge et al. (2006) suggest may be achieved either by letting a separate advance works contract or by letting the work using two-stage tendering or negotiation. From the perspectives of transparency, probity and sustainability, however, none of these suggestions would overcome the 'two camp' divide between the parties and, as Morledge et al. (2006) suggest, a strategy based on price competition is a recipe for adversarial relationships to develop.

12.2.2 Design and build

Masterman (2002) distinguishes between 'separated' and 'integrated' procurement systems and suggests that the integration achieved by adopting design and build procurement may overcome some of the problems associated with separation under traditional methods. He also explains the variants to design and build , including novated design and build, package deal, develop and construct, and turnkey methods.

Design and build takes many forms but, essentially, contractors are asked to provide a design to meet the client's requirements, along with a price, usually in a limited completion of say four or six competitors. The advantage to the client is that there is a single point responsibility for the design and construction of the project but the downside is that the contractor's offer may not be exactly what the client wants or what he would have achieved with a traditional 'architect/engineer-designed' project. **Novated design and build** can overcome such problems by arranging for the client's designers to be transferred (novated) to the preferred contractor, who can thereby be involved in the contractor's development of the design and, at the same time, ensure that the client's objectives are met. In practice, this is easier said than done as both Masterman (2002) and Turner (1997) explain.

The costs of design and build are greater for contractors than traditional procurement because designs have to be produced in order to be able to submit a tender and, as only one tenderer can win the contract, abortive design fees which cost money may be incurred. Additionally, as pointed out by Morledge *et al.* (2006), the quantity risk is greater because it is the contractor which must produce the bills of quantities and must take on board the possibility that mistakes have been made or items have been omitted in error.

Counterbalancing the downside risk is the possibility of making money during the contract, firstly from design changes ordered by the client and, secondly, by cutting corners on specifications which normally would not be feasible on a traditional contract. This, however, does little to redress the transparency and probity frailties complained of in traditional procurement methods. Additionally, clients have little or no protection from the tactical pricing strategies of contractors and, in many respects, design and build offers more possibilities for 'commercial opportunism' than traditional procurement.

For one thing, design and build does not give the client the 'transparency' of priced bills of quantities and, for another, the pricing document that is normally submitted under design and build arrangements is easier for the contractor to manipulate, in terms of early cash flow generation and the valuation of client-initiated design changes, than traditional bills of quantities. The JCT Design and Build Contract 2011, for instance, offers Alternatives A and B for arranging interim payments. Payments are determined under Alternative A by reference to a breakdown of the contract sum into stages with cumulative sums of money allocated (by the contractor!) and, under Alternative B, by reference to a contract sum analysis (again prepared by the contractor!) which enables periodic payments to be made (usually monthly).

The sustainability concerns of traditional contracting are to some extent mitigated under design and build in circumstances where repeat business can be attracted from satisfied clients, thereby enabling design and site management teams to be kept together.

12.2.3 Management contracts

Masterman (2002) classifies management contracting and construction management procurement routes as 'management-oriented procurement systems' and, interestingly, disaggregates design and manage from his 'integrated procurement systems' that covers design and build.

Management contracting is distinguished from construction management in that the former engages a management contractor who then engages, and pays, his own subcontractors whilst the latter creates separate, direct, contracts with the client (employer) to both the construction manager and each work package contractor.

The management contractor is paid a fee for organising and managing the site and coordinating the work of package contractors, who are in direct contract with the management contractor for defined packages of work, usually on a lump sum basis. The advantage to the client is that the management contractor can be appointed early in the design process and thereby contribute to design development and feasibility. This does not obviate, though, the lack of transparency between the client and the management contractor and the subcontract packages. It is conventional for the client's quantity surveyor (QS) to scrutinise package tenders before contracts are let but this is no guarantee that probity will not be undermined by the usual 'smoke and mirrors' that contractors are practiced at!

In **construction management** procurement, the construction manager is also engaged on the basis of a fee but this time the work packages are procured directly by the client, usually on a traditional basis with, perhaps, some design element involved. The construction manager hence becomes a *defacto* member of the client's design team; this brings considerable benefits at the early stages of design. The duties of the construction manager normally include the coordination and integration of work packages and management of the site and he may also be involved in the valuation and certification of the package contractors' work in progress but not in their payment.

Under **design and manage** arrangements, a single organisation is appointed to design and manage the construction project whilst the work is carried out by works package contractors directly engaged by the client. The design and management functions may be undertaken by a contractor or a firm of consultants. To all other intents and purposes, this method is similar to construction management procurement.

Construction management and design and manage methods of procurement are seemingly more transparent than either traditional, design and build or management contracting because of the removal of the 'main contractor layer' of the process. Whilst this is true, the 'smoke and mirrors' game can now be played by the works package contractors which have the opportunity to manipulate their prices and stack the bills of quantities in their favour regarding cash flow, variations and re-measurement. It could be called 'the poacher turning gamekeeper' syndrome!

12.2.4 'Pain and gain' systems

'Pain and gain' systems are also known as **target cost contracts**; these have been around for a long time. Perhaps the first authoritative reference on the subject was by Perry and Thompson (1982) who explained the process in their detailed CIRIA Report. Morledge *et al.* (2006) provide a more concise explanation and also inform us that many partnering contracts are arranged on the basis of a **target cost** and **guaranteed maximum price**.

The usual process is that the contractor is engaged, early in the design process, on the basis of a fee percentage and a 'target cost' is developed in conjunction with the client's quantity surveyor – in a similar way to the traditional cost planning process. This target cost is then firmed up later on when sufficient design information is available and the contractor commits to a guaranteed maximum price for the job. If the contractor is able to procure the work packages for less than the target, savings will be shared between the contractor and the client. On the other hand, if the target cost is exceeded, then the contractor bears the loss. Hence the expression 'pain and gain'.

The advantage to the client is early contractor involvement and, *prima facie*, greater price transparency and certainty. However, no contractor 'worth his salt' will knowingly enter into a contract where he is likely to lose money and it is questionable whether the eventual guaranteed maximum price is more transparent than a traditional tender. The price certainty comes at a cost because, although the works packages will be tendered competitively and scrutinised by the client's quantity surveyor, the contractor is always going to be paid his percentage irrespective of the prices received. Admittedly, the percentage fee will be initially tendered in competition, but it will be a fee that the contractor is happy to live with. In any event, the fee is the minimum that the contractor will make and the opportunity for additional margin is presented in the form of the shared 'gains'. An example is shown in Box 12.1.

> ## Box 12.1 Margin calculation in target cost contracts
>
> | | | | | £ |
> |---|---|---|---|---:|
> | A | Guaranteed maximum price | | | 10 000 000 |
> | B | Management fee | 10% | $B/_{100} \times A$ | 1 000 000 |
> | C | Efficiency gains | | | 500 000 |
> | D | Share | 50/50 | $C \times 50/_{100}$ | 250 000 |
> | E | Actual cost | | $A - C$ | 9 500 000 |
> | F | Total margin | | $B + D$ | 1 250 000 |
> | G | **Overall margin on turnover** | | $F/_E \times 100$ | **13.16%** |

12.2.5 Partnering

Partnering is a 'discretionary procurement system' according to Masterman (2002) because it is a voluntary arrangement that acts as a framework within which other procurement systems may work more effectively. Consequently, competitive tender, negotiation, design and build, management contracts and target contracts can all work within a partnering environment.

Ostensibly partnering agreements overcome many of the difficulties of traditional and other less transparent methods of procurement in that the interests of both the client and the contractor and, possibly, other members of the supply chain, are uppermost in a spirit of mutual trust and cooperation. However, any thoughts that this arrangement is more open than other methods are self-deluding because contractors are naturally averse to opening their accounts to the scrutiny of outsiders and professional consultants are imbued with an ingrained mistrust of contractors. It is the culture of the industry.

There are some shining examples of the reverse of this situation but they are few and far between, especially where partnering is on the basis of single project partnerships. Where partnering has gained most success, however, is in the strategic partnering deals with clients who have long-term repeat business to offer.

12.3 Conditions of contract

12.3.1 Payment mechanisms

A number of payment mechanisms are commonly used in the construction industry including:

- a single payment upon completion of the contract;
- interim payments based on:
  - a valuation of work carried out
  - an S-curve forecast;
- stage payments based on:
  - construction stage reached (e.g. substructure, first floor, eaves)
  - progress related to the construction programme
  - handover of completed work (e.g. new houses, refurbished properties, factory units);
- income generated during a concession period (e.g. tolls from bridges, tunnels etc.);

- public sector service level agreements (e.g. payment for delivery of services such as prisons and schools).

The terms of payment for the work done are normally set out in the contract agreement between the employer and contractor or between the contractor and subcontractor. This might be a verbal agreement for a small project but, more usually, will be expressed in writing either in formal correspondence or in a written order or in a contract – typically, though not necessarily, a standard contract or an amended standard contract.

Part of the payment mechanism written into standard forms of contract concerns the question as to who prepares the valuation of work in progress. The general approach in JCT contracts is that the employer's architect/contract administrator will determine the amount to be paid to the contractor. Under some JCT conditions (but not all) the payment will be supported by a valuation prepared by the employer's quantity surveyor, usually with the assistance of the contractor's quantity surveyor. Under NEC conditions the project manager assesses the amount due to the contractor whist under ICC conditions it is usual for the contractor to submit an application for payment which will be checked and verified by the engineer's representative (a quantity surveyor or measurement engineer).

As far as subcontracts are concerned, the DOM/1 Conditions of Subcontract provide that it is the main contractor which values the amount due for interim payment whereas under CECA conditions of subcontract the subcontractor is required to make an application for payment along with a written statement of how the sum claimed was arrived at.

12.3.2 Payment procedures

Arrangements for payments by the employer to the main contractor and by the main contractor to subcontractors vary considerably from contract to contract and the terms for payment may be very different depending upon whether standard or non-standard forms are used. Most standard forms of contract provide for certification of payment for work in progress usually, though not always, based on a valuation of work done.

It is normal practice that the contractor's monthly payment follows as a consequence of the issue of the interim certificate by the architect/engineer/contract administrator providing the client (employer) does not default on the contract. The issue of the interim certificate signals that **payment is due** and this starts the 'clock ticking' as regards:

1. **payment** by the employer who must pay within the period stipulated in the contract (e.g. JCT SBC/Q 2011 = 14 days);
2. the **statutory notices** that must be given in order to comply with section 110 of the Construction Act 1996 (as amended by the Construction Act 2009) regarding what is to be paid, how it has been calculated and whether there are any deductions or set-off.

Payment is normally subject to deduction of retention monies by the employer at the rate stated in the contract – Chapters 6 and 17 contain more information on retention.

Looking at payment from a cash flow point of view, the contractor does not actually receive the money until the employer's cheque or inter-bank transfer has cleared the bank. This can be up to six working days after a cheque has been paid in to the contractor's bank account but is much more immediate with an electronic transfer by BACS, CHAPS or FPS. The payment arrangements in standard subcontracts usually provide that the subcontractor will be paid some time after the main contractor is paid (there is a one week delay under the JCT Standard Building Subcontract). This should avoid any cash flow problems for the main contractor provided, of course, that the employer pays on time!

All standard conditions of contract contain specific provisions regarding payment. Whilst the JCT, NEC, ICC and other families of contracts all have slightly different payment arrangements and use different terminology, it is nevertheless true that the principles of valuation, certification and payment are well established in all main contracts and subcontracts in common use and that all standard contracts are Construction Act compliant.

The requirements of the Construction Act apply to all contracts involving 'construction operations' with a duration of more than 45 days and, consequently, impose certain rights and obligations as regards dates for payment, withholding payment, suspension of performance for non-payment and the prohibition of conditional payment provisions.

12.3.3 JCT Standard Building Contract with Quantities 2011 (SBC/Q)

Under this form of contract, interim certificates are issued by the architect/contract administrator at monthly intervals, unless the contract particulars are amended otherwise. However, a valuation is only carried out when the architect/contract administrator considers them to be necessary. Consequently, the contractor has no contractual right to a valuation but, to all intents and purposes, it is extremely difficult to determine the amount due to the contractor without one. When a valuation is done, this is carried out by the client's quantity surveyor usually in conjunction with the contractor's quantity surveyor.

A further option is available under this form of contract whereby the contractor may submit an application for payment not later than seven days before the date for issue of interim certificates, in which case the quantity surveyor is obliged to make an interim valuation. Should the quantity surveyor disagree with the amount stated on the contractor's application then a statement showing the areas of disagreement must be provided to the contractor.

Under JCT SBC/Q 2011, once an interim certificate is issued, payment is to be made no more than 14 days from date of the certificate. The amount paid by the employer to the contractor will be subject to retention in accordance with the provisions of the contract. However, it should be noted that certain items that might be included in the gross valuation, such as payments for loss and expense and insurance payments, are not subject to retention. The retention figure shall be 3% unless otherwise stated in the contract particulars.

12.3.4 Infrastructure Conditions of Contract (ICC) – Measurement Version

Under this form of contract, the contractor shall submit a statement of estimated value to the engineer or the engineer's representative at monthly intervals. Within 25 days the engineer shall certify the amount which he/she considers is due to the contractor. Within 28 days of the date of the certificate the employer shall pay the contractor the certified amount.

Retention is deducted at the rate stated in the Appendix to the Form of Tender – Part 1, which is recommended not to exceed 5%. An unusual feature of the contract is that a limit to the amount of money held under retention applies where stated in the Appendix. This is recommended not to exceed 3% of the tender total but it could be any figure or zero.

It should be noted that this is a re-measurement contract and that the quantities stated in the contract bills are, therefore, approximate. Consequently, there is no contract sum as such and the tender total is therefore only used as an indication of the 'value' of the contract and as a reference point for the calculation of the retention limit where this applies.

12.3.5   NEC Engineering and Construction Contract 3rd Edition

Under the NEC form, the project manager assesses the amount due at each assessment date but there is no contractual requirement to undertake a valuation. Any application for payment by the contractor is considered in assessing the amount due. The assessment interval is stated in Contract Data – Part 1 but it must not exceed five weeks in any event. The project manager certifies payment within one week of the assessment date and payment is due three weeks from the assessment date unless otherwise stated in the Contract Data.

The contract contains a secondary option clause (Clause P) which means that the deduction of retention is optional. If a retention provision is included in the contract, the retention percentage and retention-free amount are included in the Contract Data – Part 1. The retention-free amount is effectively a retention limit similar to that under the ICC Conditions referred to previously.

12.3.6   Other standard forms of contract

Most standard forms of contract contain very similar provisions as regards payment. Payments are usually supported by a valuation, certification intervals are usually monthly or four-weekly and retentions tend not to exceed 5% of the gross valuation. Some contracts, such as the JCT Design and Build Contract 2011 and the ICC – Design and Construct Version, allow for stage payments to be made when defined construction stages have been reached and others, such as GC/Wks/1, allow for payment based on an ogive curve. In both cases, the contractor submits a payment application but there are no prescribed intervals for doing so.

The JCT Major Project Construction Contract 2011 is interesting in that the contractor makes a payment application whenever he considers that the employer should issue a 'payment advice' and the rules for payment are contained in the 'pricing document'. This might be a contract sum analysis or bills of quantities and there are rules applicable to interim valuations, stage payments, progress payments or other preferred method.

Table 12.1 summarises the payment arrangements under some common forms of contract.

12.3.7   Conditions of subcontract

Payment arrangements under three of the principal standard conditions of contract are summarised in the following sections. There is, however, no substitute for reading the contract itself and for checking the contract documents for any amendments that

Table 12.1 Payment under some common forms of contract.

Form of contract	Payment provisions

JCT Major Project Construction Contract 2011	• When the Contractor considers that the Employer should issue a payment advice, it shall submit a detailed application not later than seven days beforehand
	• Rules for determining the way the Contractor will be paid are contained within the Pricing Document e.g.: a contract sum analysis or bills of quantities etc.:
	  ○ P3 Rule A – Interim valuation
	  ○ P4 Rule B – Stage payment
	  ○ P5 Rule C – Progress payments
	  ○ P6 Rule D – Some other method
	• The applicable rule is stated in the Contract Appendix which also states the date when any payment advice is to be issued. The default date is the 28th of the month
	• The payment advice shall include a statement of the proposed payment and how it was calculated
	• Payment becomes due upon receipt by the Employer of a VAT invoice from the Contractor. The Employer then has 14 days in which to pay. There is no retention under this Form of Contract

JCT Design and Build Contract 2011	• Interim Payments are to be made using either Alternative A or B. The agreed method shall be stated in the Contract Particulars
	• The Contractor shall make Applications for Interim Payment
	  ○ Alternative A – Stage Payments: applications shall be made on completion of stages specified in the Contract Particulars
	  ○ Alternative B – Periodic Payments: applications normally made at monthly intervals unless otherwise stated in the Contract Particulars
	• Final date for payment is 14 days from receipt of the Contractor's application
	• The Contractor's payment is subject to 3% retention

JCT Intermediate Building Contract 2011	• Interim valuations shall be made by the Quantity Surveyor whenever the Architect/Contract Administrator considers them to be necessary • Alternatively, the Contractor may submit an application for payment not later than seven days before the date for issue of the Interim Certificate • Interim certificates are to be issued at monthly intervals unless the Parties to the contract agree to stage payments • Payment is to be made by the Employer to the Contractor no more than 14 days from date of Interim Certificate • The percentage of the total value of work carried out is stated in the Contract Particulars. This is normally 95% of value of work done

(Continued)

Table 12.1 (*Continued*)

Form of contract	Payment provisions
JCT Minor Works Building Contract 2011	• Progress payments are to be certified at four-weekly intervals by the Architect/Contract Administrator • Payment is due no later than 14 days from the date of the Certificate • The Contract Particulars state that 95% of the value of work shall be paid
ICC – Design and Construct Version	• The Contractor is to submit a statement to Employer's Representative showing the amounts considered due under the Contract • Within 28 days the Employer's Representative shall certify and the Employer shall pay the amount which is considered due to the Contractor • Where a payment schedule has been included in the Contract, stage payments are made when progress has reached the required stage for payment • Retention shall be deducted at the rate stated in the Appendix to the Form of Tender - Part 1 which is recommended not to exceed 5% • A retention limit may be stated in the Appendix.
ICC – Minor Works Version	• At monthly intervals, the Contractor shall submit a statement of estimated value to the Engineer: • Within 28 days the Engineer shall certify and the Employer shall pay the amount which the Engineer considers is due to the Contractor • Retention shall be deducted at the rate stated in the Appendix to the Form of Contract. The Contract Guidance Notes recommend a rate of 5% for retention • The total amount of retention is limited to that stated in the Appendix to the Form of Contract. The Contract Guidance Notes recommend a limit of between $2\frac{1}{2}$% and 5% of the estimated contract value.

may apply to a specific project. All too often assumptions are made that payment will follow a certain pattern only to find out too late that the 'goalposts have been moved'.

Subcontractors, in particular, are notoriously slipshod at tender stage and simply do not read (or understand) the tender documents. In practice, payment does not always follow performance of a contract and it is very common to find subcontractors in difficulty when they find that the payment 'rules' are not as expected or that payment for variations carried out in good faith is disputed due to a lack of documentary evidence.

JCT subcontracts 2011

The JCT 'family' of subcontracts provides a variety of sets of conditions suitable for various ways of subcontracting parts of the main contract works. These include where the subcontractor designs the whole or part of the subcontract works, where stage payments as opposed to monthly interim payments are agreed and where the contractual arrangements are lump sum or re-measurement.

Under the JCT SBCSub/C 2011, interim payments are geared to the certification process under the main contract and payment becomes due to the subcontractor on the date when the interim certificate for the main contract is issued. The final date for payment is 21 days after payment becomes due as opposed to 14 days under the main contract (e.g. JCT SBC/Q 2011). This makes sense because the main contractor would otherwise have to pay the subcontractor on the same day that payment is received from the employer. The one week payment delay allows for the employer's payment to clear the main contractor's bank account, thereby making the funds available, and for the subcontract payment to be issued in a timely fashion.

The main contractor is responsible for ascertaining the value of the subcontract works carried out but the subcontractor also has the right to submit his own valuation. If the main contractor agrees with this valuation then this shall be the basis upon which payment is made.

The usual 'Construction Act' notices apply to the subcontract and these will be of particular interest (and perhaps concern) to the subcontractor. A withholding notice or pay less notice will signal that the expected interim payment will be reduced, perhaps due to a measurement dispute or incomplete or defective work or for contra-charges levied by the main contractor. Having said that, subcontractor payment notices are a rarity in practice but this does not prevent main contractors from paying less than the amount applied for by the subcontractor.

The full rate of retention stated in the subcontract will be deducted from the gross payment until such time that the subcontract works reach practical or partial completion, whereupon the retention percentage will be halved. The provision by the subcontractor of a retention bond is possible under these conditions of subcontract, in which case there will be no deduction of retention monies from the monthly payment.

DOM/1 (1980 Edition)

DOM/1 is a long-standing form of subcontract that has been extensively amended over the years. It is, nevertheless, still in use despite being perhaps less 'user friendly' than the more modern subcontracts available.

This subcontract provides the option for the parties to agree to stage payments as opposed to interim payments and it may be used for either a lump sum or a re-measurement contract. This requires the deletion of either Article 2.1 or 2.2 as appropriate. If this is not done, subcontractors should be aware that other documents, including tender documents, correspondence or the minutes of pre-contract meetings, that may be incorporated into the subcontract, may state whether the subcontract is to be re-measured or not.

Subcontractors should also be alive to the payment implications as between lump sum and re-measurement contracts. In the former case, the subcontract sum may only be adjusted for limited reasons (such as variations) and in the latter case the works will be completely re-measured. Re-measurement may not be the best option for the subcontractor because the re-measurement will be carried out by the main contractor's quantity surveyor (as opposed to the employer's quantity surveyor) and the re-measure may not be either accurate or in accordance with the standard method of measurement.

The valuation for payment purposes is determined by the main contractor under this form of contract. The subcontractor may also make a statement as to the amount of any valuation, together with details to substantiate the statement, but there is no express right for the subcontractor to submit a valuation nor is there any obligation on the main contractor to consider one.

The final date for payment under DOM/1 is 17 days after the due date. There is no linkage to the main contract certification process under this form of subcontract and the payment due date is given as not later than one month after commencement of the subcontract works and at intervals not exceeding one month thereafter.

Depending upon the timing of the commencement of the subcontract works in relation to the valuation dates for the main contract, it may be that the subcontractor will become entitled to payment for part of his work in progress before the main contractor is paid for this work. This would be bad news for the main contractor's cash flow but may also be bad news for the subcontractor, as the main contractor will undoubtedly find some reason for delaying payment or making an underpayment in such circumstances.

The retention percentage under DOM/1 is deemed to be 5% unless otherwise stated. This can give the main contractor a nice cash flow advantage where, for instance, JCT SBC/Q is the main contract, because the retention percentage under this form of main contract is 3%.

CECA form of subcontract

Formerly known as the FCEC 'Blue Form', the CECA forms make up a small 'family' of subcontracts used for civil engineering work in conjunction with the former ICE Conditions of Contract. There are subcontracts for both the measurement and design and construct versions of the ICE Conditions, and they are still 'blue'! Also in the 'family' are subcontracts for use with GC/Wks/1 and PC/Wks/1 government contracts which cover civil engineering, building and mechanical and electrical works.

The 'blue forms' place the onus for payment applications firmly on the shoulders of the subcontractor. There are strict time limits written into the subcontracts for making payment applications and failure to submit a valid application, on time, will result in no payment being made!

The payment application takes the form of a 'written statement' of work done under the subcontract and includes any materials on/off site for which payment is claimed. It is the main contractor who decides how the statement is to be presented including what form it should take and what details it should contain.

Submission of the subcontractor's written statement must be made not less than seven days before the date specified in the subcontract schedule. This 'specified date' represents the due date for payment and the final payment date is 35 days from this date.

Retention is deducted from interim payments in the usual way but subcontractors should carefully read the schedules to the subcontract as this is where the retention percentage and retention limit is stated. There is no set retention percentage (so the figure could be anything!!) and it is permissible to state different percentages for work in progress and for materials on site if the main contractor so desires.

NEC3 Engineering and Construction Subcontract

Where the NEC3 ECC is the main contract, the contractor is obliged to tell the project manager which form of subcontract is intended to be used for each subcontract unless the NEC3 Engineering and Construction Subcontract is to apply. Eggleston

(2006) reports that uptake of the NEC subcontract is not widespread due to the level of administrative burden that its use creates for the main contractor and for legal and commercial reasons. However, Eggleston (2006) also suggests that the NEC subcontract is nonetheless useful where the main contractor wishes to create a subcontract that is 'back-to-back' with the main contract, as the NEC3 subcontract duplicates the NEC3 main contract almost exactly.

The NEC subcontract differs with many other subcontracts in that it is the main contractor who values the subcontractor's work in progress. This has to be done to a strict timetable. The main contractor is given two weeks to certify payment counted from the 'assessment date' and has a further four weeks in which to make payment. This gives the main contractor a one week 'breathing space' for certification and payment relative to the main contract timetable.

Retention is a secondary option (option P) under the NEC3 Engineering and Construction Subcontract in common with the main contract. Subcontractors should note that where the main contract has no retention provision this does not necessarily apply to the subcontract, nor is the main contractor obliged to keep to the same retention percentage as the main contract.

12.4   Method of measurement

It is important to remember the key role that standard methods of measurement play in the measurement, valuation and eventual certification of construction work and in the allocation of risk, especially below ground risk. The method of measurement that applies to the contract works may be identified in the conditions of contract by an express term or may be included in the contract particulars, works information or a preamble.

The contractual link with the method of measurement is a crucial factor in determining the balance of risk in the contract – such as who is responsible for the provision of temporary works and for payment for working space and groundwater problems. These issues are discussed in the following sections and in Chapter 15.

12.4.1   Standard methods of measurement

The most common methods of measurement in use in the United Kingdom are:

- SMM7 (Standard Method of Measurement of Building Works 7th Edition) – for building work;
- CESMM4 (Civil Engineering Standard Method of Measurement 4th Edition) – for civil engineering work;
- the Method of Measurement for Highway Works - for major road works projects.

The RICS New Rules of Measurement are in their very early stages of publication and not yet in widespread use.

12.4.2   Basis of quantities

The quantities presented in bills of quantities (and indeed the other forms of pricing document described in Chapter 4) are usually determined in accordance with, or generally in accordance with, an appropriate standard method of measurement which explains:

- how the bills of quantities is to be set out;
- how the work should be measured;
- how the various work items should be described;
- what the work items are deemed to include (coverage);
- how particular words or expressions used in the standard method are defined.

Where the contractor produces the quantities they will often be loosely based on a standard method without following all the detailed rules.

The items in schedules of work are an exception and, being of an *ad hoc* nature, they do not normally follow any standard method of measurement. Item descriptions in some schedules of rates are based on SMM7; these include the National Schedules of Rates, which are used for measured term contracts, repair and refurbishment contracts and the like. The PSA Schedules of Rates have their own measurement rules whilst the M3NHF Schedules are largely based on a schedule of works format with no quantities given.

12.4.3 Classification systems

The common standard methods of measurement used in construction are all based on similar principles but they differ significantly in certain respects especially regarding the level of detail of the measurement rules and the extent of the work covered by each measured item. All methods of measurement are based on a **classification system** which distinguishes the various work sections:

SMM7
- A   Preliminaries/General conditions
- C   Demolition/Alteration/Renovation
- D   Groundwork
- E   *In situ* concrete/Large precast concrete
- F   Masonry
- and so on

Each item description is made up from *one* descriptive feature drawn from *each* of the first three columns in the classification table and *as many* descriptive features from the fourth column as required.

CESMM4
- Class A: General items
- Class B: Ground investigation
- Class C: Geotechnical and other specialist processes
- Class D: Demolition and site clearance
- Class E: Earthworks
- Class F: *In situ* concrete
- and so on

The 26 main classes (A–Z) given in CESMM4 provide a means to describe work commonly found in civil engineering projects. There are three divisions in each class which classify the work in progressively more detail. Each division contains up to eight descriptive features. Bills of quantities items may be made up of one feature from each division of the relevant class. The classification system facilitates a useful coding system such that a bill of quantities item may be identified alpha-numerically.

Method of Measurement for Highway Works
- Series 100 Preliminaries
- Series 200 Site Clearance
- Series 300 Fencing
- Series 400 Road Restraint Systems
- Series 500 Drainage and Service Ducts
- Series 600 Earthworks
- and so on.

Each item description is drawn from one or more of the Groups listed within each Series of Chapter IV of the Method of Measurement. An item description may contain Features from as many Groups as necessary to identify the work required, but may include only one Feature from any one Group.

12.4.4 SMM rules

Each of the above methods of measurement, albeit with different approaches, include:

- **general rules** concerning general layout, coding of items, preparation and pricing of bills of quantities, method-related charges and so on;
- **measurement rules** setting out how the work is to be measured;
- **definition rules** to distinguish the extent and limits of each class of work;
- **item coverage rules** which define the work that is deemed to be included in the measured items.

Uniquely, CESMM4 requires a list of principal quantities to be given which presents a useful overview to the contractor of the scope and scale of the project not found in other methods of measurement. Table 12.2 provides a comparative illustration of the classification systems used in the three common methods of measurement.

12.4.5 Waste

Waste is an important issue in the pricing of a contract and in the management of the site, the control of subcontractors and in statutory compliance. Waste is also recognised as an issue in the standard methods of measurement used in the industry and may be categorised as:

- Measurement waste:
  - All standard methods of measurement deem the quantities to be measured net as fixed in position.
  - This means that any additional material needed to accomplish the design of the works is deemed to be included in the contractor's rates and prices or otherwise allowed for in his tender.
  - Examples include:
    - Lapping sheets of steel mesh reinforcement for *in situ* concrete slabs.
    - Blind boring' *in situ* concrete piles through ground to be later removed.

Table 12.2 Measurement classification systems.

SMM7		1 Plain vertical	1 Height >1.00 m	m²	1 Left in
1	Sides of foundations	2 Dimensioned description	2 Height ≤250 mm	m	2 Permanent
2	Sides of ground beams and edges of beds		3 Height 250–500 mm		
3	Edges of suspended slabs		4 Height 500 mm–1.00 m		
4	Sides of upstands				
5	Steps in top surfaces				
6	Steps in soffits				
7	Machine bases and plinths				

CESMM4   CLASS I: PIPEWORK – PIPES

Includes: **Provision, laying and jointing of pipes**
Excavating and backfilling pipe trenches

Excludes: **Work included in classes J, K, L and Y**
Piped building services (included in class Z)

FIRST DIVISION

1	Clay pipes	m
2	Concrete pipes	m
3	Iron pipes	m
4	Steel pipes	m
5	Polyvinyl chloride pipes	m
6	Glass reinforced plastic pipes	m
7	High density polyethylene pipes	m
8	Medium density polyethylene pipes	m

SECOND DIVISION

1	Nominal bore:	Not exceeding 200 mm
2		200–300 mm
3		300–600 mm
4		600–900 mm
5		900–1200 mm
6		1200–1500 mm
7		1500–1800 mm
8		exceeding 1800 mm

THIRD DIVISION

1	Not in trenches	
2	In trenches, depth:	Not exceeding 1.5 m
3		1.5–2.0 m
4		2–2.5 m
5		2.5–3 m
6		3–3.5 m
7		3.5–4 m
8		exceeding 4 m

Method of Measurement for Highway Works

Group	Feature	
I	1	Disposal
II		
	1	Acceptable material excluding Class 5A.
	2	Acceptable material Class 5A.
	3	Unacceptable material Class U1A.
	4	Unacceptable material Class U1B.
	5	Unacceptable material Class U2.

  - Extra backfill materials or concrete required where the bucket width of the excavator exceeds that of the standard trench width.
- Process waste:
  - This may be defined as waste naturally occurring as a function of the production process and derived from activities such as offloading, handling, cutting and fixing in final position.
  - Most construction materials suffer from process waste and allowances are made by the contractor's estimator to cover the financial implications of such waste.
  - There are statutory duties to manage such waste at site level.
  - All standard methods of measurement deem process waste to be included in the contractor's rates and prices.
- Method waste:
  - Where the contractor decides to use a method of construction which requires the use of additional materials.
  - Examples include:
    - Extra backfill materials where a battered excavation is preferred to using an earthwork support system.
    - Extra concrete used where the contractor decides to fill a trench rather than use formwork.
    - Where the contractor uses bricks rather than cutting blocks to complete a block wall which will later be plastered.
  - Under SMM7, the cost of method waste may be compensated by a measured item, such as earthwork support or formwork, which will be paid irrespective of the contractor's chosen method.
  - Under CESMM4 and the Highways Method of Measurement earthwork support is deemed included by the respective coverage rules.
  - Under these methods of measurement formwork is not measured to the sides of concrete to be cast against an earth face and the measurement risk is, therefore, carried entirely by the contractor.

12.4.6 Working space

Working space is frequently required in excavations to enable earthwork support or formwork to be fixed in position prior to backfilling or concreting. In drainage work, drag boxes or other patented shoring systems take up space beyond the 'net' dimensions of the drain trench/pipe bedding design and this is also classed as 'working space'. Figure 12.2 illustrates the issue of working space under the following standard methods of measurement where a mass concrete foundation is to be constructed:

SMM7
Working space is a measured item for excavations, under Class D20: Excavation and filling, in circumstances where the face of the excavation is less than 600 mm from the face of formwork, rendering, tanking and so on. Working space is measured in square metres and is calculated by multiplying the girth of the formwork, rendering and so on by the depth of the excavation (measurement rule D20: M8 refers).
CESMM4
Class E: Earthworks measurement rule M6 states that the volume measured is the volume occupied by the structure or the volume vertically above the structure and

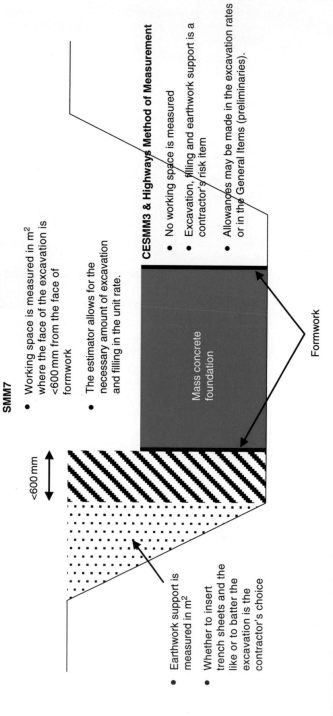

SMM7

- Working space is measured in m^2 where the face of the excavation is <600 mm from the face of formwork

- The estimator allows for the necessary amount of excavation and filling in the unit rate.

<600 mm

Mass concrete foundation

- Earthwork support is measured in m^2

- Whether to insert trench sheets and the like or to batter the excavation is the contractor's choice

CESMM3 & Highways Method of Measurement

- No working space is measured

- Excavation, filling and earthwork support is a contractor's risk item

- Allowances may be made in the excavation rates or in the General Items (preliminaries).

Formwork

Figure 12.2 Measurement of working space.

coverage rule C1 says that excavation items are deemed to include additional excavation to provide working space. Therefore, estimators need to make an allowance in their rates/prices for any additional excavation and backfilling where they think that the chosen construction method will require working space. Such allowances may be included in the unit rates or, alternatively, as a method related charge.

The CESMM and Highways Method of Measurement do not measure working space items under any circumstances and such requirements constitute a 'contractor's risk' item.

12.4.7 Temporary works

Temporary works are often required to enable the permanent works to be constructed. Such items include:

- Earthwork support
  - Timbering
  - Proprietary systems
  - Sheet piling.
- Formwork
  - A temporary support used to mould poured concrete into the required dimensions until the concrete becomes self-supporting
  - Examples include:
    - foundations
    - walls, columns and beams
    - slabs and suspended slabs
  - Usually removed but may be left in place (permanent formwork)
  - Measured under all methods of measurement where concrete requires support during casting
  - Concrete poured against earth faces is measured separately under SMM7 but not under other methods
  - Complex rules apply in the various methods of measurement in respect of the height of formwork, its inclination and the structural element involved.
- Falsework
  - A temporary structure used to support a permanent structure
  - Removed when the structure becomes self-supporting
  - Carries mainly vertical loads
  - Examples include
    - suspended slabs
    - bridge decks
  - Not measured under any standard method of measurement.

Earthwork support – and steel sheet piling, where required by the design, and specified to be used – is measurable under SMM7 but not under CESMM4 or the Highways Method of Measurement, which both deem this to be a contractor risk item linked to the choice of construction method (Figure 12.2). Formwork is measured in all the standard methods but falsework is a contractor's method issue and is not measureable.

12.5 Bills of quantities

12.5.1 The use of bills of quantities

Despite the decline in traditional procurement methods reported by the RICS (2011), bills of quantities and other pricing document formats still play a crucial role in the financial management of construction projects. Such documents provide the template for setting the contract budget and act as a control tool as well as a reference point for measuring and valuing the work carried out.

In traditional procurement, the parties are used to seeing formal bills of quantities set out in a traditional manner but, in design and build procurement and drawings and specification contracts, formal bills of quantities do not feature. Informal bills of quantities will, nevertheless, be prepared by the contractor but neither the employer nor his professional advisers will normally see this document. It is more likely that the contractor will present his price in the form of a **contract sum analysis**, which is a breakdown of the tender sum, not in quantities format, but in a series of lump sums of money related to defined elements of the design or stages of construction or activities on the contractor's master programme.

12.5.2 Structure and layout

The structure and layout of 'formal' bills of quantities will be determined strictly in accordance with the relevant standard method of measurement requirements – this creates the 'level playing field' intended by standardised measurement rules. As far as 'informal' pricing documents are concerned, the preparation of 'builders quantities' requires no observance of a particular convention but it is more than likely that standard methods of measurement will inform the way that the work is measured and presented to the estimator.

Whilst there are several methods of measurement available, and the relevant contract conditions or contract particulars will state which one is applicable, they all classify work into three main areas:

- preliminaries or general items;
- measured work items;
- prime cost and provisional sums.

The measured work items are classified into distinct subdivisions or classes reflecting the work content of the project in question. In practice, each section of the bills of quantities is brought to a total and the totals are carried to a final summary, the summation of which represents the contractor's tender figure. The final summary sometimes has an adjustment item which allows last minute alteration of the tender figure +/- depending upon the decisions made at the contractor's tender adjudication meeting. Consequently, the bills of quantities will normally consist of some or all of the following:

- preliminaries or general items;
- measured work sections/classes;
- prime cost sums;
- provisional sums (including dayworks);
- contingencies;

- final summary;
- adjustment item.

12.5.3 Preliminaries

When a contractor tenders for a construction contract, an important element of its pricing is the allowance made for running the site and supervising the work. This allowance is usually called the 'preliminaries', 'general items', 'on-costs' or 'prelims' for short. The CIOB Code of Estimating Practice (1997) provides extensive guidance as to what should be included in the preliminaries but they will include the contractor's estimated costs for:

- supervision and setting out;
- site offices and welfare facilities;
- power, water and temporary drainage;
- site transport;
- cranes and materials handling equipment;
- access roads and hardstandings;
- site security, fencing and barriers;
- and so on.

Table 12.3 shows a simple example of how preliminaries are calculated. It can be seen that some items are fixed or 'one-off' charges and others relate to the contract duration for the project. Some contractors price the cost of visiting staff, such as contracts managers and quantity surveyors, in the preliminaries (as illustrated) and some charge such costs as part of the general overhead of the company. The estimator will pay particular attention to the pre-tender programme when pricing the time-related preliminaries, as this will indicate how long major items of plant will be on site.

Fixed charges allow for costs that have no appreciable time duration including:

- erection and dismantling of the site sign board;
- the cost of transporting cabins or items of plant to and from the site;
- constructing temporary access roads and hardstandings, together with their removal at the end of the contract;
- connecting the site to the local authority sewer or to statutory utilities such as water, telephone or electricity.

Fixed charges are included in the monthly valuation as and when the item is completed. However, should the contractor include a single preliminaries item for, say, providing a site sign board, then the item would have to be apportioned for valuation purposes. The employer's quantity surveyor would decide on the proportions which might be a payment of 70% of the item when the sign board goes up and another payment of 30% when it is removed at the end of the contract.

Time-related charges are costs that are incurred each week or month and include:

- wages and salaries of site staff;
- provision of site transport such as cars and vans;
- rental costs for site cabins;
- hire charges for cranes and lifting equipment;

Table 12.3 Preliminaries calculation.

ACME CONSTRUCTION LTD
Contract   St Joseph's School
Contract value   £326000
Contract Period   12 wks

PRELIMINARIES	Quantity	Unit	Rate	Weeks	TOTAL
SUPERVISION					£
Contracts manager (visiting)	0.2	No	1200	12	2880
QS (visiting)	0.4	No	1000	12	4800
Site manager	1	No	900	12	10800
SITE ACCOMODATION					
Cabins	2	No	200	12	4800
Transport to site		Lump sum			200
Drying Room	1	No	200	12	2400
Lock-ups	1	No	100	12	1200
Remove from site		Lump sum			200
RUNNING COSTS					
Cleaning	1	No	50	12	600
Heating & lighting	1	No	100	12	1200
Telephones	2	No	10	12	240
SITE TRANSPORT					
Vans/Cars	1	No	350	12	4200
Van/Lorry visits	0	No	200		
SMALL TOOLS & TACKLE					
Labour On-cost	2	%	35000		700
WELFARE & FIRST AID					
Safety equipment	1	No	120	12	1440
PPE		Lump sum			150
SCAFFOLDING					
General Access	100	m2	12	6	7200
Mobile Towers	2	No	150	4	1200
MECHANICAL PLANT					
Mobile crane	1	No	800	6	4800
Telehandler	1	No	500	10	5000
Site dumper	1	No	120	12	1440
TOTAL PRELIMS				£	55450

● regular payments for:
  ○ telephones
  ○ heating and lighting
  ○ water supply and so on.

Time-related charges are included in the monthly valuations *pro rata* to the time that has expired on the contract. Consequently, if the contract period is 26 weeks and 12 weeks has elapsed then $^{12}/_{26}$ of the time-related charges will be paid.

Complications arise when the contractor is either ahead or behind programme or where the contractor has submitted a programme that is shorter than the contract period.

12.5.4 Measured work

The measured work represents the quantity of the permanent works as measured in the contract bills of quantities and basically comprises a list of items of work measured from the architect's or engineer's drawings by the employer's quantity surveyor. These items are priced by the contractor using either **unit rates** or **lump sums** as appropriate. Table 12.4 shows examples of the allowances and calculations that an estimator might make for materials, labour, plant, overheads and profit in the build up of the contractor's prices. The estimator's approach will depend very much on whether unit rate pricing or operational pricing is preferred for the job in hand.

Unit rate pricing methods, whether using an estimating package, spreadsheets or traditional methods, are based on calculating prices per unit of work; for example the rate per:

- cubic metre of concrete;
- square metre of brickwork;
- linear metre of drainage.

The basis of operational pricing is to calculate **lump sums** for defined operations and then to allocate the money to each of the bills of quantities items that define the work involved. To price work this way, the estimator will have to consider the:

- duration of the activity;
- resource costs as a function of time;
- total quantity of materials required for the operation including waste.

In general, unit pricing methods are conventionally used for building work whereas operational pricing is more common in civil engineering. The reasons for this largely stem from the methods of measurement used and the general tendency that 'composite' items are more commonly found in civil engineering work. However, building repair, maintenance and refurbishment work may well be described in the tender documents using composite or lump sum items in which case operational pricing may be more appropriate.

Building work:
- tends to be measured in finer detail;
- working space and temporary works items (such as sheet piling) are measured where demanded by the technical requirements of the project.

Civil engineering:
- the methods of measurement used have a wider 'item coverage' than in building work;
- the contractor largely takes the risk of temporary works items and provision for working space and so on in its prices.

Table 12.4   Estimator's allowances.

Item	Allowance	Example				
Materials	• Cost delivered to site	1000 facing bricks				800.00
	• Allowances for:					
	○ Off-loading	Waste @ 2½%				20.00
	○ Handling	Telehandler				Incl in Prelims
	○ Waste	Handling 2 hours Labour @ £15/h				30.00
		Total per 1000				**850.00**
Labour	• Basic cost of wages					£
	• Allowances for on costs:	Paid hours	1848	hours	10.30	19034.40
	○ Employers' national Insurance contributions	Incentives	1848	hours	0	0.00
		Non-productive overtime	23.1	hours	10.30	237.93
	○ Sick Pay	Travel	115.5	hours	10.30	1189.65
	○ Holidays With Pay	Public holidays	64	hours	10.30	659.20
	○ Training Levy	Subtotal				21121.18
	○ Insurances etc	Employers' NIC				2007.83
		Sick pay	1	week	401.70	401.70
		CITB levy	0.5	%	21121.18	105.61
		Holidays with pay	46.2	weeks	32.07	1481.63
		Subtotal				25117.95
		Severance pay	2	%		502.36
						25620.31
		EL+PL Insurance	2	%		512.41
		Gross Cost p/a				26132.72
		HOURLY RATE	1848	hours	£	**14.14**

Plant	Provision of plant		
	Allowances for:	Hire of excavator per day	100.00
	○ Fuel	Fuel 20 litres/day @ £1.20/litre	24.00
	○ Maintenance	Maintenance	Incl in hire rate
	○ Operator (if required)	Operator 8h @ £20/h	160.00
		Total per day	**284.00**
Overheads	Allowance for head office running costs: ○ Salaries ○ Rents ○ Utility bills ○ Company cars ○ Office equipment etc.	The annual cost of running the head office is calculated and expressed as a percentage of annual turnover	Varies from firm to firm but typically 7-10%
Profit	A margin to provide a return to shareholders and funds to reinvest in the business	The annual profit required to provide a satisfactory return on capital invested in the business. Expressed as a percentage of annual turnover	Varies according to the type of work undertaken. Could be in the range 1-25%. Typically 2½-5%

Table 12.5 Bills of quantities.

Ref	Description	Quant	Unit	Rate	£

A	Excavating
	   To reduce levels
	      1m maximum depth
		5290	m³	7.50
				6.52
					39675.00

B	For pile caps and ground beams between piles 1m maximum depth; commencing from reduced level	1049	m³	18.95 11.26	19878.55
	Page Total to Summary				59553.55

NB: Prices inserted by contractor at tender stage.
7.50 is the gross rate.
6.52 is the net rate or 'allowance'.

Table 12.5 shows a typical bill of quantities item where it can be seen that there is an item description, quantity, unit and rate with a column for extending the quantity multiplied by the rate. The highlighted area is priced by the contractor and extended to a total. Some contractors price the items 'net'– that is excluding overheads and profit – and then 'gross up' the bills of quantities after the tender figure is established, the profit margin agreed and any strategic pricing or rate loading decisions are taken. The 'net' and 'gross' rates are shown in Table 12.5. The contractor's 'net' rate is confidential and it is only the gross rates that are submitted to the employer in the contractor's 'fair copy' of the priced bills of quantities.

It is important to remember that some methods of measurement deem that the contractor's rates and prices are 'gross' (e.g. SMM7 – General rule 4.6), especially when the time comes to value variations to the contract.

12.5.5 Prime cost sums

Prime cost sums represent allowances made by the design team for items of work which have not been fully designed at tender stage. They may be included in a contract to allow the architect/engineer to spend money on specialist items of work or on the supply of materials which cannot be precisely identified at tender stage. These sums are stated in the bills of quantities and the contractor is given the opportunity to price any attendances and profit in the space provided. An example is given in Table 12.6. The highlighted area shown is priced by the contractor at tender stage and therefore represents part of the competitive element of the tender.

Attendances are facilities or services that the contractor must provide in respect of prime cost items and include:

- offloading of materials;
- the provision of storage facilities;
- the provision of water, power or lighting;

Table 12.6 Pricing prime cost sums.

Ref	Item	Quant	Unit	Rate	£	p
	PRIME COST SUMS Work by Nominated Subcontractors Provide the following prime cost sums for:					
A	Bored Piling				750 000	00
B	Profit		%	2½	18 750	00
C	General attendance:					
	Fixed charge			}	1500	00
	Time-related charge					
D	Special attendance:					
	Guide walls				37 500	00
	Piling mat				2 500	00
	Removal of spoil				3 750	00

1. Item A is the architect's/engineer's tender stage allowance.
2. Other prices and rates are inserted by contractor at tender stage.
3. Prime cost sum will be adjusted at final account stage.
4. The contractor's allowances may also be adjusted but not the profit percentage.

Attendances may be classed as 'general', such as the use of the contractor's standing scaffolding, or 'special', which might include the provision of individual access or craneage facilities.

Under the ICC Conditions – Measurement Version, the engineer may nominate a company to carry out work of a specialist nature which is beyond the expertise of the main contractor or may order the supply of materials which are to be fixed in place by the main contractor.

The nomination of subcontractors is a facility which is no longer available under JCT conditions following widespread revisions of the 2005 and 2011 versions of the JCT family of contracts.

12.5.6 Provisional sums

Provisional sums are included in the contract bills to allow the architect/engineer to spend money on items of work which may prove to be necessary to complete the contract works. A typical example is where a new factory is to be built and an electrical substation will be needed. At tender stage there will probably be no design for the substation but past experience will indicate the sort of money that it will cost. If such work can be described, but not in sufficient detail to be measured, the contractor can at least be alerted to the possibility that the work may arise at some stage in the job. This is more satisfactory than simply adding a broad contingency sum to the contract price.

For building projects, where the Standard Method of Measurement for Building Works 7th Edition (SMM7) applies, a distinction is made between 'defined' and 'undefined' provisional sums. Where **defined provisional sums** are included in the contract, the contractor is expected to allow for this work in its programme but this is not the case for **undefined provisional sums** where the scope of works cannot be described in adequate detail. When provisional sums are expended, they are usually valued using the rates in the bills of quantities as a basis for valuation. Provisional sums are often regarded by contractors as potential turnover or earned value on a contract, as there is a good chance that the money will be spent at some stage of the contract.

12.5.7 Daywork

Daywork is a type of provisional sum that allows for undefined work which will be paid for on the basis of the time spent and materials used in completing the job. The contractor's daywork rates are calculated by reference to a 'definition of prime cost of daywork'. There are several definitions and two common ones – one for building work and one for civil engineering (Chapter 17). There are four main ways of including a provision for daywork in the tender documents:

1. Include a schedule of provisional sums of money in the bills of quantities for labour, plant and materials and ask the tendering contractors to price their percentage additions to these items (Table 12.7).
2. Include a schedule of a provisional number of labour hours for different sorts of labour in the bills of quantities and ask tenderers to price their hourly daywork rates against these hours.
3. Ask tenderers to quote their daywork percentage additions for labour, plant and materials.
4. Ask tenderers to quote their hourly daywork rates for labour and percentage additions for plant and materials.

Table 12.7 Pricing of provisional sums for daywork.

Ref	Item	Quant	Unit	Rate	£	p
A	Allow the provisional sum of £5000 for the net cost of labour expended on a daywork basis				5000	00
	Add Percentage addition		%	160	8000	00
B	Allow the provisional sum of £10 000 for the net cost of materials expended on a daywork basis				10 000	00
	Add Percentage addition		%	25	2500	00
C	Allow the provisional sum of £2000 for the net cost of plant expended on a daywork basis				2000	00
	Add Percentage addition		%	75	1500	00
	Page Total to Summary				29 000	00

1. Provisional sums inserted by architect/engineer/PQS.
2. Rates inserted by contractor at tender stage.
3. Provisional sums deducted at final account stage.
4. Actual daywork authorised and carried out will be added back.

Traditionally, daywork is seen by contractors as a 'nice little earner' and a convenient way to boost profits whilst paying site operatives the normal 'flat rate' for the job. From the employer's perspective, therefore, the contract documents should be structured in such a way that the daywork rates are priced competitively at tender stage. Consequently, methods 3 and 4 should be avoided as they will not ensure competitive daywork rates.

12.5.8 Contingencies

Contingency sums are 'global' allowances to provide against the unexpected and are intended to give the designer a degree of latitude in the overall spend for the project. Contingencies are particularly useful where the cost plan allowance for an element of the design is overspent and savings cannot be made elsewhere.

In an ideal world, the contingency sum will not be spent and will, therefore, represent a saving for the client. If the unexpected does happen, however, the contingency sum acts as a 'buffer' to protect the client from a nasty surprise when the final account is prepared. Contractors regard contingencies as potential turnover and sometimes take the view that savings can be made in the tender figure in respect of the 'commercial opportunity' presented by a 'juicy' contingency sum! (Chapter 15).

Table 12.8   Adjustment item.

Adjustment item at tender stage						
Ref	Item	Quant	Unit	Rate	£	p
	GRAND SUMMARY					
A	Class E: Earthworks				146 973	96
B	Class F: *In situ* concrete				49 334	72
C	Class G: Concrete ancillaries				9874	21
D	Class P: Piles				98 652	98
E	Class Q: Piling ancillaries				13 788	67
F	Class U: Brickwork, blockwork and masonry				17 463	22
G	Class W: Waterproofing				6998	44
	Bills of Quantities Total				343 086	20
	Adjustment item				*12 196*	*20*
	TENDER TOTAL				**330 890**	**00**

Adjustment of interim valuation					
		£	p	£	p
• Paid or deducted in instalments	Gross interim valuation			104 746	96
• Adjusted in proportion to the total of the bills of quantities before the addition/deduction of the Adjustment Item	Adjustment: £12 196.20 × £343 086.20	104 746	96	(3723	60)
• Adjusted before deduction of retention monies				101 023	36
• Adjustments shall not exceed the total of the Adjustment Item	Retention 3% **Certified value**			3030 **97 992**	70 **66**

12.5.9　Final summary

Bills of quantities culminate in a final summary page, which brings together the totals of all the preceding work sections and preliminaries or general items. The total amount of money in the summary represents the contractor's tender total.

12.5.10　Adjustment item

Adjustment items enable contractors to make last minute changes to the priced bills of quantities, before submission, without having to alter all the rates and prices already entered or recalculating the totals. In building work, the contractor usually makes adjustments in the preliminaries section of the bills of quantities but, in the civil engineering sector of the industry, it is usual to provide the contractor with an adjustment items on the final summary page of the bills.

Where an adjustment item is priced this represents an increase or reduction in the unit rates in the bills of quantities as a proportion of the bills of quantities total. Table 12.8 illustrates how an adjustment item may appear in the grand summary of the bills of quantities for a civil engineering project.

Most civil engineering contracts are re-measurement contracts and, as such, there is no contract sum – just a tender total. When the time comes to value work in progress, rather than painstakingly increasing or reducing all the contractor's rates to cater for the adjustment figure, an adjustment is made to the gross valuation, before deduction of retention money, in the proportion that the adjustment item represents to the bills of quantities total (not the tender total). Table 12.8 illustrates how this works with a gross interim valuation of £104746.96 and explains CESMM4 rule 6.4 – Adjustment Item.

References

CIOB (2008) *Code of Estimating Practice*, 6th edn. Chartered Institute of Building (CIOB), London.

Cooke, B. and Williams P. (2009) *Construction Planning, Programming and Control*, 3rd edn. John Wiley & Sons Ltd, Chichester.

Eggleston, B. (2006) *NEC3 Engineering and Construction Contract: A Commentary*. Blackwell Publishing, Oxford.

FRC (2010) The UK Corporate Governance Code. Financial Reporting Council (FRC), London.

Harris, F. and McCaffer, R. (2006) Modern Construction Management, 6th edn. Blackwell Publishing, Oxford.

ICAEW (1999) The Turnbull Working Party Report. Institute of Chartered Accountants in England and Wales (ICAEW), London.

Masterman, J.W.E. (2002) *An Introduction to Building Procurement Systems*, 2nd edn. Spon Press, London.

Morledge, R., Smith, A. and Kashiwagi, D. (2006) *Building Procurement*. Blackwell Publishing, Oxford.

Perry, J.G. and Thompson, P.A. (1982) Target and cost-reimbursible contracts. Report No. 85, Construction Industry Research and Information Association (CIRIA), London.

RICS (2011) *Contracts in Use, A survey of building contracts in use*. The Royal Institution of Chartered Surveyors (RICS), London.

Turner A.E. (1997) *Building Procurement*, 2nd edn. Macmillan Press, London.

13

Budgets

Financial Management in Construction Contracting, First Edition. Andrew Ross and Peter Williams.
© 2013 Andrew Ross and Peter Williams. Published 2013 by John Wiley & Sons, Ltd.

13.1 Developing and monitoring budgets

The establishment of organisational goals for the year is informed by the firm's strategic plan and its ability to meet historic goals in previous years. All organisations, irrespective of their size or function will plan ahead, the plan may be for periods as short as a month or as long as five years. In all cases a financial plan will produced which will identify goals or targets to be achieved. As the activities over the period of the plan are undertaken, measurements of the outcomes are taken and then compared against the original targets. If a variance is identified that requires corrective action, it will usually be undertaken and the goals within the plan updated.

The budgets are usually developed by individuals who have the responsibility of achieving the goals set; a turnover budget for a region within a firm would have a contribution from the regional director whereas a project turnover budget would rely upon a discussion with the project manager. Most organisations will have the budgets shown in Table 13.1.

Table 13.1   Types of strategic and project level budgets.

Strategic	Project level
Turnover	Turnover
Cash	Cash
Overhead	Production
Profit and Loss	Profit
Balance sheet	

13.2 Types of budget

13.2.1 Strategic budgets

The development of the strategic budgets for the year is a lengthy and complex process. It will be informed by the strategic business plan, which may have the aim to increase turnover or to improve profitability or both. It may aim, as in the present economic circumstances, to reduce turnover and maintain profitability and improve cash flow. All of these considerations will eventually have to be condensed into a turnover budget for the year.

13.2.2 Turnover budget

The accounting year varies from firm to firm and so the turnover budget will follow the accounting reference period (Chapter 2) of the company. Turnover can be considered as value of production, this is different to what has been paid or what has been costed, as will be considered in later chapters. The organisations budget, therefore, is the aggregation of the value of production of all the current projects that are either completed or continuing or are to start in the forthcoming financial year. It should also make a projection on those projects that are currently being tendered and those for which tender enquiries may be expected and likely to commence in the year.

The turnover budget, sometimes considered as the sales budget, is based continuing projects and future projects. Turnover contributions from continuing projects are relatively straightforward to calculate, each project quantity surveyor will have a turnover projection to completion which is updated each month. Assumptions need to be made with regard to the contribution made by future work, as some of the work will have been secured, some projects will be at tender stage and others only speculative enquires highlighted by the business development manager and in are the pipeline. They may be subdivided into:

- definite
- probable
- possible.

Some firms take a percentage of categories to future budgets of 100%, 70% and 20% respectively. These obviously vary with the size of firm, the relationship with the clients and the bullish or bearish nature of the management team and, most importantly, the extent of competition. The turnover budget will allow the senior management to set the business development unit future targets to bring in new enquiries and the estimating and planning department associated targets for tender submissions.

13.2.3 Overhead budget

Normally head office overheads are added to a project costs as a percentage at tender stage. This percentage is derived from the projected annual turnover of the firm and the anticipated head office overhead costs (offices, service department costs etc.). If an organisation suffers from falling turnover during the year it has a dilemma, as the recovery of the overhead is less than projected and could lead to reduced profitability.

A reduction in costs is usually to be avoided as the experience of the staff is often difficult to replace, an increase in the percentage to be applied to future projects could lead to them being uncompetitive and reduce the likelihood of the project being won. It is for this reason that the overhead and turnover budget is closely linked. Project-based overheads, (staff, cabins, utilities etc.) are normally costed against the project, so do not become a problem unless the company has too few projects to deploy the project-based staff. In this case they become a general overhead to be allocated to current projects.

Table 13.2 illustrates a turnover and overhead summary budget which identifies that a company had a projected head office overhead cost of 29.5 at the start of the year. A projected turnover budget of 295 was established based on current work and future work. The actual turnover dropped from what was expected at month 3 and continued to be under budget until month 7. If this trend stopped and the turnover went back to that predicted, the resultant 40% reduction in turnover in the year would equal a shortfall of overhead recovery of 5.5. In order to recover this shortfall the percentage overhead addition to the remaining projects from period 7 onwards would increase from 10 to 21. Figure 13.1 illustrates the turnover and overhead recovery situation graphically.

The above example is somewhat simplistic as overheads, like all costs, can be fixed, variable or semi-variable and related to the volume of throughput of potential

Table 13.2 Overheads and sales budget.

Month	Sales	Overhead recovery	Actual sales	Actual overhead recovery	Variance
1	10	1	10	1	−0.9
2	10	1	10	1	−0.9
3	20	2	15	1.5	−1.8
4	30	3	15	1.5	−2.7
5	20	2	10	1	−1.8
6	10	1	5	0.5	−0.9
7	40	4	20	2	−3.6
8	30	3			
9	45	4.5			
10	23	2.3			
11	34	3.4			
12	23	2.3			
Projected turnover	295	29.5			

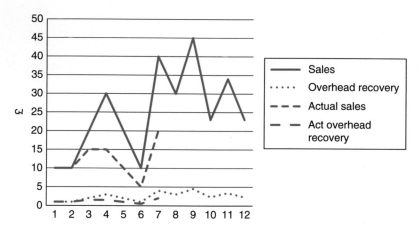

Figure 13.1 Representation of overhead recovery.

and completed projects. The example does, however, illustrate the importance of strategically monitoring the organisation's turnover budget at regular periods. A shortfall in the firm's turnover budget can have an impact at a project level. Conventional measures, such as ensuring correct valuations are completed, can be a focus for management action, less conventional methods, such as acceleration of works, early and enhanced valuations and early deliveries of materials on sites, have also been witnessed. The impact at a project level can appear quite minimal. However, when the project outcomes are aggregated together they can have a significant impact.

Cash budgets, profit and loss and balance sheet budgets have been considered in Chapters 8 and 11 respectively.

13.3 Project level budgets

The financial value of a project is usually determined by the estimator who combines in-house production standards with quotations for materials, plant and subcontracts into an estimate that when combined with the project overheads and margin becomes a tender. The project budget is made up of a series of associated budgets for turnover, cash flow, production allowances, overhead, commercial opportunity and profit.

A simple definition of project budgetary control is a system in which actual income and spending are compared with planned income and spending, so that management can determine if the pre-project plans are being followed and whether those plans need to be changed in order to make a profit. This definition is easy to understand but the implementation of budgetary control in construction projects is quite complex.

The client's quantity surveyor undertakes budgetary control at the design stage by comparing projected expenditure with a cost plan and provides advice on any remedial actions to be taken to bring the project back in line with the cost plan. This management process is arguably a lot easier than that which is undertaken by a contractor's quantity surveyor who is responsible for commercially managing a project, reporting on resource budget variances, advising on remedial action in both client and supply chain directions and monitoring and reporting a profit.

The starting point for project control is the budget. This, as discussed in Chapter 3, is developed by a department whose specific role is to secure work. The tender document is often in a form that requires it to be adapted to support budgetary control; for example, an elemental cost plan has budgets allocated to elements that will be carried out by resources that overlap with other elements. The de-scoping of the tender documentation to provide a basis for control is discussed in Section 13.5.

13.3.1 Project turnover budgets

The level of detail a project turnover budget can be expressed in is in direct proportion to the amount of design information available. If little design information is available, the budget may be based on an S-curve formula or $\frac{1}{4} : \frac{1}{3}$ approach discussed in Chapter 11. In contrast, an estimate for a traditionally procured project, which has been tendered on a completed design, will be supported by a construction programme, a schedule of major subcontractors and supplier quotation, and a costed preliminaries schedule. At the early stage of the project it is relatively straightforward to use this information to complete a more accurate turnover budget than the model-based budgets.

At the commencement of the project, the turnover budget can be fairly uncertain, as there are many assumptions that the pre-contract team has made regarding production, precedence of activities and so on. As the project progresses and as built programme emerges, the turnover budget starts to become more certain.

13.3.2 Project cash budgets

Each project has specific factors that influence the money in and out of the company. Some relate to the client's procurement approach, such as payment terms, some relate to the technological nature of the project, for example whether the project has a condensed programme with a lot of value generated in the early months, and some factors relate to the supply chain and their credit terms. The modelling of project cash budgets and how a contractor seeks to minimise the aggregated impact of its cash requirements is considered in Chapter 11.

13.3.3 Production budgets

Production budgets can fall into two categories, those that are used to procure resources, such as materials, plant or subcontractors, and those that are used to consider the activity. The procurement or buying budgets are developed during the estimating phase of the project and are based on quotations received from tendering suppliers and subcontractors; an adjustment to the prices received is often taken at tender stage and the adjusted price becomes the buying budget. The activity-based budgets are based on output standards, often contained within the estimating software 'library' of items or estimator's notes and are adjusted for factors such as access, difficulty, repetition and so on. Both buying budgets and activity budgets can be combined to create a work package budget.

13.3.4 Procurement budgets

Materials

It is important for the project team to be able to untangle the losses or gains that are a result of procurement of resources from those that are a result of production. During the estimating stage, the estimator will compare the quotations received from various suppliers and base the materials component of the estimate on this information. Some organisations have strategic alliances with materials suppliers which will provide a schedule of prices for commonly used items; these can be accessed via a look-up system and reduce the requirement for sending for quotations for every material item. The development of the budgets for materials is based on the supplier prices and also includes an allowance for materials that are wasted; these can be divided into two categories, that which is as a result of design and specification and that which is a result of construction activity.

Examples of design wastage include:

● Drylining: cutting of plasterboard sheets and metal studs to fit wall heights and openings.
● Flooring: cuttings of floor tiles to fit room layouts.
● Ceilings: cuttings of ceiling tiles and fixings to fit room layouts.
● Insulation: cutting of insulation boards to fit openings.
● Tiling: cutting of floor and wall tiles to suit design and room shapes.
● Paving: cutting of paving slabs to fit layout.
● Brickwork and blockwork: cutting of bricks and blocks to suit building dimensions and building services.

Examples of waste as a result of construction activity include:

● Inaccurate or surplus ordering of materials that are not used.
● Damage through handling errors.
● Damage through inadequate storage.
● Damage generated by poor coordination with other trades.

- Rework due to low quality of work.
- Inefficient use of materials.
- Temporary works materials (e.g. formwork, hoarding etc.).

To make an accurate assessment of the wastage that may result from design activities the estimator must cross reference the specification, bills of quantities and drawings and visualise how the construction will be undertaken. A detailed knowledge of materials sizes and packaging quantities is obviously required.

When a competitive design and build tender is being produced, often the estimator will be required to assess the acceptability to the client of two or more different alternatives. This requires knowledge of the technology and the performance once incorporated into the project as well as an understanding of the price of the interdependent and consequent designs related to the alternatives – a demanding task given the short tender periods demanded by clients! Occasionally, a tender will be submitted including qualifications that identify that a price has been offered using a cheaper although similarly performing alternative.

Almost all construction activities involve materials wastage; if possible, main contractors will try to offset the risk of this cost to their subcontractors by requesting quotations which include labour, materials and plant. Where this is not the case and the estimator has to include an allowance for materials, often a percentage will be added for waste which is based on historic records. If a work package is to be undertaken by a labour and plant subcontractor with the main contractor supplying the materials, a clause in the subcontract may be included to allow the contractor to contra-charge the subcontractor if materials wastage exceeds a given percentage. This penalty is often levied against groundwork contractors which have either over-dug trenches or have produced poorly stoned up formation levels for slabs and the activity concrete wastage percentages have been exceeded.

It is becoming mandatory for contractors and subcontractors to have a site waste minimisation plan. Some recent research indicated that the true cost of filling and disposing of a skip full of mixed construction waste was £1343; this consisted of skip hire £85, labour to fill the skip £163 and the costs of the unused materials £1095. If the considerable environmental implication of disposing of the unused materials is overlooked, the cost of the materials can be considered as a loss and has a direct impact upon the project's margin.

The outcome of the estimator's calculations for the materials budget will be a materials allowances for each quantified item, which is usually based on a number of materials quotations, often 'muscled' down to anticipate additional discounts offered if the tender was successful. This allowance will form the basis for the standard to be used for post-contract budgetary control. Figure 13.2 shows an example of the development of a materials procurement budget.

Subcontractor procurement budgets

A similar process is used to develop a subcontract budget. During the estimate stage the subcontract quotations will be compared and the most favourable prices will be included within the estimate. At the tender stage, the committee will consider the 'safeness' of the prices by reference to the comparable prices, the reliability and history of the subcontractor and will agree upon the extent of commercial opportunity and muscle that are applied to each trade package. Once the project has been won the quantity surveyor will develop a subcontract procurement schedule and often

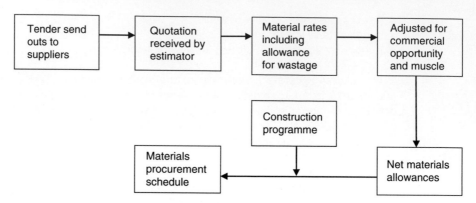

Figure 13.2 Development of a materials procurement budget.

will re-send out for new quotations. The enquiry process, comparison and analysis of subcontract prices are considered in more detail in Chapter 3.

13.4 Activity level budgets

The estimator has to visualise an activity and develop a set of allowances that represent the resources required to undertake this activity. Imagine building a brickwork wall up to 4 m high; the activities required to undertake this will include transportation of the materials to the place of construction (this may involve plant and man hours), providing a secure access for the bricklayers to lay the bricks (this may involve plant hire – erection and alteration as the wall increases in height), the brick, mortar and brick ties and associated wastage and the labour resource, which may be a combination of craftsmen and labour. The activity will have an on-cost which will represent site overheads, head office overheads and profit margin. The activity will have a planned duration and will be preceded and followed by linked activities.

The contractor's programme would indicate a period of time for the construction of the wall, the designer's specification would identify the type and quality of materials to be used and the expected tolerances. The time allowed for the activity will be calculated from the quantity to be carried out, the resource levels to be deployed and the expected outputs from the planned resource. Consequently, the activity level budget can have resource level budgets supporting it. This is illustrated in Figure 13.3, which shows budget hierarchies for a composite activity.

At tender stage, a reconciliation of the resource allowances included by the planner within the construction programme and those made by the estimator and included within the estimate will be undertaken to ensure that the logic of programme and estimate are consistent.

The above simple example allows the high level activity to be broken down into a number of resource budgets. These can be used for procurement of the labour, materials and plant, used to monitor the outputs of the resources, used to reconcile material wastage, used for the procurement of the scaffolding subcontract and can also provide valuable feedback to the estimator and planner about the assumptions they made at estimation stage.

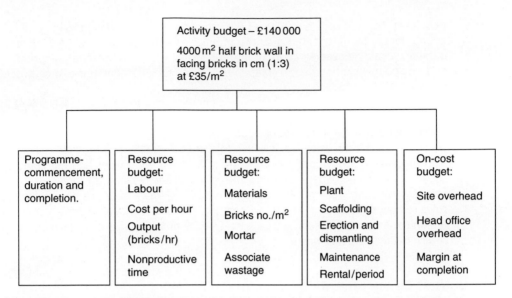

Figure 13.3   Hierarchy of budgets at activity level.

13.5   De-scoping bills of quantities

As can be seen from Section 13.4, an activity can have multiple cost bearing aspects that can be used for resource procurement, monitoring and prediction. The Royal Institution of Chartered Surveyors (RICS) undertook extensive consultation with contractors during the development of methods of measurement with the aim of attempting to reflect the construction process within the measurement rules. The common arrangement of work sections was introduced in SMM6 and refined slightly for SMM7. The New Rules of Measurement (NRM) committee recognised that the bill of quantities were used as a basis for procuring prices from specialist subcontractors and the new rules recognise this and ensured that work can be subdivided using work breakdown structures and then procured and monitored as work packages.

The **New Rules of Measurement** define a work package contractor as 'a specialist contractor which undertakes particular identifiable aspects of work within the building project; e.g. ground works, cladding, mechanical engineering services, electrical engineering services, lifts, soft landscape works or labour only. Depending on the contract strategy, works contractors can be employed directly by the employer or by the main contractor'.

The NRM support the breakdown of the bill of quantity into three different structures:

- Elemental – this is consistent with the logic of the cost planning.
- Work section – this relates to the 41 NRM tables from preliminaries to builders work in connection.
- Work package – Measurement and description are divided into employer or contractor defined work packages. Works packages can be based on either a specific trade package or a single package comprising a number of different trades.

The measured items within the work sections are measured net in place and the NRM committee has not attempted to provide a level of coding beneath the measured item level. The new coding reinforces the use of a database style approach that has been adopted by most bill of quantity production software such as CATO. Such a database could be used as an input into a contractor's work package procurement system and the addition of new codes supports the procurement and cost control of some activities.

Most estimating software systems use a similar approach to the development of work package send outs. The estimator will have a clear idea of how risk for the different sections of the work is to be apportioned and will put together work packages which reflect this; for example, if a high tolerance floor slab and finish is required a work package that included the concrete slab laying and power floating could be separated from the general groundwork work package, which may include excavation and concrete work to pile caps and ground beams. There are no established rules to how an estimator packages items together and although the main contract preliminaries would be 'sent out' to specialised contractors it is rare for them to be return detailed breakdowns.

13.6   Budget development

The nature of estimating is that it is a process of arriving at a net price which is then adjusted by a tender process to create a price which will win work in a market condition. As such the estimate is rarely thought of as a definitive accurate standard by which to control the costs of the construction, it can be thought of as a point of reference by which to identify the need for remedial action if variance is discovered. However, the reference point needs to have a logical basis.

The standards used for developing the estimate are usually developed from data that have been captured in a work study or other standards that are published in price books and are expressed in hours/unit of measured work. Few organisations have formalised systems for capturing and analysing data from site to refresh these standards. Consequently, they are often perceived by the estimator as rough guides that are subject to adjustment using a range of quantitative and qualitative factors. In fact, many organisations discourage estimators to seek feedback on the efficacy of the assumptions made at tender stage. The reason behind this is that if estimators were to include the 'true' costs of undertaking a project the net price would be too high to win the job when compared with the company's competitors. This leads to a well established attitude by estimators that their role is to win the work and the construction team's role is to build the project within the price they have been given. Feedback about the estimate that comes from the site team is usually negative and it would be unhelpful to the company if the estimator attached too much credence to it.

13.6.1   Labour

The need for accurate output standards is further weakened as the responsibility for labour budgetary control is passed to others down the supply chain. A high proportion of labour is employed as labour only subcontracting, either directly to a main contractor or to another subcontract organisation. Often the size and nature of these subcontractors means that they do not have the resources or expertise to invest in systems for capturing and analysing work study data. It is perhaps as a consequence of the remoteness of control, the fragmented nature of construction teams and the variable nature of the construction process itself that the more sophisticated systems of **parametric estimating** used by some manufacturing organisations have not found a place in construction.

An estimator, whether working for an organisation that is undertaking the work or working for an organisation that is pricing work that will be undertaken by others, will have to calculate a budgetary allowance for the labour element of the work to be undertaken. When developing the unit rate for a measured work item, the estimator has to refer to historic data on performance achieved by labour on previous projects. Usually the estimator will have a standards database that can be referred to. The standards can be expressed in hours per resource per activity with a range of adjustments for differing circumstances. For example, an estimating labour standard for installing cable trays would identify the height to soffit, the type of soffit materials to be drilled, the type of cable tray and the fixing itself.

A budgetary allowance for an item would consequently be generated and this allowance would be used as a basis to procure labour resources, monitor the costs and provide a basis for payment. It can also be used to support a contractor's claim for additional money as a consequence of a variation or disruption claim.

The labour budgetary allowance also supports the establishment of incentive or bonus systems. A **bonus system** in construction is based on the principle of incentivising labour to achieve outputs that lead to a saving. These can be categorised into four different approaches

1. Piecework
2. Time saved
3. Performance related
4. Job and finish.

A **piecework scheme** is the easiest to develop, if a rate is agreed with a craftsmen. If bricklayers lay 1000 bricks they can be expected to earn £300, from which a notional hourly cost can be deducted (Table 13.3).

Table 13.3   Gang bonus calculation.

	Output	Rate £/1000		Total
Earned	7000 bricks	300		£2100
Cost	h/man	Cost/h(£)		
2 bricklayers	35	15	1050	
1 labourer	35	10	350	(£1400)
	Bonus earned for gang			£700

Another example which is also output related with a base line of expected output and a bonus paid for work exceeding expected output is shown below. The costs of administration of bonus schemes are measuring and agreeing work undertaken. Agreeing the bonus itself can be problematic, as often labour can be disrupted during production and can rightly claim that non-productive time should be discounted against the bonus (Box 13.1).

The most common scheme in the construction industry is the **time saved system** – output targets are compared with the actual time taken and any hours saved are paid at a predetermined rate per hour. The advantages of this scheme are obvious; a lot of

Box 13.1   Agreeing a bonus

Assume a 102 mm brickwork wall (i.e. 60 bricks/m²), overall area of 280 m². A 2:1 gang could typically lay 220 bricks per hour. A gang cost is £40/h (bricklayer cost £15/h, labourer cost £10/h), this is equivalent to £3.67/m². This would give a budget of £10.91/m² or 0.27 h/m².

There would be other costs associated with this item, which could be access scaffolding, protection, supervision. If these are assumed to cost £300/week and the overall activity was programmed for a two-week period, the allowance bill would include £600 for plant.

The aim of a bonus system is to incentivise production by paying a weekly amount that will provide a saving; the time-related cost above is £8.57/h. Industry norms dictate that approximately 60% of the saving is passed on, this would equate to £5.14/h. This would mean that a bonus of approximately £5.00/1000 bricks in excess of the weekly production target of £7700 would be paid.

Table 13.4   Estimation of time saved.

			hours
Target hours	7000	50 bricks/h	140
Actual hours	3 men	35 h	105
		Time saved	35

Table 13.5   Time based gang bonus.

Gang rate	Craftsmen	2	15	30
	Labourer	1	10	10
				40
	Bonus (at 25% rate)			10
	Bonus			350

preliminary costs are time related and if the time taken for activities can be reduced there are knock on savings that can be achieved. An example is given in Table 13.4.

Most time-related schemes have a notional rate or a percentage of the employed rate rather than the full hourly cost. If each hour saved is paid as 25% of the gang rate, this will give a gang bonus of £350 (Table 13.5).

Some construction companies use standard times and pay bonuses according to a predetermined scale which may increase for each point of performance for **performance-related** schemes. The scheme can be difficult to implement as non-productive time needs to be considered and the scales often need adaptation for work of differing complexity.

Using the example above, the target hours is 140, the gang took 105 hours to complete the work, a simple ratio calculation of target/actual gives a value of 1.33. This can be then used against a standard scale to determine an hourly bonus for the gang (Table 13.6).

The gang bonus would therefore be 105 hours at £3.00/h = £315.00.

Table 13.6 Performance ratio for bonus calculation.

Performance ratio	Gang bonus (£/h)
<0.8	Nil
0.8–0.9	0.50
0.9–1.0	1.00
1.0–1.1	1.50
1.1–1.2	2.00
1.2–1.3	2.50
1.3–1.4	3.00

It is beyond the scope of this text to consider bonus schemes in great detail; most are operated within subcontractor and small companies and are tailored to the specific technology. The approach can have a dramatic effect upon the approaches taken by the various operatives to their working day. A plastering subcontractor may pay an all-in rate of £3 per m^2 of production. A plasterer can only 'lay up' approximately 25 m^2 at any one time; as this first section is going off another 25 m^2 is applied. By the time the second section has been applied the first section is ready for finishing and once finished the second section can be completed. 50 m^2 of production has been achieved, giving a payment of £150 for the day's work. This may have been achieved by 3.00 p.m. and any production over and above this would be difficult to achieve as the productivity output for any additional sections within the day would be low and unattractive to the craftsman.

13.6.2 Materials

Tender documentation measures construction work net, there is no allowance made within the measurement of an item for the costs associated with producing the item. The cost allowances must be made within the pricing rate. The industry briefly considered **operational bills of quantity** as a means of measuring the resources that are used in production but dropped the idea as it was too complex to introduce in practice. The contractor will be paid on the basis of net output, and will have to make an allowance for wastage whether it is from theft, design or production. The estimator will take into account wastage in the build up to the net rates and these will form part of his notes; it is not usual for these notes to be communicated to site. The site team will be provided with a materials allowance which is used as a basis for the budget for the procurement of the materials, their transportation to site, their storage and eventual placement into position.

Most organisations centralise the material procurement function under the control of a **buying department**. The advantages of this are that it provides the company with an opportunity to develop relationships with a few suppliers for their materials and by aggregating the demand from all the sites it will achieve a better price and credit terms. It also provides a level of control, as sites have to go through well defined and auditable procedures to order materials. The buyer will use the budget allowances to set a target for the materials procurement. If the materials are purchased for less than this target a buying gain has been achieved. If over target a buying loss is recorded. It is important for the site to be aware of this when monitoring the profitability of the materials component of a project to allow them to disentangle procurement gains from production losses.

Materials are ordered by the site through the completion of requisitions which identify quantities, dates and locations. The nature of the construction requires daily and sometimes hourly communication between the site and the suppliers. Materials that are delivered are recorded by the site staff on materials received sheets; these form the basis of the cost records, which are discussed in Chapter 18. It is good practice for the site quantity surveyor to undertake **bulk reconciliations** after the completion of an activity to establish the amount of wastage that has occurred. This is of particular importance if the materials are costly and subject to high percentages of waste, such as *in situ* concrete. It will also allow for the early identification of problems that require attention. In one case that the author had experience of, the supplier of ready mixed mortar had submitted invoices, supported by delivery tickets for mortar used in the production of facing brickwork. Reconciliation with what has been undertaken indicated that the wastage was 350%. This was excessive even for ready mixed mortar. Further investigation by the site team uncovered a scam whereby the labour only brickwork subcontractor was diverting some of the tubs of mortar to a nearby housing site!

The regular re-measurement of works and reconciliation of actual materials used against budget allowances give invaluable information that can identify where the problems lie and support the site team in taking remedial action. Not all cases are as dramatic as the one above, often problems of poor storage or inaccurate preparation are highlighted which early action can have an impact on. Materials reconciliation also flags up situations where the conditions of undertaking the work may have been different from those envisaged by the estimator at tender stage. If this change is a result of a variation the costs may be recoverable and the reconciliation can be used as justification. The completion of the **project cost value reconciliation** will often highlight macro variances which require explanation; the buying report along with the materials reconciliations completed should provide information which will support such an explanation.

13.6.3 Plant

Few general contractors own their own plant, the costs of owning, operating and disposing of plant has been passed on to subcontractors. Specialised subcontractors which own plant develop hourly rates which include for the costs of depreciation, maintenance and transportation to and from site. These hourly rates are then used to build up unit rates. There are many books on estimating to which the reader can refer if they wish to discover more about this approach.

The plant that the main contractor uses is hired and, as with materials supply discussed above, strategic alliances are formed with plant hire companies which agree a schedule of hire rates for popularly used plant. The plant is called off in a similar fashion using a plant requisition schedule, where the buying department places the order and the plant is delivered to site. The hire company keeps an inventory of what plant has been delivered and bases its monthly invoice on an itemised list applying the agreed rate. The costs of plant are calculated on the basis of the plant returns, the buying department's inventory and the invoices received by the plant hire company. The costs are reported monthly and the site team is required to monitor the costs against the budget allowances established by the estimator.

The budget is often within different parts of the tender document; for example, site cabins and small tools may be priced within the preliminary section, ground pumps may have been allowed for in the excavation unit rates and access towers may be included within a subcontract allowance for painting. It is important to establish where the tender allowances are and use them to compare actual with expected. As with the buying of materials, gains and losses can be made in agreeing rates for plant and these need to be disaggregated from those which are a result of plant being hired for too long or in too high a quantity. Construction managers have a deserved reputation for keeping plant on hire 'just in case', which can add considerably to the costs. Regular reconciliations with the allowances and the as built programme can identify opportunities to recover some of the additional costs from either the client if the work has been varied or by contra-charging subcontractors if they have delayed the works.

13.6.4 Preliminaries

The preliminaries budget will consist of time-related items, such as hire of cabins, site staff, and attendant labour, and fixed costs, such as delivery of cabins to site, establishment of telephones, water and other utilities. The tender committee often trims the preliminaries budget to reduce the price to win the work and it is often the same staff which allocate the costs of the overheads once the job is underway!

The allowances should have a full breakdown, for example staff costs will identify the number, the grade and the percentage of time allowed on the contract. This can be used to develop a preliminaries allowance budget for the project which can be used to monitor the costs against.

The quantity surveyor will agree a schedule of payments against preliminaries when discussing valuation principles at the commencement of the contract. The client's quantity surveyor will only have the priced document and will not have access to the allowances. The priced preliminaries are often front loaded to enhance cash flow and it is important than the allowances are adjusted to take account of this loading when calculating the variances. This principle is explored in Chapter 18. A typical schedule of monthly preliminary costs used as a basis for interim valuation is shown in Table 13.7.

The cost value reconciliation (CVR) also will only usually report on the preliminaries heading of staff cost, the other allowances and costs subsumed within the headings of labour, plant and subcontractors. The consequence of this is that the site quantity surveyor will need to monitor the adequacy of the preliminary allowances for these

Table 13.7 An example of how cost items are expressed within the method of measurement.

	Months					
	1	2	3	4	5	6
Allowances						
Staff	6000	4000	4000	5000	5000	4000
Site accommodation	2000	2000	2000	2000	2000	2000
Services and facilities	7000	1500	1500	1500	1500	2000
General attendance	3000	3000	4000	4000	2500	2500
Temporary works	8000	1000				
Prelims	26000	11500	11500	12500	11000	10500
Cumulative	26000	37500	49000	61500	72500	83000

heading as insufficient allowance for, say, scaffolding may lead to a variance in the subcontracting heading of the CVR which will require explanation.

13.7 Variance analysis

An analysis of variance from budgets requires information from a range of sources; the valuation provides information regarding what has been paid, the internal measure identifies what has been completed, the cost reports indicate the recorded costs and the buying report will identify any losses or gains which have been the result of the procurement approach. The site team will often have to assess individual work package activities to identify variances. A simple example of a productivity assessment is given in the following section; the example commences with a typical item within a bill of quantities and demonstrates how the item needs to be subdivided into budget allowances before assessing the position now and projecting into the future. The example also illustrates how the adjustment of over and undermeasure should be taken into account to ensure the correct comparison of allowances with costs is being made.

The following item is priced in a bill of quantities and the profitability of this item needs to be calculated up to the end of valuation 1 and projected until completion.

13.7.1 Productivity assessment: an example

'102.5-mm thick wall in facing bricks in cement mortar (1:3), pointed one side 5000 m2 at £49.60.'

The estimator's pricing notes indicate the following:

- Labour – a 2:1 gang, bricklayer at £15.00/h, labourer at £ 10.00/h, the output of the gang has been calculated using the estimator's standards as 50 bricks per bricklayer per hour.
- Materials – at tender stage a quotation from the brick suppliers was used, the supplied cost was £300/1000 bricks. The estimator included an additional 10% for wastage at tender stage. The mortar supply was allowed for within the preliminaries.
- The estimator allowed for an on-cost of 15% to include for site and head office overheads and profit.

The first valuation for this item has been carried out and the details are:

300 m² have been completed.
350 m² have been claimed.
330 m² have been paid.

The cost report indicates that the materials are being purchased by the buying department for £280/1000 and the labour returns indicate 200 gang hours have been recorded for this activity. to undertake the assessment of allowance against cost, an allowance for labour and materials is required (Table 13.8).

The allowance bill for the item can then be developed (Table 13.9).

The next stage is to calculate the **net allowance** for valuation number 1 for labour and materials. This is required to compare income generated with costs incurred. The net allowances are given below, labour is £7101, materials £5841 and on-cost £1941 as illustrated in Table 13.10.

Table 13.8 Labour and materials allowance.

All-in rate build up			
Labour		**£**	
2:1 gang	Bricklayer	15	
	Labourer	10	£40/h
The output of the gang bricks/hour	100		
Assuming 59 bricks/m²			
Labour allowance rate per m²		**23.67**	
Materials			
Facings £/1000	300		
Cost per m²	17.7		
Waste 10%	1.77		
		19.47	
Subtotal		43.13	
Overheads and Profit 15%		6.47	
External rate		49.60	

Table 13.9 Establishment of Allowance bill.

Allowance Bill				
		Labour	Materials	Overheads and profit	Total
102.5 brickwork	5000 m²	118 350	97 350.00	32 350	248 050

Table 13.10 Aspects of a valuation.

Valuation 1					
		Labour	Materials	On-cost	Total
Claimed	350 m²	8 284.00	6 814.50	2 264.50	17 363.00
Complete	300 m²	7 101.00	5 841.00	1 941.00	148 883.00
Overmeasure	50 m²	1183.50	973.50	3 23.50	2 480.50

The costs incurred can then be compared with these allowances. This has been simplified, as in reality direct and accrued costs would need to be considered. The materials costs are £5451.60 (Table 13.11) and labour costs £8000.

This **buying gain** is £1.30/m². If projected until completion of this activity would generate a surplus total of £6500. This is in contrast to the labour position (Table 13.12).

The production loss is £3.00/m². If this is projected to completion it would give an additional loss of £7000, a total labour loss of £15 000.

This example demonstrates the need to disaggregate buying gains from production losses, as, if taken as a total, a buying gain can mask the problem of a production loss.

Table 13.11   Calculation of a buying gain.

Materials					
300 m²		59	17 700.00	280.00	4956.00
Waste	10%				495.60
					5451.60
Buying gain					
300 m²		59	17 700.00	300.00	5310.00
Waste					531.00
					5841.00
Buying gain at Valn 1					389.40

Table 13.12   Assessment of labour budget variance.

Labour				
200	hours	40/h	Cost	8000.00
Included in estimate				
300 m²	23.67		Allowance	7101.00
Loss at Valuation nr 1				899.00

13.8   Control procedures

For effective financial governance of projects a system of reporting of variances from budget is required. The level and frequency of reporting will depend upon the control cycle required for the resource considered. A high cost resource which has a tight control cycle, such as concrete pours, will require reconciliations to be undertaken after each pour; labour budgets are often reported upon weekly in correlation with the bonus calculations; subcontract and materials budget variances are usually reported monthly. This provides an analysis of the to-date situation as well as a projection to completion.

For effective control to be effected, the remedial measures need to be considered before the variances are identified. This allows actions to be planned beforehand and supports timely interventions, if the materials budget is exceeded and this is to have an impact upon profit margins this should inform the CVR process. If the plant budget has been exceeded due delays caused by design variations, the client should be made aware of potential future costs.

Most of the control procedures are brought together in the monthly contract performance meeting, where the CVR and programme are considered by the senior management of the company.

13.9   Earned value analysis

In earlier chapters this book has considered approaches to capture of costs, reconciliation of value and accounting policies and regulations. In this section the use of earned value analysis (EVA) as a project control tool is explored. Unlike the CVR approach which uses the project estimate and associated allowances, EVA uses the project programme as a source of reference. Earned value (EV) is not a new concept. It has been around since the 1960s, originating in the United States when the Department of Defence used it as a method of measuring project performance. It was known as C/SCSC (Cost Schedule Control System Criteria).

13.9.1 Definition of EVA

EVA can be defined as a method for measuring project performance. It compares the amount of work that was planned with what was actually accomplished to determine if cost and schedule performance are as planned. It can be said that the best indicator of how a project is doing is from current performance and the data that are available for this. Earned value is able to monitor the budgeted work to date, the amount of work done to date and also the costs to date for which that work has been done. It relies on information from the programme, with regards to the works which have been completed and also the cost data for the works done to date. The allocation of this cost data is crucial to getting a true picture of the project at any given moment. It has only recently become more widely adopted within the construction industry, although some consider that the time and effort involved in making the system work are too great for the benefits that can be achieved from it. The wider acceptance of this technique will depend on a better understanding of its capabilities and how it can help the performance of a project.

EVA has its own set of jargon. However, the fundamental concepts are relatively straightforward. A project is to be divided into a rational set of activities, measured and phased with the financial flow of aligned sequence and resources. A sequence of milestones can then be tracked and evaluated based upon a set of integrated metrics including cost and time (the schedule), both budgeted (forecasted or planned) and actual. There are three measures which are ultimately compared: planned work accomplished or earned work, and actual costs of work accomplished. The metrics can be compared using graphical reports and calculated ratios. Tracking these metrics allows a manager to identify problems and then focus upon the cause and, hopefully, a solution.

Earned value uses three data values, which are computed each week, month or whatever other period is being used. The term 'analysis date' is used to refer to the date when the three values are analysed. The three values are:

- **Budgeted Cost of Work Performed (BCWP).**
- **Budgeted Cost of Work Scheduled (BCWS).**
- **Actual Cost of Work Performed (ACWP).**

The project programme can be used to supply a baseline against which any future alterations to the programme can be monitored. The resources that are required for each individual works process are identified on the programme. This is first done as a baseline profile for the entire project which is then changed, usually on a weekly or monthly basis, once the construction process has started.

The programme usually adopts a **work breakdown structure**, which should be agreed before the commencement of the construction phase, this then allows the programme to show progress against individual activities as the progress figures are input into the software.

13.9.2 The components of EVA

The first basic requirement for the application of EVA principles is that a planned budget of the works that are scheduled on the project is established; this is the Budgeted Cost of Work Scheduled (BCWS). There are a number of standard derived metrics that are used in the monitoring of a project using EVA, a metric being defined as a means of deriving a quantitative measurement. The metrics used by EVA are shown in Figure 13.4.

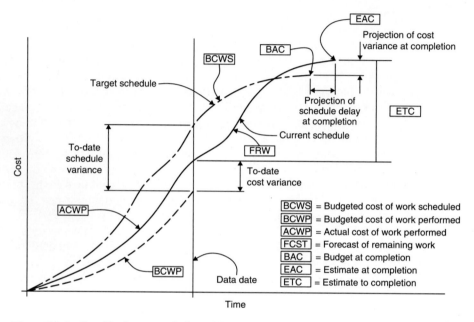

Figure 13.4 Graphical representation of the metrics.

These metrics are used to support assessment of the cost performance of a project. Currently the estimated cost to complete (ECC) and forecast cost to complete (FCC) are based on past performance on the project and, as such, forecasting may be incorrect as future works may be different from what has been done before. This approach to projection was used in the CVR process when looking to establish the security of the future margin. When carrying out EVA it is important to understand the difference between the earned value of work performed and the cost of work performed.

The derived metrics are standards used for the monitoring and reporting of EVA on projects. The most common of these that are used are the cost performance index (CPI) and scheduled performance index (SPI) figures. These are calculated as follows:

CPI = BCWP / ACWP

SPI = BCWP / BCWS

For both of these values, anything greater than one means that the project is being completed at less than that was budgeted and that the project is ahead of schedule, respectively.

Figure 13.4 above is a graphical representation of where the various metrics are taken from when the data for cost and time are presented in this way. For the SPI and

CPI calculations, these take into account the BCWS, BCWP and ACWP. The BCWS is simply the baseline of costs over time, which is set out at the beginning of the project. ACWP is the cost line which shows the profile of spend over time, often done on a month by month basis. This should show accurate cost allocations and be as up to date as possible. BCWP is the earned value of what has been done to date. This is the profile of the budget that has been earned based on the progress made on the project to a specified date. All of these factors are able to help calculate the CPI and SPI to give an overall view of the project. If the three points are considered along the data line, the following can be seen, the budget cost of work performed is less than the actual cost of work performed (i.e. a loss is being made), the situation is even worse as the budgeted cost of work scheduled (i.e. that originally considered) is higher than the budget costs of work performed (the project is behind programme). The to-date schedule variance and to-date cost variance indicate this on the graph. A forecast of remaining work (FRW) can be developed based on the performance to date; this will provide an estimate at completion (EAC), which can be compared with the budget at completion (BAC) to give a projection of the cost variance. The difference in time between the end of the original programme as defined by the BCWS and the FCW gives a projection of the delay to the project. The estimate budget to completion (ETC) gives an indication of the 'wriggle' room available to management to bring the project back to budget.

13.9.3   EVA in practice

The main tasks required to create a generic methodology are:

- construct the work breakdown structure (WBS);
- estimate the cost for each activity;
- construct a programme for the project based on the WBS;
- identify project milestones;
- monitor the progress of the activities and keep the programme up to date;
- identify and record the actual costs incurred at each milestone;
- calculate the earned value metrics at each milestone.

Lukas (2008) noted that there were ten mistakes that could be made when trying to implement EVA. These are:

5. no documented requirements;
6. incomplete requirements;
7. WBS not used or not accepted;
8. WBS incomplete;
9. plan not integrated;
10. schedule and/or budget incorrect;
11. change management not used or ineffective;
12. cost collection system inadequate;
13. incorrect progress;
14. management influence and/or control.

These 10 challenges are not inconsiderable and are one of the reasons why EVA has struggled to be accepted by construction. One of the major difficulties is establishing a robust work breakdown structure which can be used on different projects.

13.9.4 Work breakdown structures

The work breakdown structure (WBS) allows a project to be broken down into smaller works packages that can be managed and monitored on a smaller scale, which in turn allows greater control of a project for the project manager.

'It is a hierarchical representation of the work contents, whereby the project is progressively subdivided into smaller units' (Ibrahim et al., 2009:389) or, according to the PMBOK (APM, 2000), is 'a deliverable orientated hierarchical decomposition of the work to be executed by the project team to accomplish the project objectives and create the required deliverables'.

The WBS is a fundamental building block that gives control in the initiating, planning, executing, monitoring and controlling processes used in order to manage a project. The upper levels of any WBS usually outline the main work areas or packages of the project, narrowed down into logical work processes. The lower elements provide more detail relating to each and focus on support of project management processes, such as schedule development, cost estimating, resource allocation and risk assessment.

A WBS also helps with assigning responsibilities, resource allocation, monitoring the project, and controlling the project. It makes the deliverables more precise and concrete, so that the project team knows exactly what has to be accomplished within each deliverable. This also allows for better estimating of cost, risk and time because it is possible to work from the smaller tasks back up to the level of the entire project. Finally, it allows a double check of all the deliverables specifics with the stakeholders to make sure there is nothing missing or overlapping.

The first stage in developing a WBS is to make a list of all of the project deliverables. These are the main items of work that lead to the overall project objective, or outcome. The second step is the decomposition of the deliverables into successively smaller chunks of work to be completed in order to achieve a level of work that can be both realistically managed by the project manager and completed within a given time frame by one or more team members. The use of data from past projects can be useful in determining the WBS for the various elements within a project.

13.9.5 Establishment of an EVA management system

For EVM to be beneficial to a project there are a number of factors that need to be adhered to and put into place. Firstly, it is imperative that the full scope of works involved in the project is considered. The total value and costs should be input into the system. Howes (2000) suggested that a Work Package Method could be used in order to breakdown how EVA looks at project. Instead of using EVA to look at the whole project, it was to be split into a series of works packages that interrelate by time and sequence.

Ibrahim et al. (2007) found that the most frequently used criteria in the formulation of a WBS for building projects are, elements, work sections, physical location and construction aids. For any given project it is essential that the criteria for which the WBS is formed are based on packages or areas that are not too broad, and yet not split down too far in that the costs cannot be attributed to the relevant areas.

EVA has to be able to combine the WBS along with programme and the costs associated with each package and its timescale. Its strength is that it can be used to identify trends or problems in the various works packages based on cost and progress. It can also be used as an early warning system that can be acted upon in order to change the way in which the job is progressing.

Once the WBS has been set out and budgets allocated to each works package, the project can be executed. Once the works have started the process of EVA requires continuous monitoring, and updating, at least monthly, although this may be more frequent depending on the size of the project. As Earned Value Management is a useful early warning tool, it is sometimes thought of by some practitioners that the updating and reporting of earned value should be done weekly if possible, as this can allow for the addressing of potential problems as early as possible. Regular monitoring is also needed to compare the current progress of the project with the original targets set out in the programme and against the budgets allocated to each works package.

The work breakdown structure will identify the various works packages that are to be monitored on an individual weekly basis and those that are to be aggregated and monitored monthly. The EVA system allows the works to be broken down into various cost centres, which in turn allows the costs to be allocated in an accurate manner.

The one main limitation of EVA is that is unable to take into account the quality aspect of a construction project. A project could be on time and on or below budget but could have costs and delays in the future as a result of low quality.

13.9.6 Benchmarking project performance using EVA

To assess whether a project is progressing in a similar fashion to previous projects there is a need for standards to assess against. Few organisations systematically store data on the performance of their projects, the view usually taken is that the CVR data is so project specific and deals with the here and now that there is little value in storing it for future analysis.

EVA data, however, may be in a form that could produce derived metrics that could be stored and used for the development of standards against which current project work packages could be compared. This form of analysis is also known as benchmarking, which is a process that helps to identify any failures in the performance and allow management to take early remedial action.

13.9.7 Predicting performance

The forecasted cost to completion of a project is a figure produced from where the project is currently at, based on the costs and progress to date. EVA can be used to produce these figures more accurately by using current and forecasted costs. There are three elements that look at this:

1. BAC - Budget at completion
2. ETC - Estimate to Completion
3. EAC - Estimate at Completion

Budget at Completion (BAC) is simply the outturn planned cost for the particular project or subelement of the project or the cumulative total for the BCWS for the project or subelement.

The **Estimate at Completion** (EAC) figure allows the user to analyse and determine what the project or the particular subelement of the projects final outturn cost may be.

The calculations for ETC and EAC are:

$$EAC = AC + ETC$$

where AC is the actual cost to date plus the estimated costs to completion.

$$ETC = BAC - EV$$

This is the budget for the whole project less the value of work done to date. Putting these two equations together gives:

$$EAC = AC + BAC - EV$$

This will allow the forecast of how the job is likely to end up at its completion to be made more accurately. Although this is often something requested by the employer, it is also something that has to be carried out by contractors before a profit at completion can be established and taken.

A contract such as the NEC3 Option C (target cost with activity schedule) may well use this process on a monthly basis as part of the commercial reporting, as part of the profit margin of this contract is the gain share, should the works be done for less than the target cost or overall budget. Any differences between costs and budget are often shared based on a percentage basis, agreed between the parties at the beginning of a project. With this is mind these tools are extremely useful for any project manager and can be used to monitor project performance for its duration.

13.9.8 EVA in action: an example

To illustrate how EVA can be used on a project the following is a simple example of a work package that uses the various measures described above. The schedule in Table 13.13 indicates the budgeted cost of work scheduled (BCWS) for a work package. Figure 13.5 illustrates this graphically.

Table 13.13 Budgeted cost of work scheduled (BCWS) on time line.

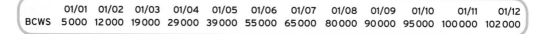

	01/01	01/02	01/03	01/04	01/05	01/06	01/07	01/08	01/09	01/10	01/11	01/12
BCWS	5000	12000	19000	29000	39000	55000	65000	80000	90000	95000	100000	102000

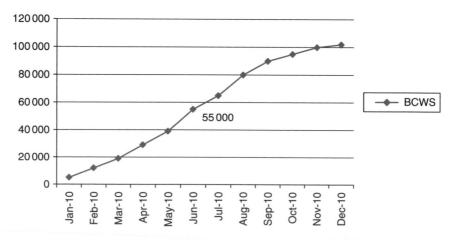

Figure 13.5 Graph of BCWS against time.

Table 13.14 Budgeted cost of work performed (BCWP) and Actual cost of work performed (ACWP) on time line.

	01/01	01/02	01/03	01/04	01/05	01/06
BCWP	5000	11000	17500	26200	35000	49000
ACWP	5000	12500	19600	29900	39900	56000

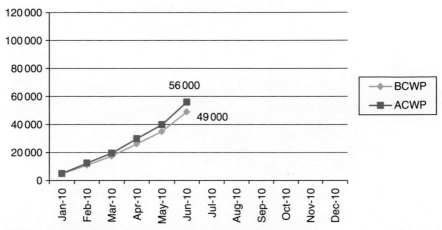

Figure 13.6 Budgeted cost of work performed and actual cost.

Definition: Budget cost of work scheduled – *Planned cost of the total amount of work **scheduled** to be performed by the milestone date.*

If the work package performance is considered on the first of June, it can be seen from both the table and the graph that £55 000 is budgeted to be completed.

If the data are expanded to also consider the performance of the work package to include the budgeted cost of the work performed and the actual cost of the work performed, as shown below, a richer picture of the work package performance is obtained (Table 13.14).

Definition of budgeted cost of work performed: *The **planned** (not actual) cost to complete the work **that has been done**.*

Definition of actual cost of work performed: Cost incurred to accomplish the work ***that has been done*** to date.

It can be seen in Figure 13.6 that the budgeted cost of work performed is £49 000, that is below the scheduled amount, and that the actual cost of the work performed, £56 000, exceeds that budgeted.

The third comparison that can be made is that between the budgeted work performed and the budgeted cost of the work scheduled. This can help give an idea of progress against programme. The data are shown in Table 13.15 and Figure 13.7 is a graphical representation of BCWP and BCWS.

The summary of the data is represented in Table 13.16 and Figure 13.8.

The strength of the EVA approach is that it can be used to benchmark a work package against others by using a series of derived metrics. These are illustrated below.

The schedule variance metric, SV: Schedule Variance (BCWP – BCWS):

- A comparison of amount of work performed during a given period to what was scheduled to be performed.
- A negative variance means the project is behind schedule.

Table 13.15 Budgeted cost of work performed (BCWP) and Budgeted cost of work scheduled (BCWS).

	01/01	01/02	01/03	01/04	01/05	01/06
BCWP	5000	11000	17500	26200	35000	49000
BCWS	5000	12000	19000	29000	39000	55000

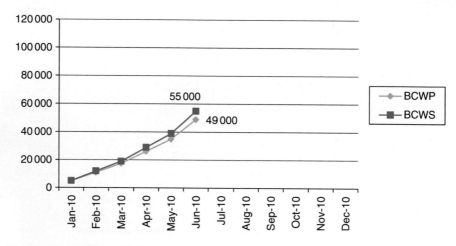

Figure 13.7 BCWP and BCWS.

Table 13.16 Performance metrics on time line.

	01/01	01/02	01/03	01/04	01/05	01/06
BCWS	5000	12000	19000	29000	39000	55000
BCWP	5000	11000	17500	26200	35000	49000
ACWP	5000	12500	19600	29900	39900	56000

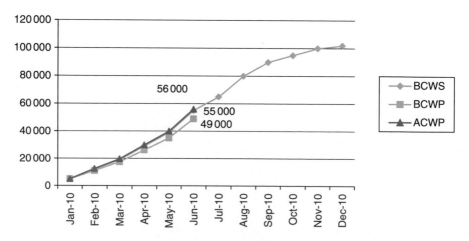

Figure 13.8 Comparison of BCWS, BCWP and ACWP.

Table 13.17 Scheduled performance index at project milestones or time line.

Schedule Variance (£)	0	-1000	-1500	-2800	-4000	-6000
Schedule Variance (%)	0.0	-8.3	-7.9	-9.7	-10.3	-10.9
Schedule Performance Index	1.000	0.917	0.921	0.903	0.897	0.891

Table 13.18 Cost performance index at project milestones or time line.

Cost Variance (£)	0	-1500	-2100	-3700	-4900	-7000
Cost Variance (%)	0.0%	-13.6%	-12.0%	-14.1%	-14.0%	-14.3%
Cost Performance Index	1.000	0.880	0.893	0.876	0.877	0.875

The schedule variance for June is £49 000–55 000, which is -£6000. If this is calculated as a percentage scheduled variance and scheduled performance index (BCWP/BCWS) it can be seen that the project has been slipping behind since month two, this is shown in Table 13.17. A negative variance and index less than one indicates a delay. The project is 10.9% behind schedule.

The cost variance metric, CV: Cost Variance (BCWP–ACWP)

- A comparison of the budgeted cost of work performed with actual cost.
- A negative variance means the project is over budget.

The schedule variance for June is £49 000–56 000, that is -£7000. If this is calculated as a percentage cost variance and cost performance index in a similar manner to that used above, the cost variance is -14.3 % and cost performance index is 0.875. The project is 14.3% over cost. This is shown in Table 13.18.

One other metric that can be derived from this data is the cost schedule index, which is calculated as cost performance index×schedule performance index. In the above example this would be $0.875 \times 0.891 = 0.780$.

13.9.9 Making predictions based on the derived metrics

The cost performance index and schedule performance index can be used to predict the performance as measured in June until the end of the project. Table 13.19 indicates how the data can be used to project the performance of the work package until the end of the year.

Budgeted cost work scheduled and performed and actual cost of work performed for work package until year end are shown graphically in Figure 13.9.

Table 13.19 Projecting the performance of the work package.

	01/01	01/02	01/03	01/04	01/05	01/06	01/07	01/08	01/09	01/10	01/11	01/12
BCWS	5000	12000	19000	29000	39000	55000	65000	80000	90000	95000	100000	102000
BCWP	5000	11000	17500	26200	35000	49000	57915	71280	80190	84645	89100	90882
ACWP	5000	12500	19600	29900	39900	56000	66189	81463	91646	96737	101829	103865

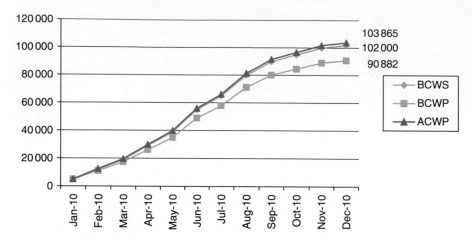

Figure 13.9 Projection to completion based on the derived metrics.

13.9.10 Benefits of EVMS

One of the benefits of an Earned Value Management System (EVMS) is that an organisation can implement a single management control system using reliable data. This has not been achieved by any construction organisation.Some have attempted to integrate the EVA and CVR concepts but have found that the WBS requirements are the main stumbling block. If this can be overcome a database of completed projects with integrated work package allowances, programme and out-turn costs could be developed which could provide a basis for comparative analysis. The system could provide historic data for the various functional departments, estimating, planning and so on as a basis for lessons learned for future projects. The metrics of CPI and SPI are useful early warning indicators that highlight the need for management action and can be used to monitor the effectiveness of management interventions. If company's collected this data they could establish upper and lower parameters which could be used to support a traffic light system of intervention (red-urgent, amber-moderate, green-no action), the interventions could also be measured to assess their impact.

The derived metric 'to complete performance index' (TCPI), which is the ratio of remaining work to remaining budget, can be calculated and may be useful, as it gives stakeholders an opportunity to assess the reasonableness of a contractors estimate to complete. It indicates the level of performance that the contractor must achieve to reach a financial goal. Thus, this earned value metric can help the manager assess the reasonableness of critical financial goals, such as completing the remaining work within the targeted cost.

References

APM (2000) *Project Management Body of Knowledge*, 4th edn. Association for Project Management (APM), London.

Howes, R. (2000) Improving the Performance of Earned Value Analysis as a Construction Project Management Tool. *Engineering, Construction and Architectural Management*, **7** (4), 399–411.

Ibrahim, Y.M., Kaka, A.P., Trucco, E., Kagiouglou, M. and Ghassan, A. (2007) Semi-Automatic Development of the Work Breakdown Structure (WBS) for Construction Projects. In: *Proceedings of the 4th International Research Symposium (SCRI)*, 26–27 March, Salford, UK, 133–145.

Ibrahim, Y.M., Kaka, A., Aouad, G. and Kagioglou, M. (2009) Framework for a generic work breakdown structure for building projects. *Construction Innovation: Information, Process, Management*, **9** (4), 388–405.

Lukas, J.A. (2008) Earned Value Analysis – Why it Doesn't Work. *AACE International Transactions*, EVM 01.

14

Resource procurement

14.1   Introduction

All the resources used on a construction site from a nail to an air handling unit will have been procured by one of the organisations working on the project. The resource will have been specified, priced, ordered and delivered. These resources can be categorised as:

- labour
- plant
- materials
- subcontractors.

Financial Management in Construction Contracting, First Edition. Andrew Ross and Peter Williams.
© 2013 Andrew Ross and Peter Williams. Published 2013 by John Wiley & Sons, Ltd.

This chapter considers the approaches contractors use to procure the resources needed.

14.2 The resource budget

The approach to procuring the project's resource is influenced by whether the client or the contractor take the risk for cost and time overruns from the project's original budget and programme. This is influenced in a large part by the procurement arrangement. For example, a client may procure a project using management contracting procurement. The management contractor will be selected on its expertise and be paid a fee to cover the management infrastructure for the project. Part of the criteria for selection of the management contractor would be the strength of its supply chain and the ability of the supply chain to meet the client's requirements.

In this case, it is likely that a design team will have provided the client's quantity surveyor with enough information to establish a cost plan. This cost plan would form the budget for the work package procurement. A programme for the project would be prepared; it will indicate the commencement on site for each work package and this will establish a procurement programme.

If the work package excludes design, the design team will complete the design and the client's quantity surveyor will assemble a tender document that will be sent out to the contractors supply chain. The send out will be coordinated by the management contractor and will include design information, main contract and subcontract conditions, the outline programme for the work package indicating commencement and completion dates and the format for the returned quotation (this could be as detailed as a bill of quantities or as simple as a schedule of rates and lump sum).

The quotations are returned, analysed by the client's quantity surveyor and management contractor and, if acceptable, a contract is placed. Any risk of the work package budget being exceeded rests with the client, although some arrangements do have a gain/pain arrangement which incentivises the contractor to achieve the budget figure.

To contrast this approach, if the project was a competitive design and build lump sum, the contractor would have won the project on the basis of an estimate which would have been developed over a limited time and used pre-contract quotations from the company's supply chain partners. In this case, the onus would be upon the project team to procure the resources within the budget set by the estimator. If the budget was exceeded this would be considered as a buying loss and reduce the contractor's profits.

14.3 Resource procurement programme: subcontractors

The contractor's site manager usually works in close association with the planner and quantity surveyor to produce a procurement programme. The contractor's project programme will be the basis for establishing the key processes that are required before the resource is required on site. The approach to subcontract procurement will be influenced by whether the subcontractor has quoted at tender stage and is to undertake the works post-contract; this is often referred to as a 'one to one arrangement'.

The reasons why a main contractor uses one to one arrangements relate to the attributes of the subcontractor or the project. The subcontractor may have invested a lot of resources in developing a design at tender stage, may have a particularly specialised technology that the project requires or it may be that there is simply not enough time to go back to the market for alternative quotations.

If the main contractor has some flexibility about which subcontractor can undertake the work and also has the time to go back to the market, a procurement programme would be developed for each work package.

The tender enquiries to be send to the subcontractors may be compiled by the buying department or by the site quantity surveyor. This would depend upon the size of the company, project and complexity of the work package. It is important at this stage to ensure that the tender documentation includes all the necessary information to allow for an unambiguous price to be returned by the subcontractor.

Often, main contractors will employ consultant quantity surveyors to quantify the work packages into bills of quantities or schedules of builder's quantities (not usually in accordance with SMM7). This reduces the subcontractors cost of tendering and increases the likelihood of receiving a *bona fide* tender. It also provides a uniform basis for comparison of the different quotations when received.

The send out would include the following:

- drawings, specification, schedule of works;
- timescale for the project;
- main contract conditions, including any LADs that would apply to the subcontractor if delayed;
- subcontract conditions;
- health and safety requirements;
- attendances provided by the main contractor.

Before the tender enquiries are sent out it is usual to confirm that the subcontractor is interested in pricing the works. Most site teams have a good rapport with a number of subcontractors and will encourage them to price the works. As the cost of tendering can be significant, the subcontractors will hope to reduce their abortive pricing works to a minimum.

14.4 Tender assessment

It is not usual for main contractors to follow the same tender assessment procedures as clients when considering work package tenders, the imperative is to secure the services of a competent subcontractor who has the resources to do the work and the right price. This means that after initial assessment there is usually a period where qualifications are negotiated away, prices discussed and whether any additional discounts can be offered are determined.

The initial assessment will be similar to that done by the estimator at tender stage; a table indicating the price from each subcontractor will be developed with rows for each item. The idiosyncrasies of the subcontractors pricing are highlighted at this stage; some may only price some items as labour and plant, and some may bracket a number of items together and provide a lump sum.

It is important to ensure that the quotations are compared on a like for like basis and that any omissions from one quotation are highlighted for potential further discussion with the subcontractor.

This is an exciting aspect of the project, the subcontract prices are often less than those quoted at tender stage as the main contractor has won the project and the likelihood of the contract being placed is a lot higher. The quantity surveyor gets a better idea of whether the estimator's budget is a realistic one or not and also can start to firm up the projected costs to complete the project and the projected final margin.

It can also be at this stage that errors made at tender stage are highlighted. Often subcontractors which priced at tender stage discover errors in their quotations and withdraw their initial quotation. The consequence of this is that the project work package budget is no longer founded on a quotation and it is with some trepidation that the new quotations are compared with the original budgets. A project's profit margin can be wiped out at this stage.

14.5 Tender negotiation

Once the comparisons of the subcontract prices have been made the contractor opens a dialog with the most favourable subcontractors. This may not be solely on those who have priced the work package the lowest, other criteria such as quality, ability to complete on time and technical expertise are also considered. There is no point appointing the lowest priced subcontractor if it cannot complete the job to specification. This aspect leads to a great deal of discussion between the contractor's quantity surveyor and the site manager; one view could be that the work has been specified and priced and that it is the site management's job to ensure the subcontractor complies. Another view is that the management resources to chase a poor subcontractor may be better deployed elsewhere. Usually a compromise is reached.

The price work the works will eventually be agreed after a well rehearsed set of stages to the negotiation – initial confirmation of the price and establishing the basis for any qualifications or amendments to the original tender enquiry, exploration of opportunities to reduce the price by requesting additional discounts or by promising advantageous payment terms. The quantity surveyor has to be careful not to seek to muscle the subcontractor down to a price which is uneconomic as it may cause difficulties in the future.

The extent of discounts offered by the subcontractor will depend upon its workload and forward order book and also the perceived level of competition for the work package. It is not unusual for levels of discount of 20% to be offered. Newer internet reverse auctions are used by some contractors to procure their subcontractors which allow this discounting to occur over a set period and to allow the subcontractor to see how its discounted price compares with other competitors. This is discussed later in this chapter.

14.6 Buying gains and losses

Once the subcontract has been let the quantity surveyor has a clear indication of the buying gain or loss that has been achieved. This is calculated by comparing the subcontract contract rates with the allowance rates for the work package. This will also highlight any allowance rates which may be lower that the subcontract rates. This may cause problems in the future if variations to contract cause the quantities of these items to rise. The converse of this situation can also occur if the rates are lower.

14.7 Newer approaches to subcontract procurement

As internet technology becomes more sophisticated and the ability to communicate with many parties in real time improves, newer approaches to subcontract procurement have developed. These have not been universally welcomed as they are seen to encourage subcontractors to enter into cut-throat competitions for work.

The approaches vary with different companies but the basic stages are:

- once the main contractor has won the work, a work package procurement programme is established;
- tender enquiries are then sent out as normal, usually on a CD or can be downloaded from a website;
- the tenders are priced and returned to the main contractor as usual;
- the main contractor now has a number of subcontract tenders for the project;
- the key difference to a normal arrangement is that the tenderers then enter into a reverse e-auction to win the work.

14.7.1 Reverse e-auction

A specialist external supply chain organisation, such as Bravo Solutions, is used to host the competition. The competition rules are made explicit at the tender stage, an auction start and completion date and time will be given as well as the incremental amounts the tender can be reduced by, as indicated in Table 14.1.

The subcontractor is then invited to discount its initial tender by incremental amounts until the auction closes. The main contractor then enters into discussion with the most favourable bidders prior to entering into a contract.

14.7.2 Pre-auction stage

Before the auction opens the contractor will have considered the tenders received and weighted the attractiveness of each company against a set of criteria that include:

Table 14.1 A reverse auction call.

Project information		
Project Code	**Project Title**	**Project Reference**
Tender; LJMU	e-auction March 2012	

Auction settings			
Auction Code	**Auction Title**	**Auction Description**	
LJMU 123	Byrom Street	Windows and external doors	
Auction Start Date and Time:	**Auction End Date and Time:**	**Overtime Interval**	
19/03/2011 10:30	19/03/2011 11:00	Restart from 3 minutes	
Base Price	**Reserve Price**	**Minimum Bid Decrement**	
1700 000.00	1500 000.00	2500.000	
Currency	**Awarding**	**Supplier to improve on own bid only**	**Auction Lot Status**
GBP	Discretional	Yes	live

Table 14.2   Subcontract price weighting criteria.

Criteria	Weighting	Supplier A		Supplier B	
		Score	Weighted Score	Score	Weighted Score
		/10		/10	
Price*	30	2	60	1	30
Quality	20	4	80	5	100
Programme	10	7	70	8	80
Location	5	4	20	3	15
Buyer relationship	20	8	160	9	180
Financial stability	15	5	75	4	60
Total	100		465		465

*Price = discount from base price, 10% = 1, 15% = 1.5.

- quality
- reliability
- previous experience
- price.

The example shown in Table 14.2 illustrates how non-price criteria can be used to develop a weighted scoring system which incentivises subcontractors to discount their price from the base price. The main contractor would have assessed the technical bids prior to the action and ascribed scores to factors such as financial stability, ability to meet the subcontract programme, previous quality of work, health and safety record.

14.7.3   The auction stage

The system provides the tendering subcontractors with feedback on the attractiveness of their bids each time they enter a bid. This relates to their relative position in the competition and takes into account the other non-price dependent criteria

14.7.4   Post-auction stage

The information the main contractor receives from the e-sourcing provider shows the respective bidder's prices over the auction period. Table 14.3 is an example taken

Table 14.3   The results of the auction.

Supplier ranking				
Rank	Supplier	Bid Date/Time	Bid issued	Total Score
1	Easyglaze Windows	11:02:03 a.m	1485000.000	99.000
2	See Thru Windows	11:10:26 a.m	1554198.000	94.770
3	Quikfix Windows	11:07:42 a.m	1544624.150	94.333
4	Trans Windows	10:55:09 a.m	1590000.000	92.726

from a company that uses this approach. It has indicated that it has achieved savings of up to 30% on the procurement of materials such as kitchens and bathrooms and 15% for subcontractors.

A comparison of bidding trends can also provide interesting feedback to the main contractor's procurement department and provide insightful information on the approaches taken by the different subcontractors (Figures 14.1 and 14.2).

14.7.5 Live e-auction results

The data shown in Tables 14.1 to 14.3 are taken from a live auction. It is interesting to note the subcontractor's behaviour over the bidding period:

- The winner of the auction, EasyGlaze, only place four bids and won the contract with a weighted score of 99 at 11.02 a.m.
- The second placed bidder, See Thru Windows, made further three bids after this time in an attempt to improve its score. However, the final bid of £1.54 m gave a score of 94.77.

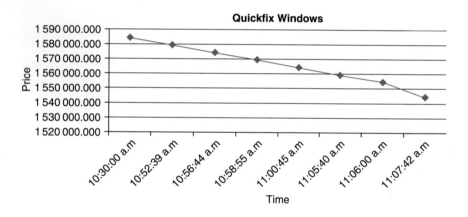

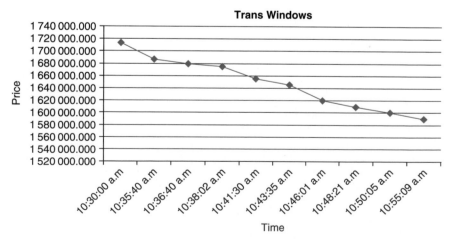

Figure 14.1   Bidding over time.

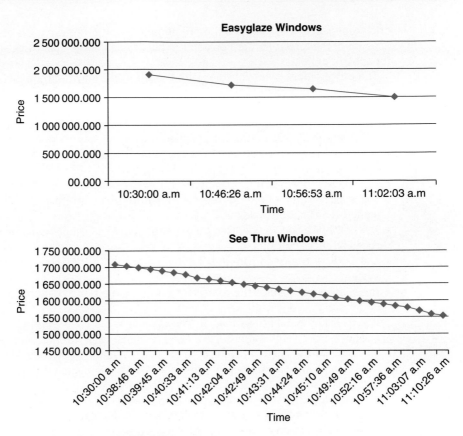

Figure 14.2 Subcontract e-bidding trends.

- It would appear that in this case either the scoring system was not transparent or that See Thru Windows did not understand the system as it would have ceased bidding as soon as it was apparent that its price reduction was insufficient to win the job.

14.7.6 Reflections

The approach is in its infancy for the procurement of subcontract packages. Obvious differences are evident from the purchase of commodities such as skips or printer cartridges to complex construction work packages.

Bidders dislike the system and feel that it reduces the buyer/seller relationship to one purely on price and have expressed concern regarding the subjectivity of the scoring system of the qualitative criteria.

14.8 Materials procurement

Materials can form up to 40% of a project's value and suppliers prices can win or lose a company a project. Most organisations will have a strategic alliance with local or national suppliers for the commonly used materials. The supplier will quote prices that are fixed for up to four months and these will be used by the estimator when pricing projects. These rates will be re-negotiated on a regular basis and will take account of the forward order book and, in particular, if large quantities are required for specific projects.

The procurement process begins with the site staff, which will complete a materials requisition programme. This will be sent to the buying department indicating quantities and dates. The following will then be initiated:

- an order will be placed with the supplier;
- a copy of the order will be held by the accounts department, the site office and the buying department;
- the site will then 'call off' the materials from the supplier;
- the supplier will deliver the required quantity and confirm the delivery with a signed confirmation (a delivery ticket);
- the delivery ticket will be used by the site team to a complete materials (goods) received sheet (MRS);
- the MRSs are passed to either buying or accounts for entry into the costing system;
- the supplier will invoice the contractor based on the signed delivery tickets.

It is usually the quantity surveyor who is responsible for ensuring that the costs are reported correctly and often will need to check back through the MRS records to ensure that the accruals are accurately calculated. It is not unusual for delivery tickets to go astray from the point of delivery to the accounts office, whereas it is unusual for a supplier to mislay a delivery ticket.

Materials logistics is a key part of the site manager's role. Often site compounds for storage are small and careful handling and storage can reduced costly wastage. Contractors are often assessed by the extent of waste generated on sites and produce, as part of their method statements, site waste management plans.

14.9   Plant procurement

A similar arrangement will be entered into with local plant hire companies. Rates for popularly used equipment will be negotiated and supplied to site. The site manager will call off plant via the buying department, the plant hire company will deliver the plant to site and the hire period will commence. The arrangements with the hire company vary. Sometimes plant will be hired with an operative; alternatively, it may be hired on a maintenance basis.

Hire periods can vary from hourly, daily, weekly and monthly and the definition of plant encompasses mechanical plant such as cranes and excavators, small tools, accommodation such as cabins and even leased company cars.

The control of costs of hired plant is a complex issue and the site team have to balance the convenience of keeping plant on hand to avoid delays against the cost of hire.

14.10   Labour procurement

All projects require skilled and unskilled labour to complete them. The main difference between labour on site now to that forty years ago is their employer. Most large contractors employ few tradesmen. The reasons for this were briefly examined in Chapter 1.

Most of the labour on site will be employed by subcontractors and with this employment goes the risk for continuity of work, cost control and payment. There are a few instances where it is advantageous for the contractor to employ direct labour:

- in a maintenance capacity;
- to complete snagging and defects on completed projects.

14.11 Labour-only subcontractors

If labour is required, it is often supplied on a labour-only basis by a subcontractor. The two most popular labour-only trades are the bricklayers and joiners. The budget for the labour element of a project is often established by the estimator using output rates which have been adjusted to take account of the specific job requirements. The approach to seeking labour-only rates at tender stage often fails as the subcontractors which undertake labour-only work normally keep their overheads to a minimum and only price work once it has been won by a contractor and which they feel they have a chance of winning. The reverse auction approach discussed above has been used to procure site labour by offering work to pre-registered craftsmen and labourers on agreed piecemeal rates.

A send out to labour-only subcontractors would be undertaken in a similar fashion to labour, materials and plant subcontractors. A comparison would be made between the bidding subcontractors and the contract place with the most suitable company. A company often has a number of preferred labour-only subcontractors. Often projects require flexible labour requirements and the labour-only subcontractor may be called upon to find additional resources with limited notice. Often, where work is unscheduled and often not measured or measurable, the work will be completed on a daywork basis. The hourly daywork rate would have been agreed prior to entering into the contract.

15

Project risk and control

15.1 Introduction

Project risk begins with the contractor's tender and ends with agreement of the final account. This is when the consequences of all the gambles taken and all the 'known-knowns', 'known-unknowns' and 'unknown-unknowns' become a harsh

Financial Management in Construction Contracting, First Edition. Andrew Ross and Peter Williams.
© 2013 Andrew Ross and Peter Williams. Published 2013 by John Wiley & Sons, Ltd.

reality. It is only then that the true profit or loss situation is established and the contractor knows whether or not all the effort and worry has been worthwhile.

Project risk comes in many forms and mostly has financial outcomes – some good, some bad. Sometimes project risk cannot be measured in monetary terms and the cost is counted in human suffering and distress following a serious accident or fatality. This book takes a somewhat narrow view of 'risk' but that is not to minimise the risks taken with people's health and safety as a consequence of construction activity or of the technical risks that construction projects pose which are invariably overcome by the ingenuity and inventiveness of all those involved at the 'sharp-end' of the construction process.

Consequently, attention is focused in this chapter on the following classification of risk as it concerns the subject matter of this book:

- tender risk
- contract risk
- claims
- insolvency risk.

15.2   Tender risk

15.2.1   Programme and method

Tender documents normally state the commencement and completion dates for the contract or the period within which the works are required to be completed. Alternatively, the tendering contractors may be asked to submit their own duration for completion of the works which then becomes an additional competitive element to the tendering competition.

Employers (clients) generally have a good idea of when they require a contract to be finished (usually yesterday!!) and this thinking is often linked to the rate of liquidated and ascertained damages (LADs) for failure on the part of the contractor to complete on time. Where the employer dictates the construction period, this is usually based on the knowledge and experience of the employer or its professional team (architect, engineer, quantity surveyor etc.) and by reference to past projects of a similar nature. There is no science to such empirically-based decisions and the duration given may equally be incorrect as correct. Indeed, even with the latest project management software, years of experience and competent project planners, contractors cannot be certain how long a project will take with any real degree of accuracy. Contractors, nevertheless, habitually tender on the basis of a shorter contract period than that stated in the tender documents with no certainty that the programmed duration is correct or indeed achievable.

Keane and Caletka (2008) suggest that the inherent uncertainty in determining accurate task durations may be improved by using PERT methodology which introduces three-time probabilistic estimates into the process. Lowsley and Linnett (2006) cast some doubt on this theory, however, reflecting that construction projects exhibit a degree of 'uniqueness' such that there is always an element of uncertainty and risk in project timings.

Consequently, financial decisions that contractors might make based either on a stated contract duration or a programme using project management software are likely to be less than accurate and, therefore, subject to a degree of risk. This means that a tender based on a shorter contract period may not, in reality, generate the savings in time or the reduction in time-related costs anticipated. This could, therefore, jeopardise expected profitability on the contract or even result in the contract making a loss. The upside to this risk is that, with tight control of the programme and the careful selection and management of subcontractors, a better than expected result may be achieved.

Additional profit contribution may also be achieved by adopting different construction methods to those planned at the tender stage. Indeed, the construction manager's judgement as to construction method may well differ from that of the estimator; this is quite usual. The estimated allowance for an activity purely represents a budget for the work required and it is good practice for contractors to consider the time and cost implications of alternative solutions. This is particularly useful for important activities on the programme, such as large concrete pours, multistorey construction, alternative crane and lifting arrangements and so on.

Cooke and Williams (2009) suggest that **sequence studies** are a useful means of determining the time and cost implications of alternative construction methods, of ensuring continuity of work and ensuring that the best use is made of available resources.

15.2.2 Ground conditions

Responsibility for the risk of unfavourable ground conditions is a complex legal issue and there is a significant body of case law on the subject. Delving into such complexities is clearly beyond the confines of this book but the subject is nevertheless significant both from a contractual risk point of view and in terms of the resulting financial implications.

Basically speaking, in common law, the below ground risk is borne by the contractor unless the contract says otherwise. Therefore, if the contract is silent on the matter, the contractor carries the risk. In this context, 'the contract' refers to the various documents that make up the contract documentation including the form of contract itself , the drawings, specifications, bills of quantities and so on. 'The contract' also includes other documents that might be referred to in the contract, such as methods of measurement.

Matters are further complicated when borehole information is included in the tender documents. Borehole logs are not normally included as contract documents as such and are usually marked 'for information only', with a disclaimer that the employer does not warrant the accuracy of the information. It is, therefore, up to the contractor to draw his own conclusions about the prevailing ground conditions. Keating (Furst and Ramsey, 2012) suggests, however, that a court is unlikely to uphold such a disclaimer where the information is given knowing that reliance will be placed on it by the contractor.

The common law complexities of below ground risk are to some extent clarified by reference to some of the commonly used standard forms of contract. In the ICC – Measurement Version, for instance, the contractor is entitled to claim for *'physical conditions and artificial obstructions'* encountered if such conditions could not *'reasonably have been foreseen by an experienced contractor'*.

Clause 60 of the NEC3 ECC allows a somewhat less generous relief to the contractor, whereby compensation is granted should the contractor encounter *'physical conditions ... which an experienced contractor would have judged ... to have such a small chance of occurring that it would have been unreasonable for him to have allowed for them'* in its tender.

The JCT SBC/Q 2011 is less clear on the issue but, should the contractor encounter adverse ground conditions that require a variation instruction from the architect/ contract administrator, this would qualify as a *'relevant event'* and could lead to a claim for loss and expense.

Further clarification of how ground conditions are dealt with under 'the contract' may be found in circumstances where the contract works are measured in accordance with a standard method of measurement. The JCT SBC/Q 2011, for example, states that the contract bills have been prepared in accordance with the Standard Method of Measurement which is defined, in clause 1.1, as the **Standard Method of Measurement of Building Works 7th Edition** (SMM7). The ICC – Measurement Version similarly defines the method of measurement for preparing the bills of quantities as the **Civil Engineering Standard Method of Measurement Fourth Edition** (CESMM4).

The importance of the method of measurement relates to how work below ground is measured and how, if at all, adverse ground conditions are dealt with. If, for instance, the contractor encounters water-bearing ground, this is a measurable item under SMM7 but not under CESMM4. Under SMM7 working space is a measured item where Measurement Rule M7 applies whereas there is no similar provision under CESMM4. Rock, as defined in the contract or the method of measurement, or other artificial hard materials are measured under both methods of measurement. Table 15.1 compares SMM7 with CESMM4 as regards the measurement of work below ground.

Consequently, the risk attached to ground conditions encountered on site depends upon the express terms of the contract, which method of measurement applies and the common law situation should there be a dispute over borehole information supplied by the employer. In most cases there is a reasonably fair balance of risk as between the employer and the contractor but the issue of 'below ground risk' has nevertheless appeared more than once before the Courts!

In conclusion, Figure 15.1 illustrates how a below ground risk problem may be dealt with under a JCT SBC/Q 2011 contract using SMM7 as the method of measurement.

Table 15.1 Measurement of work below ground.

Condition	SMM7		CESMM	
Earthwork support	D20: Excavation and filling	Measured item		No provision
Sheet piling	D32: Sheet piling	Measured where ground conditions warrant its use		No provision
Changes in the water table	D20: Excavation and filling	Subject to re-measure	Class E: Earthworks	No provision
Excavation below the water table	D20: Excavation and filling	Measured item		No provision

Stage A

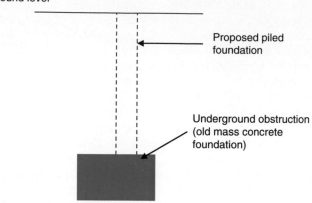

Stage B

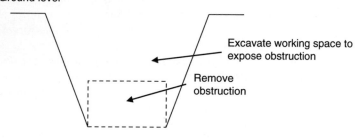

Stage C

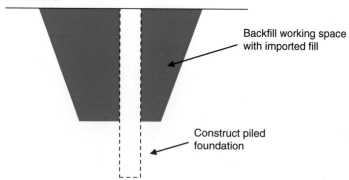

Figure 15.1   Below ground risk.

Figure 15.1 relates to a site is being developed for a new factory and the foundations are to be piled using CFA piles. An underground obstruction – an old concrete foundation – is encountered and the piling rig cannot continue.

Because the contract will have been tendered on the basis of measured bills of quantities, the piling work will have been quantified according to the standard method of measurement. However, because the project risk has changed, both in terms of its nature and balance, the work resulting from the presence of the obstruction will need to valued according to the provisions of the contract. To understand how the work in progress would be valued, it is first necessary to consider how the contractor would deal with the physical problem of the concrete obstruction. One way would be to:

- excavate down;
- stockpile the excavated material or remove it to tip;
- safeguard the excavation;
- bring special plant on to site to break out the concrete;
- fill the hole with suitable material (stockpiled or imported);
- start piling again.

In the meantime, the piling rig is either standing or, to mitigate loss, is working somewhere else on the site. Either way, the main contractor can expect a claim from the subcontractor which will create a liability in the project accounts. This will need to be passed on to the employer via the main contract if at all possible.

As far as valuing the work in progress is concerned, there are several ways of going about this. The following is suggested as one possible approach:

- any standing time will paid for on the basis of time – either on the basis of an hourly rate priced in the bills of quantities or 'daywork' – and records will be kept of the delay to the rig and crew and the contractor will be paid accordingly;
- breaking out the concrete could be measured and valued at bills of quantities rates or, failing an appropriate item in the bills, may be paid for on fair rate basis or on 'daywork';
- excavating and filling the hole would be a further variation and valued similarly;
- any consequential delay and/or disruption to the contractor's programme would lead to a loss and expense claim and, possibly, an extension of time.

There is no guarantee in this process that the contractor will recover all his costs or make any money out of the situation and much depends on the professional quantity surveyor's interpretation of events and the contract. Whilst the main contractor may well come out of it all 'smelling of roses', the lack of certainty in the process is indicative of the risks in construction, the problems of valuing work in progress and the difficulties of reporting a 'true value' in the annual accounts.

15.2.3 Subcontractors

Subcontracting – or the *vicarious performance* of part of the contract works – is an everyday feature of the vast majority of construction projects and the number of main contractors capable of undertaking all, or even the majority, of the 'trades' required to complete a construction project is relatively small. Most main contractors, therefore, largely plan, organise and control the project with most of the physical work being carried out by specialist trades.

The task of managing a battery of subcontractors should not be underestimated and Cooke and Williams (2009) suggest that it is common practice for a subcontract coordinator to be engaged by the main contractor to ensure that the various subcontract 'packages' run smoothly.

Subcontracting might be seen as a convenient way of saving costs and reducing contractual risk by transferring risk to others and there is no doubt that there is potential 'upside risk' to engaging subcontractors. There is, of course, a 'flip-side' to this expectation and Figure 15.2 indicates some examples of upside and downside risk in subcontracting. The 'bottom line' for the main contractor is that the benefit of the contract can be subcontracted but not the burden and the performance of his subcontractors reflects directly on his own performance and reputation.

Subcontract quotes at tender stage can create problems for the main contractor; for example:

- subcontractors may not quote for all the items listed in the enquiry;
- the validity of the quote may be subject to a time limit for acceptance (e.g. 30 days from the date of the quotation);
- there may be qualifications to the quotation which will limit risk for the subcontractor and add cost for the main contractor, such as:
  - the provision of attendances such as lock-ups, lighting, water and power
  - the use of standing scaffolding
  - limitations on work at height (e.g. 3.50 m for painting work)
  - the use of a particular brand or supply source for materials.

Subcontractors are fully aware that the successful tenderer for a project will come back to them for better rates or additional discounts should their initial quotation be in the right 'ballpark'. They also know that some main contractors will use their quotation to get better quotes from other subcontractors and that this could lead to a 'Dutch auction', whereby prices could spiral downward to such an extent that it would be impossible to make any money out of the job. Desperation for work can lead subcontractors to take on such contracts sometimes with disastrous results. However, losing a subcontractor part-way through a contract can cause delay and additional costs to the main contractor, as it may not be easy to engage another subcontractor at such preferential rates.

Main contractors have a simple choice:

- squeeze subcontractors to such an extent that they go out of business or become disillusioned with the contractor and refuse to tender in the future;
- accept a reasonable price for a good service but take the risk that the preferred subcontractor will not be available for a specific contract;
- partner with subcontractors and build long-term relationships with the reward of keen prices in return for continuity of work.

15.2.4 Suppliers and materials

At tender stage the main contractor will send out enquiries for the supply of materials for work that the contractor will carry out. Such work may be done by the contractor's own labour force or it may be sublet to a *bona fide* subcontractor on a 'labour-only' basis, with the main contractor supplying the materials. In both cases, the 'materials risk' will remain with the contractor.

Upside risk	Downside risk
Vicarious performance of work package: • Specialist expertise • Access to skilled resource base • Availability of specialist plant & equipment • Partnering gains	**Main contractor retains contractual obligations to:** • Complete the works • Finish on time • Rectify defects • Rectify latent defects within statutory limitation period
Transfer of risk: • Market risk • Working capital • Health and safety	• Ability of subcontractor to carry risk • Risk of subcontractor default or insolvency • Cost of completing work using other subcontractors
Single point responsibility for each work package	• Poor quality supervision • Poor quality record keeping • Potential for subcontractor insolvency
Price competition: • Greater price certainty • Lower prices per unit of work • Higher discounts	• Cutting corners on quality • Poor quality health and safety provision • Under resourcing of the work • Delays and problems with other subcontractors
Time savings on master programme: • Willing to work flexibly • Motivated workforce	• Defects • Incomplete work requiring additional visits • Interface problems with following trades • Potential contractual claims from other subcontractors

Figure 15.2 Subcontract risk.

The major risk issue that the contractor faces at tender stage is whether the materials allowances in the tender will be achievable once the contract has been awarded. Suppliers never quote 'fixed' prices but the contractor is, nevertheless, bound by the tender figure and cannot ask the employer for more money if prices go up in the meantime.

During the time lapse between tender submission and contract award it is possible that suppliers' prices may increase as a function of inflation or as a function of supply and demand. Materials' price inflation is determined by the costs of production and, where imported from abroad, exchange rate fluctuations. Supply and demand issues are a function of the market for construction materials and this can be a real problem when the construction economy is buoyant. In either case, the contractor's tender allowance is at risk.

There is, of course, the possibility that the contractor may be able to negotiate more favourable prices once the contract has been awarded, as materials suppliers may be just as keen for new business as anyone else. On the other hand, it is equally possible that materials prices will 'firm up' when it is known that the contractor has been awarded the contract, especially where there are a limited number of suppliers of the materials in question. Some suppliers quote their keenest prices at tender stage whilst others allow a margin for negotiation should the main contractor's tender be accepted by the employer.

It is normal practice for the contractor to obtain several quotes from different suppliers for each individual material and the prices quoted may vary considerably. The 'gamble' for the contractor is to decide which price to base his tender on:

- the lowest overall price quoted (to submit a 'keen' tender);
- the second lowest quote (allowing for a margin or 'comfort' factor);
- the arithmetic mean price (to be on the safe side);
- the lowest individual rates from each of the quotes received (to be 'ultra keen').

Table 15.2 illustrates how a contractor may compare the various quotes received for a variety of designed concrete mixes. The individual designed mixes are indicated with their associated quantities, which are then multiplied by the rates quoted by each supplier in order to give a total overall quote.

It can be seen that Acme Readymix has submitted the lowest overall quote after deducting the cash discount. However, only one of its rates is the lowest in the competition, as shown in Acme's list of net rates. Whilst not the lowest quote overall, Betamix has quoted the lowest rates for C15 and C20 concrete and the main contractor could possibly use this as a lever with this supplier to negotiate a reduction in its other rates.

An alternative course of action for the contractor is to use the lowest rates quoted across the board and this comparison is shown in the far right column of Table 15.2. Using the lowest rates would achieve an overall net saving on the Acme quote of £90 220.85 - £89 175.62 = **£1045.23**.

The other main issue with materials is, of course, waste. Normally the estimator will make a waste allowance in his rates to cover for losses, misuse, breakages, laps and so on, usually by adding a percentage to the net cost of the materials in question. Whether this is sufficient is a matter for site management, which must also attend to the contractor's statutory waste management obligations.

Where the main contractor supplies materials to labour-only subcontractors, however, this can lead to gratuitous waste way beyond the estimator's allowances.

Table 15.2 Comparison of suppliers' quotations (Ready Mixed Concrete Comparison).

Design mix	Quantity m³	Acme Readymix		Betamix		Newmix		Sitemix		Lowest rates	
		Rate (£)	Total (£)	Rate (£)	Total (£)	Rate (£)	Total (£)	Rate (£)	Total (£)	Rate (£)	Total (£)
C10	249	49.60	12 350.40	48.50	12 076.50	46.00	11454.00	47.00	11703.00	46.00	11454.00
C15	226	52.35	11831.10	51.00	11526.00	52.50	11865.00	52.00	11752.00	51.00	11526.00
C20	187	61.00	11 407.00	59.00	11033.00	62.50	11687.50	60.50	11313.50	59.00	11033.00
C25	414	73.20	30304.80	72.00	29808.00	74.10	30677.40	70.50	29187.00	70.50	29187.00
C30	297	89.70	26640.90	88.00	26136.00	88.90	26403.30	89.50	26581.50	87.46	25975.62
Subtotals			92534.20		90579.50		92087.20		90537.00		89175.62
Discount	2½%		2 313.36	Net	0.00	Net	0.00	Net	0.00		0.00
TOTALS			90220.85		90579.50		92087.20		90537.00		89175.62

Net rates
48.36
51.04
59.48
71.37
87.46

Not only can this lead to a significant loss on the materials budget but it can also create additional costs for keeping the site clean and tidy and for segregation and disposal of waste. To mitigate this risk, main contractors often 'contra-charge' sub-contractors for the costs of dealing with site waste or, on large sites, they may ask each subcontractor to contribute (via a contra-charge) to the overall costs of keeping the site tidy.

Labour-only subcontractors can be made responsible for materials waste by asking them to quote for the job on a 'labour and materials' basis with the main contractor supplying the materials. Here, the labour-only subcontractor includes an allowance in his rates using the main contractor's list of materials prices. This effectively creates a subcontractor's budget for materials which, if exceeded, will result in additional contra-charges by the main contractor. The risk for the main contractor is whether the labour-only subcontractor is capable of doing the calculations in the first place and whether it is financially able to 'take the hit' should the budget be exceeded.

15.2.5 Commercial opportunity

Contractors' tenders are made up of two basic elements:

- A **cost element** comprising an estimate of the net cost of the labour, materials, subcontractors, plant hire and site on-costs required for carrying out the construction work as described in the tender documents taking into account the particular constraints of the project in question.
- A **commercial element** which determines the profit margin and overhead contribution required on the contract taking into account turnover targets, competition for the work and the technical and contractual risks of undertaking the work.

By 'drilling down' into the cost element of the tender, the commercial element can be brought into sharper focus and the competitiveness of the tender enhanced without undermining the inherent profitability of the contract. This is achieved by taking advantage of the 'commercial opportunity' offered by the contract in question. 'Commercial opportunity' is sometimes referred to as 'scoping allowance'.

Commercial opportunity is an aspect of risk. It is not a risk allowance but is rather a 'calculated risk' based on perceived savings that might accrue should the contract be awarded. It is a strategy for increasing competitiveness without compromising profitability by adopting a particular attitude to the way that the contract is priced.

When the commercial opportunity or 'scope' for cost reduction has been determined, the resultant calculation is a net deduction from the tender total. Such decisions are usually taken at the tender adjudication stage and, therefore, the cost reduction is effected by an adjustment to the tender total rather than a reduction in the rates for individual units of work. The balancing deduction will normally be taken out of the preliminaries total for ease of calculation unless an adjustment item has been provided in the bills of quantities. As a consequence, should the contract be awarded, the unit rates must be regarded as having a lower intrinsic value and this needs to be reflected in the monthly cost value reconciliation.

Table 15.3 illustrates a simple example of how a tender can be 'scoped' for commercial opportunity without reducing anticipated margin. At tender stage, a contractor will typically take the view that the award of a contract will put him in a strong bargaining position with subcontractors and suppliers, meaning that preferential terms and additional discounts may be negotiated as a result. As a consequence, and

Table 15.3 Tender scoped for commercial opportunity.

Contract	Wiggleton factory project
Employer	Wiggleton Ltd
Stated contract period	20 weeks
Proposed construction period	16 weeks
Normal tender margin	10%

	£	£
Preliminaries		
Fixed		10 000
Time related		90 000
Own work		
Materials	64 000	
Labour	96 000	
Net total own work		160 000
Net total subcontracts		750 000
Provisional sums		100 000
Subtotal		1 110 000
Normal margin 10%		111 000
Tender total		1 221 000
Less Commercial Opportunity		97 950
Revised Tender Total	£	1 123 050

Commercial opportunity

Scope allowance £		Notes
18 000	4 weeks	Saving on programme duration
3 200	5%	Muscle on materials
16 000	10%	Scope on variations
56 250	7.5%	Additional discounts
4 500	5%	Take scope on £100 000 less 10% margin
97 950		

depending upon the contractor's attitude to risk, commercial opportunity may be taken on his tender calculations by 'gambling' on the amount of money that negotiations could generate and this sum is then deducted from the tender figure in order to reduce the level of the bid and thereby increase competitiveness. In this example, the contractor is anticipating a buying margin of 5% on materials on the main contractor's own work and additional discounts on subcontractors of 7½%.

A contractor may further speculate that it can complete the contract works in a shorter time than is stipulated in the tender documents or in the pre-tender programme. Table 15.3 shows that completion of a 20-week contract in 16 weeks would result in a significant sum in time-related cost savings. This sum is calculated by dividing the net total of time-related preliminaries allowed in the tender (£90 000) by 20 weeks and multiplying by the four weeks time saving. This makes an overall net saving on the tender allowance for time-related preliminaries of £18 000.

Where bills of quantities contain provisional sums, this indicates to the contractor that the architect or engineer has made an allowance for work additional to that which has been measured and billed. Consequently, the contractor might regard this as additional turnover which will be carried out at the rates and prices stated in the contract. 'Commercial opportunity' may thus be taken on the value of this work which can be added to the scoping allowance in the tender. In Table 15.3, the contractor has taken a 5% scoping allowance on the value of provisional sums. The saving is taken on their net value (£90 000) after deducting the contractor's normal 10% margin. When the provisional sums are expended, the contractor must make savings of 5% on net costs but the 10% margin will be preserved.

Tender documents frequently contain errors, omissions and inconsistencies. Where the contractor suspects this to be the case, it is a fair bet that the contract administrator will have to issue variation instructions to put matters right. This can give rise to additional profit to the contractor due to the way that variations are measured and valued and in Table 15.3 the contractor is anticipating an additional 10% return on the value of the work concerned (i.e. 10% of £160 000).

Errors, omissions and inconsistencies may also result in the tender being 'loaded' in such a way as to take commercial advantage of the situation. When it is suspected that a quantity will have to be increased due to a mistake, for instance, the contractor can insert an enhanced rate in the bills of quantities so that additional income will be generated when the work is re-measured.

Commercial opportunity may be taken by 'moving money' in the bills of quantities to take advantage of undermeasured items. This is done by moving 'margin' (i.e. overheads and profit) from one billed item to another so that some rates will be increased and others reduced in value. If the quantities are then increased when the contract is underway, the contractor may capitalise through increased revenue. Alternatively, cash flow may be improved, and financing costs reduced, by increasing 'margin' on items of work to be carried out early in the contract, or by loading money into early preliminaries items, thereby generating additional funds during the early stages of the project.

It must be remembered that enhancing rates in the bills of quantities, moving money around or loading certain items must be achieved without increasing the tender figure. Therefore, if rates are enhanced the money must come from other items in the priced bills of quantities. Equally, when margin is moved from items to be carried out later in the contract in favour of earlier items then the later items of work will have no margin. Consequently, the contractor's cost value reconciliation will have to reflect any profit and overhead taken in advance when determining the profitability or otherwise of the contract at any given point in time.

15.3 Contract risk

15.3.1 Delay and disruption

Lowsley and Linnett (2006) refer to **delay** both in the context of delay to the planned completion date for a project and also with respect to the delay of a particular activity or sequence of activities and they remind us that an activity, or sequence of activities, may be delayed without impacting on overall completion of the works. Keane and Caletka (2008), on the other hand, argue that **disruption**, once established, has a direct measurable financial consequence but that delay does not.

Cooke and Williams (2009) explain that delay and disruption, whilst commonly regarded as synonymous, are not the same and have separate and distinct meanings. They further indicate that a project may be delayed without any disruption having taken place or that site activities may be disrupted without causing any delay to contract completion. Keane and Caletka (2008) concur with this view.

Notwithstanding the above, when project completion is delayed, both client and contractor can suffer financially to a greater or lesser extent:

- Occupation of the completed building will be delayed, thereby resulting in loss of income or utility.
- The contractor will be on site longer and will incur additional expenditure in time-related costs (preliminaries).
- Subcontractors may be delayed in completing their work and may have a claim against the main contractor.
- Delay to practical completion will result in retention monies being outstanding for longer, with the consequential effect on cash flow and the reduction of working capital.

Redress for delay is often sought by either or both parties through the mechanisms provided in the construction contract. Disruption, on the other hand, is unilaterally experienced by the contractor who equally may look to the conditions of contract for recompense.

Delay may be the fault of:

- **the contractor**, for example slow progress;
- **the employer**, for example late supply of information to the contractor;
- **both** contractor and employer, that is where each party contributes to the delay but not necessarily equally;
- **no-one** – where neither party is at fault but where one or the other carries the risk for the delay (e.g. delay by subcontractor = contractor risk; delay due to exceptionally inclement weather = employer risk);

The standard conditions of contract provide mechanisms whereby the employer can be compensated for delay to the completion of his project and the contractor can receive time and, possibly, money for delay and/or disruption.

Under the JCT SBC/Q 2011, for example, the employer may be recompensed in the form of liquidated and ascertained damages whereas the contractor may be entitled to an extension of time for completion and, possibly, for loss and expense. The Infrastructure Conditions of Contract (ICC) similarly recognise that the employer may be entitled to liquidated and ascertained damages for delay and that the contractor may suffer **delay and extra cost** for specified reasons. Under the NEC3 ECC, the inclusion of LADs in the contract is an option (Option R) whereas time and money claims for delay and/or loss and expense are dealt with under Section 6 (**Compensation events**).

Where there is delay and/or disruption, the role of the architect or engineer is to determine the reasons why and come to a balanced judgement to establish entitlement under the contract, if appropriate. Establishing the reasons for delay is not done scientifically and none of the standard contracts requires any special approach or technique to be employed. Under normal circumstances, however, and especially where the causes and effects of delay and disruption are straightforward, the architect's/engineer's assessment of delay will be acceptable to the contractor and any financial recompense due under the contract will be paid by the employer.

However, when any degree of complexity is involved in the delay causation, the assessment of delay, and the financial recompense that follows, is more difficult and may lead to disagreement or even a dispute between the parties to the contract. The problem is that the causes of delay can be complex and there may be many reasons for a delay. Some of these may be the fault of the employer (**employer risk event**), some the fault of the contractor (**contractor risk event**) and some the fault of neither party (**neutral risk event**) but where one party or the other nevertheless carries the risk.

Apart from relief from having to suffer liquidated and ascertained damages, extensions of time are of no benefit to a contractor without associated costs. Consequently, where there is delay, a contractor will be looking for an extension of time **relevant event** that has a direct loss and expense entitlement attached. Figure 15.3 distinguishes between those relevant events that do and those that do not give rise to time and money claims under the JCT SBC/Q 2011.

15.3.2   Delay analysis

When there is delay and/or disruption to the programme, the contractor will notify the architect/engineer/project manager who will then assess the situation according to the rules contained in the contract and make a decision regarding an extension of time with/without costs and/or the award of loss and expense. This decision will then be communicated to the contractor. The NEC3 ECC (clause 61.6) adds the further provision that, where the effects of a compensation event are too uncertain to be forecast reasonably, the project manager shall state his assumptions and, if later found to be wrong, will correct his decision.

When complex delay/disruption arises, the normal 'intuitive' method is not good enough to distinguish the reasons for the delay/disruption, or to evaluate the money flowing from it, and thus a more forensic type of analysis is needed. However, whilst a court may conclude that an architect should make a calculated rather than impressionistic assessment (*John Barker Construction Ltd v London Portman Hotel Ltd, 1996*), it is unlikely that a great deal of science, time or effort will go into it. The effort is more likely to come from the contractor who will want to justify his time and money claim in the best way possible. This will involve using delay analysis.

Keane and Caletka (2008) explain that delay analysis is concerned with the identification and assessment of delay/disruption entitlement by establishing causation and liability and by demonstrating the extent of time-related damages or disruption costs arising as a direct result of the events relied upon. They further explain that delay analysis methods fall into three categories – additive, subtractive and analytical.

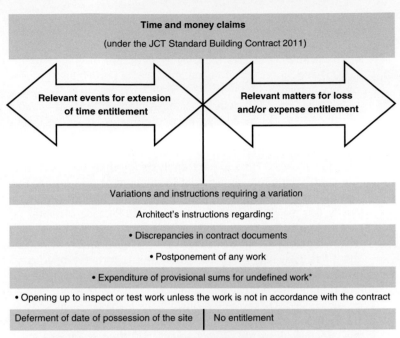

Time and money claims

(under the JCT Standard Building Contract 2011)

Relevant events for extension of time entitlement | **Relevant matters for loss and/or expense entitlement**

Variations and instructions requiring a variation

Architect's instructions regarding:

• Discrepancies in contract documents

• Postponement of any work

• Expenditure of provisional sums for undefined work*

• Opening up to inspect or test work unless the work is not in accordance with the contract

Deferment of date of possession of the site	No entitlement

The execution of work where an approximate quantity in the contract bills is not a reasonably accurate forecast of the quantity of work required*

Suspension of work by the contractor where the employer fails to pay in full the amount of an interim certificate

Any act or omission by the employer, architect, quantity surveyor or other of the employer's persons

Work carried out or not carried out by a statutory undertaker	No entitlement
Exceptionally adverse weather conditions	No entitlement
Loss or damage caused by fire, lightning, explosion, storm, flood etc. (Specified Perils)	No entitlement
Civil commotion or terrorism	No entitlement
Strike etc. affecting:	No entitlement
• Trades employed on the works	No entitlement
• Trades employed in the preparation, manufacture or transportation of goods and materials	No entitlement
• Persons engaged in the preparation of designs for which the contractor is responsible (contractor designed portion)	No entitlement
Exercise of Government statutory powers	No entitlement
Force majeure	No entitlement

* See SMM7 General rule 10.

Figure 15.3   Time and money claims.

The primary methods of delay analysis are explained by Lowsley and Linnett (2006) and others whilst Keane and Caletka (2008) also propose a number of secondary derivative methods.

Additive methods include impacted as-planned and time impact analysis. **Impacted as-planned** works by adding (impacting) delays onto the planned programme. By accounting for variations and delays, a revised programme is produced that shows the impact of the changes. It is thought to be the simplest and most popular method of delay analysis. **Time impact analysis** looks at the effect of the delay events on the contractor's future plans but in the light of progress at the time of the delay event. It is considered the most thorough method but can be time consuming and costly. It is the preferred method of the SCL Protocol (Chapter 16).

As-planned vs. as-built is an analytical method and is popular if basic. The planned programme is compared to the as-built programme but, whilst the method identifies delay in individual activities, this is normally insufficient to determine why overall completion was late. Consequently, Lowsley and Linnett (2006) prefer methods that identify critical delays and use critical path analysis to determine the effect of such delays. These methods include the impacted as-planned (additive), **collapsed as-built** (subtractive) and time-impact analysis (additive) as well as **windows analysis**, which reviews the project at specific times (windows) and builds a picture of progress during the window based on the records available.

The collapsed as-built method removes changes to the as-built programme in order to produce a programme that shows what would have happened but for the delay events. This method is simple but suffers from the defect that it is often difficult to construct an as-built programme from available records – simply because the records are often not good enough.

The types of delay in question are explained by several authorities including Lowsley and Linnett (2006) and Cooke and Williams (2009). Keane and Caletka (2008) categorise them as:

Compensable delay	An employer risk event where the contractor is entitled to an extension of time and time related costs
Concurrent or parallel delay	Where there are two or more delays and at least one is an employer risk event and another a contractor risk event
Critical delay	Delay to a critical activity which causes delay to overall completion of the project
Excusable delay	Where the contractor is relieved from delay damages and where there may be time and money entitlement under the contract
Non-excusable delay	Delay caused by the contractor for which there is no relief from delay damages and no time or money entitlement
Global delay	Where actual completion exceeds planned completion but individual employer risk events are not identified
Local delay	Delay to non-critical activities which do not impact the contract completion date.

Any successful delay analysis begins with a reliable **as-planned (baseline) programme** and an accurate **as-built programme**. This sounds simple but is difficult to achieve in practice. Contractor's programmes are inherently unreliable because they are frequently based on poor logic, they contain mistakes and omissions and, if there is a critical path analysis, it is frequently wrong because the logic links are wrong. The as-built programme relies on accurate records of events but, in practice, these are often incomplete, missing or inaccurate.

15.4 Claims

15.4.1 Extensions of time

Under the standard conditions of contract, the contractor is entitled to an extension of the time for completion if there is a qualifying delay and the contractor is not at fault. If it is likely that completion will be delayed (JCT conditions) or if a cause of delay has arisen (ICC) or if an event has happened or is expected to happen (NEC3 ECC), then the contractor is obliged to notify the architect/engineer/project manager in writing.

Where there is a **relevant event**, as specified in the JCT SBC/Q 2011, which qualifies the contractor to an extension of time, the architect is required to estimate a fair and reasonable later date for completion of the works. Under other conditions the principles are the same but the terminology may be different. For example, the NEC3 ECC refers to **compensation event** whereas the ICC – Measurement Version prefers the phrase **cause of delay**.

After due consideration, and if the architect/engineer/project manager considers that the works are likely to be delayed beyond the contract completion date, the contractor will be granted an extension of time for completion for the relevant event in question. The standard contracts stipulate the procedures necessary for the exercise of these duties and specify time limits where appropriate. Decisions may, under certain circumstances, be reviewed after completion of the works but the contract administrator may not revise the completion date to one earlier than that stated in the contract documents.

The underlying purpose of the extension of time provisions in the contract is twofold:

1. to determine the aggregate time in which the works are to be completed bearing in mind the contract completion date and
2. to preserve the employer's right to deduct liquidated damages for subsequent delay which is the fault of the contractor.

Consequently, where the fault for late completion lies with the employer (or its agents) then an extension of time for **non-culpable delay** would be granted and the contractor would be relieved from paying liquidated and ascertained damages. Judgement as to the contractor's entitlement to an extension lies with the architect or contract administrator. This does not deny the employer the right to charge LADs if the contractor should be at fault for any further delay (**culpable delay**) in which case the employer may deduct damages at the rate stipulated in the contract.

Extensions of time can be granted at any stage during a contract and the reasons for granting an extension of time are well defined in the standard contracts. Establishing the contractor's exact entitlement, however, is not defined and relies very much upon the judgement, expertise and fairness of the architect, engineer or project manager. The courts have been critical of such judgements (*John Barker Construction Ltd v London Portman Hotel Ltd, 1996*) and much will depend upon the quality of back-up information supplied by the contractor and the contract administrator's awareness of events on site and progress to date.

When the (extended) contract completion date has passed but the works are not finished, the contractor is in culpable delay and may face paying liquidated and ascertained damages. In these circumstances, should instructions be issued for a variation to the contract, an extension of time can still be given following the decision in the case of *Balfour Beatty Building Ltd v Chestermount Properties Ltd* (1993). In this case, the court decided that the contractor would be entitled to an extension of time for the time taken to carry out the variation but not for the period of culpable delay (the net method or 'dot-on' principle) but added words of caution that judgements should be carefully considered and fair and reasonable in the circumstances.

Extensions of time are important with regard to the financial reporting of contracts if a true and fair view is to be established. A contractor's reporting procedures should ensure that suitable provision is made in the accounts where:

1. An extension of time is granted without costs

   The cost of preliminaries in the additional time period
   The cost of financing the work in progress
   Loss due to inflation in the costs of labour or materials
   Claims for delay/disruption from subcontractors

2. An extension of time is granted with costs

   Under-recovery of preliminaries in the additional time period
   Finance costs not reimbursed under the contract

3. No extension of time is granted

   Liquidated and ascertained damages payable
   The cost of preliminaries in the additional time period
   The cost of financing the work in progress
   Loss due to inflation in the costs of labour or materials
   Claims for delay/disruption from sub-contractors

15.4.2 Loss and expense

Where the contractor suffers loss and expense for which it would not be reimbursed under other clauses in the contract, such as the variations clause, the standard contracts provide that the employer will pay him additional monies. To qualify for such payments, the contractor must refer to a list of events or matters which quality for reimbursement. These include postponement of the works, late instructions from the architect/engineer and failure to give access to the works or part thereof. Payments will be valued by the engineer or quantity surveyor and paid in the relevant interim certificate.

Nothing is guaranteed to raise emotions more within the employer's camp than a claim from a contractor because it often implies some error, misjudgement or oversight by the employer's team when the contract was set up. This is, to an extent, understandable as some contractors are particularly claims conscious and unduly aggressive in pursuing entitlement. There is nothing wrong with the pursuit of legitimate claims however, indeed it is perfectly legitimate where provided in the contract. **Loss and expense** is the JCT term for 'claim' whereas the ICC conditions use the word **claim** and the NEC3 ECC refers to **compensation event**.

The contractor is obliged to give written notice and apply for entitlement whichever contract is used. The key to any successful claim, however, is to keep perspicuous records of what happens on the project as these records will be vital evidence when it comes to justifying entitlement – subcontractors please note! Such records include:

- labour allocation sheets;
- plant and plant hire returns;
- letters, faxes and emails;
- requests for information;
- site diary entries and daily management reports;
- photographs;
- minutes of meetings;
- variation instructions;
- revised and updated programmes.

Many contractors keep a **programme of the day** – literally depicting daily events – to record the impact of delay events on the contractor's progress.

Contractors' claims are frequently viewed with suspicion by employers and their agents and have a reputation for being over-inflated, even spurious. There is probably a great deal of truth in such anecdotal views and most people would acknowledge that asking for more than one is prepared to accept is only human nature after all.

Keating (Furst and Ramsey, 2012) reflects that careful consideration should be given to the legal principles upon which a claim is based and that claims may be viewed as **legally enforceable** or *ex gratia*. Keating also explains that legally enforceable claims have their basis in rights to payment under the contract whereas *ex gratia* claims depend upon a statement of principle, accepted by the employer, which, if not paid, will lead to litigation and the associated costs and interest if the court case is lost.

Much depends, of course, on the way that a claim is prepared, upon the evidence available, upon the strength of the argument and the terms of the contract relied upon. It is also the case that the nature of construction projects is such that contractors are regularly caused delay (and disruption) due to the occurrence of large numbers of variations to the design and the frequency of delays that occur in the issue of design information and other instructions.

Construction contracts nevertheless contain express mechanisms for legitimate contractors' claims and a contractor would be foolish not to pursue a genuine claim for fear of upsetting the employer or the architect. Delay to site operations and disruption to the order and sequencing of the work costs money and the contract's profitability will be severely dented by costs not anticipated in the contract budget.

Costs resulting in a loss and expense claim will, therefore, be debited to the contract account but any 'claim', whether submitted or not or agreed or not, cannot strictly be taken into value in the accounts until such time that a sum is included in a certified payment due to the contractor. This is due to the high degree of uncertainty that a claim will actually be settled, in whole or in part, and to the requirement of normally accepted accounting standards that requires prudence in the reporting of work in progress. In this regard, the value of claims should only be recognised in the accounts when negotiations have reached an advanced stage and when the employer has accepted the claim in principle with an indication being given of the amount available (SSAP9 Appendix 1 paragraph 27).

15.4.3 Evaluating prolongation expenses

The recovery of prolongation expenses depends upon the terms of the contract between the employer and the contractor and the cause of the employer risk event. Usually, the cause of delay/disruption will be due to a variation or a breach of contract (e.g. delayed possession or late instructions) or some other provision within the contract (such as the discovery of unforeseen ground conditions).

Once the cause has been established, it is for the contractor to prove that he has actually suffered a financial loss or delay or both before becoming eligible for compensation under the contract.

Because the contractor's tender is based on rates/prices, with allowances for preliminaries (including time-related sums), it is commonly misunderstood that prolongation costs will be based on the tender allowances. Whilst the valuation of variations is made on the basis of the contract rates, this is not the case when it comes to evaluating a prolongation claim. The SCL Delay Protocol (2002) considers that the contractor's tender allowances for preliminaries have little bearing when calculating the amount of compensation due for prolongation or disruption.

Consequently, if the contractor has made insufficient or no allowance for site overheads in his tender, this does not reduce or remove his entitlement due under the contract. This is because the contractor's recovery normally depends upon the actual cost of prolongation (i.e. the actual cost of remaining on site for additional time) or the actual cost of work necessitated by disruption to the contractor's planned work. Keating (Furst and Ramsey, 2012) concurs with this view because the contractor's damages are to compensate for *actual* loss and must be proved (in court if necessary). Although many contractors would not agree, Keating adds authority to the view that damages for breach of contract are not intended to put the contractor in a better financial position than if the breach had not occurred.

To underline this reasoning, the ICC conditions define 'cost' as *all expenditure properly incurred … including overhead finances and other charges … but does not include an allowance for profit*. The JCT Conditions refer to *direct loss and/or expense*. Under the NEC3 ECC there are a lot more rules attached to the evaluation of compensation events but *assessments are based upon the assumption that … the additional actual cost and time due to the delay event are reasonably incurred*.

As far as variations are concerned, the SCL Protocol considers that tender allowances may have some relevance as a baseline for calculating delay/disruption consequences but only so far as the circumstances under which the variations are carried out are sufficiently different so as to make use of the contract rates/prices invalid.

Where there is a concurrent delay, the contractor must be able to disaggregate the costs associated with the contractor risk event from those of the employer risk event in order to be able to claim prolongation compensation. In practical terms this will usually mean that the contractor will only recover costs for the period by which the employer risk event exceeds the contractor risk event (i.e. the net method of calculation).

The contractor is under a general duty to mitigate the loss flowing from an employer risk event. This may be an express condition of the contract or, if not, may be implied into the contract under common law. The duty is to take reasonable steps to minimise the loss flowing from the event.

15.5 Insolvency risk

15.5.1 Risk rating

Insolvency is an ever-present danger in contracting that can be seriously damaging for the project and its participants. However, whilst the industry is certainly at or near the top of the insolvency 'premier league', the risk of an insolvency on any particular project is probably very small. At a very rough estimate, there are something like 200 000 construction projects on the go at any one time and, therefore, the 'gut instinct' is that an industry-wide insolvency risk factor of 'D' – 'low impact-low likelihood' – is about right (Figure 3.2). Having said that, it is difficult to generalise regarding the likely risk of insolvency because there is such a wide variety of type and size of contractor and subcontractor in the industry, all of which have different financial structures and problems.

When a big insolvency event occurs, however, the fall-out has far reaching implications for the injured parties and the impact will be in the 'high' category. The vast majority of those affected will be unsecured creditors in the insolvency – likely to receive little or nothing at the end of the day – but they will undoubtedly have their own creditors to pay for goods or services rendered. If the main contractor is insolvent, as happened spectacularly in 2010, a string of suppliers and subcontractors will be owed money but subcontractors will still have their own suppliers and plant hire firms to pay.

The most serious eventuality is, of course, the insolvency of the employer because this can expose the main contractor and its subcontractors to financial problems that could test their own solvency. Happily, employer insolvency is a relatively rare event that might be regarded as a 'high impact-low likelihood' risk with a 'B' risk rating. If the employer is insolvent, the main contractor will have suppliers, plant hire firms and subcontractors to pay and subcontractors will be similarly indebted. The risk of a large main contractor becoming insolvent would carry a similar rating but smaller main contractors are often less financially stable, and therefore more prone to insolvency, and a rating of between 'B' and 'A' might be more appropriate. If a subcontractor becomes insolvent, the main contractor will have to find a replacement subcontractor with all the delay, disruption and extra cost that this entails.

Therefore, the more likely insolvency event would be that of a subcontractor and the risk rating in this case could be between 'C' and 'D'. This is because the impact on the overall project would be relatively low and the likelihood of the event might be considered 'low-medium' depending upon the size and quality of the subcontractors engaged. This is not to minimise the disruptive effect of a subcontractor failure nor the expense that the main contractor will undoubtedly suffer. However, the procurement of a replacement subcontractor would be relatively straightforward to arrange and this can usually be done quickly with the minimum of inconvenience to all concerned.

15.5.2 Definitions

Insolvency is a complex area of law best left to professionals. However, quantity surveyors are regularly engaged in insolvencies, either because a subcontractor has 'gone bust' or because of a specific engagement for professional services. The quantity surveyor is ideally placed to contribute to the process because a large part of construction insolvencies involves the valuation contracts and work in progress.

'Going bust' is an idiomatic term or colloquialism and the correct term is 'insolvency'. This relates to both individuals and companies that fail the solvency tests as prescribed by the 1986 Insolvency Act. There are several formal procedures available to insolvent individuals and companies, each with a specific objective, depending upon whether the company is a sole trader, partnership or limited company:

Bankruptcy applies to individuals (including sole traders and individual members of a partnership) where an individual or partnership is insolvent and involves the evaluation of the debtors assets which are used to satisfy creditors as far as is possible.

Voluntary arrangement is a procedure that allows an individual or a company to reach a binding agreement with creditors with the purpose of enabling debts to be repaid in whole or in part over a period of time.

Administrative receivership is where an administrative receiver is appointed whose purpose is to secure the assets of the business. 'Receiver' is the commonly used name for an administrative receiver and the procedure can have a rescue function which can enable a company to be sold on a 'going concern' basis, thereby securing jobs and completing contracts.

Administration is where an administrator is put into a company in difficulty to protect the company from its creditors. An administrator may be appointed either by application to the court or by a new process known as 'an appointment out of court'. The administrator runs the business and will try to save it and the procedure gives the company 'breathing space' to consider its options and the chance to trade through its difficulties, subject to a time limit - normally 12 months.

Winding up is where an insolvent company is wound up or liquidated (compulsorily or voluntarily) by order of a court.

Voluntary liquidation is a method of liquidation by consensus (agreement), not involving the courts, for solvent companies - members' voluntary liquidation and for insolvent companies - creditors' voluntary liquidation.

Compulsory liquidation is the winding up of a company or a partnership by means of a court order. A petition (application) is made to the court, usually by a creditor, which may issue a winding up order. A liquidator is appointed to sell the company's assets, recover amounts owed to the company (e.g. from contracts), pay all fees and charges and pay the creditors. Creditors are paid, as far as funds allow, in a strict order of priority.

15.5.3 Legislation

The primary legislation concerned with insolvency in the United Kingdom is the Insolvency Act 1986 which was informed by the findings of the Cork Report (Cork, 1982). The Act introduced, *inter alia*, the concept of the Insolvency Practitioner and set out their powers and duties. The Act also sets out procedural rules, including the rights and duties of company directors and individuals and provides a definition of formal insolvency.

Since the Act was introduced significant changes have been made to the way it operates through a plethora of Amendment Rules and via primary and secondary legislation including the Insolvency Act 1986 (Amendment) Regulations 2002 and the Insolvency Acts of 1994 and 2000 (The Insolvency Service, www.insolvency.gov.uk). Further changes and consolidation is planned and the Enterprise Act 2002 made important changes to insolvency law through significant amendments to the administration procedures for companies in distress and by the introduction of the 'out-of-court' administration route.

15.5.4 Termination of main contracts and subcontracts

The possibility of the insolvency of the contractor or the employer or of a subcontractor is contemplated in the standard conditions of contract used in the industry and in the standard conditions of subcontract as well. Standard conditions of contract define 'insolvency' in the narrow sense of the contract itself and set out explicit procedures to be followed. Most standard contracts are similar but the NEC, ICC and JCT families use somewhat different phraseology. Each contract should be read carefully for the fine detail.

The JCT family of contracts provides a definition of the meaning of insolvency and prescribes the formal notices to be given if the worst happens. A definition of what constitutes a default is also given and protocols in the event of insolvency are provided. Insolvency is defined in a pragmatic rather than a technical way (i.e. by applying the financial tests prescribed by the 1986 Act) and an insolvency event happens where either party to the contract:

- makes an arrangement with creditors,
- decides to wind up his business, or
- a winding up order or bankruptcy order is made against him, or
- an administrator or administrative receiver is appointed.

An insolvency event triggers 'termination'. Therefore, where the **main contractor** is insolvent, the employer must be notified immediately in writing and it may give a written **notice of termination** at any time. If the **employer** is insolvent the employer shall immediately notify the contractor in writing and the contractor may give **notice to terminate** its own employment under the contract. If a **subcontractor** is insolvent the main contractor may give written notice to terminate the subcontractor's employment under the subcontract. In some contracts and subcontracts, termination is automatic.

15.5.5 The effect of termination

If the main contractor is insolvent, the effect of the employer's notice to terminate is to suspend the contractor's obligation to carry out and complete the works, to permit the employer to safeguard the site, the works and the site materials and to require the contractor not to hinder such measures. Where the employer is insolvent, the contractor's obligation to carry out and complete the works is similarly suspended. If a subcontractor fails, the subcontractor's obligations under the subcontract are suspended and the main contractor shall take measures to protect the subcontract works and any materials on site.

If either the employer or the main contractor is insolvent, this leads to **termination of the contractor's employment under the contract**. This means that the contractor's obligations under the contract are suspended whilst at the same time keeping 'alive' the terms and conditions of the contract. Termination therefore preserves a contractual right to deal with the insolvency as it applies to the contract, to administer the financial conclusion of the contract and, possibly, to facilitate completion of the contract if the main contractor is the insolvent party. Similar provisions apply where it is a subcontractor that is insolvent.

Contracts usually state that termination notices must not be issued unreasonably or vexatiously (i.e. out of anger or spite) – a sensible precaution because emotions can run very high in an insolvency – and termination shall take effect upon receipt of the notice. The JCT standard contracts also provide that the main contractor's employment may be reinstated at any time and on such terms as the parties shall agree. If the contractor is insolvent, the employer has the option not to continue with the project.

15.5.6 Subcontractor's insolvency

The insolvency of a subcontractor is a private matter for the main contractor as there is no privity of contract with the employer – unless third party rights have been created by virtue of Part 2 of the JCT SBC/Q 2011 Contract Particulars. In other words, the main contractor retains responsibility for the performance of the contract and subletting does not absolve him from the burden of his contract with the employer. Additionally, the main contractor may have no right to claim an extension of time or loss and expense. Under JCT SBC/Q 2011, the insolvency of a subcontractor (whether a listed subcontractor or not) is not a 'relevant event'. Under the ICC – Measurement Version, there is a possibility that a subcontractor insolvency might qualify for an extension of time under 'special circumstances' – it might be worth a try!

Subcontractors usually undergo a pre-qualification process before being allowed to tender for work with a main contractor and part of this process should be to verify, as far as possible, the financial stability of the subcontractor. Investigations can be made in the trade and bank references can be obtained but formal trade references (e.g. from suppliers) are problematic due to the privacy laws in the United Kingdom (e.g. the Data Protection Act 1998). The subcontractor's ability to provide a bond (e.g. performance bond) is a good test of a subcontractor's financial standing (Chapter 6).

If things should go wrong, the main contractor has a number of problems of immediate concern; some are practical problems and some are financial. Assuming the insolvent subcontractor is engaged on the plastering package:

- Practical problems:
  - The subcontract work will stop immediately on entering insolvency because:
    - the subcontractor will be unable to pay the wages
    - credit for materials will be withdrawn by suppliers and merchants
    - termination by the main contractor will be automatic in most cases
  - The plastering work will be part-way through and:
    - other trades may be delayed or unable to continue (e.g. joinery, painting)
    - the work continuity of other trades may be disrupted;
- Programme:
  - if the plastering (and associated trades) are on the critical path the project completion date may be in danger
  - time will be required to find and procure a replacement subcontractor and mobilise the work on site.
- Financial problems:
  - delay and disruption to other trades will lead to claims against the main contractor
  - a new subcontractor may be unable or unwilling to undertake the work at the same rates and prices as the original subcontractor
  - an incomplete package of work will be less attractive to other subcontractors and the rates may be considerably higher as a consequence

o defective work may have to be re-done which will cost the main contractor money and further delay the project

o additional quantity surveyor time will be needed to resolve the original subcontract

o the main contractor may incur liquidated and ascertained damages if the contract completion date is not met.

The first thing that the main contractor must do is try to engage another subcontractor to complete the work. The quicker this is done the less will be the disruptive effect on the project as a whole and on other subcontractors individually. The main contractor has the right to use the materials, plant and equipment of the insolvent subcontractor if so desired, subject to retention of title and ownership issues, and this enables the new subcontractor to 'hit the ground running'.

There is no hiding from the fact, however, that this will be a costly process for the main contractor – with limited possibilities for recovery of sums due from the insolvent party – and there will undoubtedly be claims from other subcontractors for delay and disruption to face as well as the possibility of delay to contract completion and liquidated and ascertained damages to pay the employer. As far as the main contractor's accounts are concerned, the insolvency creates a liability that must be recognised in the cost value reconciliation for the project. The main contractor will prepare a subcontract account of the value of work done and the reasonable cost and loss/damage incurred in completing the subcontract. A simple example is given in Tables 15.4–15.6.

Table 15.4 shows J Brown & Sons' tender for a plastering subcontract with White Contracting Ltd and Table 15.5 shows the remaining quantities of work to be done at the point where the subcontractor notified the main contractor of its insolvency. These quantities have been priced by a replacement subcontractor and a completion contract has been agreed.

Table 15.6 sets out a suggested approach to the insolvency in order to calculate the financial position of J Brown & Sons upon completion of the plastering subcontract by AB Plastering. The usual procedure is to prepare a 'notional final account' for J Brown & Sons which sets out:

● the subcontract sum adjusted for variations and other changes;
● the costs incurred by the main contractor (Whites) in connection with completion of the plastering subcontract including:
  o the completion contract (with AB Plastering)
  o rectification of defects in Brown's work
  o claims from subcontractors for disruption to their work
  o the consequential costs of Whites for delays to the main contract
  o legal fees incurred in dealing with the new subcontract.

The basic idea is to adjust the subcontract sum to arrive at the figure that would have resulted if J Brown & Sons had not become insolvent before completing the subcontract (notional final account) and then to compare this with the actual cost of completing the work together with any consequential costs arising. The difference between the two figures gives the extra cost of completing the subcontract following the insolvency. Because the insolvent subcontractor had received interim payments before termination, these have to be taken into account in determining the money owed by the insolvent subcontractor to the injured party (i.e. the main contractor).

Table 15.4   J Brown & Sons tender/contract sum.

White Contracting Ltd is main contractor.
J Brown & Sons has a subcontract in the sum of £149 883.94 for a plastering contract.
The form of subcontract is DOM/1.
Brown's tender is shown below.
A main contractor's cash discount of 2.5% has been agreed.

Ref	Item	Quant	Unit	Rate (£)	Cost (£)
	Floor, wall and ceiling finishes *Thistle plaster* Plaster; Thistle; 10 mm cement and sand (1:3); 3 mm finish; 13 mm work to concrete, brick or block base:				
A	over 300 mm wide to walls	3022[a]	m²	13.06[b]	39 467.32
	Plaster beads and the like Catnic galvanised steel beads; fixed with plaster dabs:				
B	Supasave angle bead	115	m	1.90	218.50
	Dry linings and partitions Thistle baseboard; 9.5 mm work; fixed to timber base with galvanised nails; taped butt joints:				
C	over 300 mm wide; to ceilings	7440	m²	5.69	42 333.60
D	not exceeding 300 mm wide; to walls	218	m²	6.36	1386.48
	Thistle plaster Plaster; Thistle; 3 mm one coat board finish; steel trowelled; internal; to plasterboard base:				
E	over 300 mm wide; to ceilings	7440	m²	8.76	65 174.40
F	not exceeding 300 mm wide to walls	218	m²	5.98	1303.64
	Subcontract sum				149 883.94
	Less main contractor's discount	2½	%		3747.10
	Net value of subcontract				146 136.84[c]

[a] 3022 = original quantities; [b] 13.06 = J Brown original rates; [c] 149 883.94 = J Brown tender figure.

It can be seen that the cost of completing the subcontract exceeds Brown's original contract figure (as adjusted). This creates a debt of £17 970.17 owed by Browns to the main contractor (Whites). As Browns had carried out some uncertified work in progress, the value of this work is credited to its account and the debt is reduced to £5328.61 accordingly. Previous payments of £46 665.35 made to Browns are then factored in to the calculation, which effectively reduces the adjusted subcontract sum and increases the final debt owed by Browns to the main contractor to the sum of £51 993.96.

This sum represents the main contractor's liability for the insolvency because, although Browns is effectively a debtor, it is unlikely that any money will be seen

Table 15.5   Original contract sum and completion contract sum.

J Brown & Sons' subcontract is terminated part way through due to the insolvency of the subcontractor.

AB Plastering has been engaged to complete the plastering work.

J Brown & Sons has written to White Contracting Ltd to notify it that they have entered administration.

Ref	Item	Quant	Unit	Rate (£)	Cost (£)
	Floor, wall and ceiling finishes **Thistle plaster** Plaster; Thistle; 10 mm cement and sand (1:3); 3 mm finish; 13 mm work to concrete, brick or block base:				
A	over 300 mm wide to walls	1796[a]	m²	14.20[b]	25 503.20
	Plaster beads and the like Catnic galvanised steel beads; fixed with plaster dabs:				
B	Supasave angle bead	59	m	2.25[b]	132.75
	Dry linings and partitions Thistle baseboard; 9.5 mm work; fixed to timber base with galvanised nails; taped butt joints:				
C	over 300 mm wide; to ceilings	4398	m²	6.50[b]	28 587.00
D	not exceeding 300 mm wide; to walls	157	m²	7.50[b]	1177.50
	Thistle plaster Plaster; Thistle; 3 mm one coat board finish; steel trowelled; internal; to plasterboard base:				
E	over 300 mm wide; to ceilings	4398	m²	9.50[b]	41781.00
F	not exceeding 300 mm wide to walls	157	m²	6.75[b]	1059.75
	Completion contract As per Messrs AB Plastering Ltd tender				**£98 241.20**

[a] 1796 = remaining quantities at termination; [b] AB Plastering rates.

and, if there is, it will not be much. The main contractor can offset the loss to some extent by trying to do a deal with the completion subcontractor for early settlement of the account and by accelerating the programme to save on the time-related costs (preliminaries and liquidated and ascertained damages).

As far as the employer is concerned, the rates in the bills of quantities for the trade in question (i.e. plastering) will remain unchanged. If, as in the worked example above, the main contractor has to pay a higher price to complete the work this will be taken into account when the main contractor calculates the cost of completing the insolvent subcontractor's work. It will be a matter for Brown's receiver/administrator

Table 15.6 Notional final account.

Variations of omission and addition have been valued and agreed.
J Brown & Sons has carried out authorised daywork.
Previous payments have been made to Browns but nothing is owing at the
termination date.
Browns has uncertified work in progress at the termination date.
The completion subcontractor has rectified defects in Brown's work.
Other subcontractors have made claims to the main contractor for disruption to
their work.
The main contract completion date has been delayed by three weeks.
Legal fees have been incurred by White Contracting Ltd to set up the completion
contract and deal with the administrator.

Ref	Item	Quant	Unit	Rate (£)	Cost (£)
	Notional final account				
	Subcontract sum (net)				146136.84
			Omit	*Add*	
	Variations		2896.28	267.22	
	Daywork			799.33	
			2896.28	1066.55	(1829.73)
	Adjusted subcontract sum				**144307.11**
	Costs incurred				
	Paid to AB Plastering – Completion Contract (net)				98241.20
	Remedial work				3452.89
	Subcontractors' disruption claims				17449.56
	Additional preliminaries	3	weeks	8976	26928.00
	LADs	3	weeks	5000	15000.00
	Legal fees				1205.63
	Total cost to complete				162277.28
	Less adjusted subcontract sum (net)				144307.11
	Balance owed by J Brown & Sons				(17970.17)
	Add value of Brown's work in progress at termination				12641.56
					(5328.61)
	Less Previous payments to J Brown & Sons:				(46665.35)
	Balance owed by J Brown & Sons				**(51993.96)**

to argue whether these costs are reasonable and whether the main contractor has
mitigated his loss as is required under English law.

The debt that might be owed by the subcontractor to the main contractor is to
some extent academic, as the main contractor will be an unsecured creditor in the
subcontractor's insolvency. In practical terms it is unlikely that much money, if
any, will be forthcoming and the main contractor will endeavour to get another

subcontractor on-board at the least cost possible. Any resulting loss on the subcontract package must be posted as a potential liability in the contractor's accounts.

15.5.7 Employer's insolvency

Any competent contractor will, as far as it is able, check the financial standing of the client before entering into a construction contract. This is important because, if the employer becomes insolvent during the contract, the main contractor will undoubtedly be owed money for any or all of the following:

- the value of work carried out and certified but not paid;
- the value of work done but not invoiced (work in progress);
- the value of variations ordered but not invoiced or paid;
- the value of claims for loss and expense agreed but not paid;
- the value of materials brought to site but not incorporated in the works (materials on site);
- the value of retention monies held.

If the employer becomes insolvent part-way through a contract, there are a number of possible scenarios:

- the next valuation date may not have been reached but work is in progress;
- the work in progress may have been valued but no certificate issued;
- a certificate may have been issued but no payment received;
- further work will have been carried out prior to payment;
- payment may have been received with retention deducted.

As far as subcontractors are concerned, there will be:

- completed subcontract packages;
- incomplete subcontract packages;
- subcontractors ready to start;
- the contractor will have:
  - paid some subcontractors accounts in full (less retention)
  - partially paid some accounts with outstanding issues to resolve
  - not paid the latest subcontractor applications
  - not yet received applications for work in progress.

In any event, as the majority of construction work is carried out by subcontractors, the main contractor will have problems if the employer becomes insolvent. The majority of work in progress, or the majority of the amount due on a certificate, or the majority of amounts included in an interim payment, will be destined for subcontractors with the main contractor only being due payment for preliminaries, any work carried out by its own employees and any margin it may have on subcontract work – possibly amounting to 20–30% of the total.

Different subcontracts deal with this situation differently. In some cases the risk lies with the main contractor and in others it is the subcontractors which carry the risk. Under the JCT Standard Subcontract, the subcontractor is required to submit an account detailing the amount owed by the main contractor and this is payable within 28 days without deduction of retention. The CECA Blue forms have similar provisions. Under DOM/1, however, the risk swings completely against the subcontractor and no further payment is due to the subcontractor unless the contractor has received payment.

The contractor's duty to pay subcontractors arises under the terms of the subcontract and is not contingent upon the contractor being paid by the employer. However, conditional payment ('pay when paid') is allowable under section 113(1) of the 2009 Construction Act where the employer is insolvent. Therefore, despite the conditions of subcontract prevailing, subcontractor's are 'at risk' if the employer is insolvent and they may have to wait a long time before receiving any payment, if at all. Irrespective of this, the main contractor's financial accounts will have to reflect a liability if the subcontractors' accounts are not settled and, likewise, the subcontractor's accounts must reflect the potential loss on the subcontract.

In the event of termination due to the employer's insolvency, the JCT SBC/Q 2011 requires the contractor, as soon as reasonably practical, to prepare an account of:

- the total value of work executed;
- any other amounts due under the conditions of contract;
- any sums ascertained in respect of direct loss and/or expense;
- the reasonable cost of removal of temporary buildings, plant, tools and equipment;
- the reasonable cost of removal of goods and materials including site materials intended for use in the works;
- the cost of materials or goods ordered for the works which the contractor has paid for or is legally bound to pay for;
- any direct loss and/or damage caused to the contractor by the termination.

The amount payable by the employer is due within 28 days of submission of the account (some hope!), less amounts previously paid, with no deduction of retention. If the account is settled, the goods and materials paid for by the employer shall become the property of the employer.

The contractor will, in the normal course of events, become an unsecured creditor if the employer is insolvent and will not necessarily receive all or any of the money due. Davis (2011), however, suggests that the question of possession of the site may be an issue especially if the contractor has not yet issued a notice of termination. The contractor does not own the site but is in legal possession thereof until completion of the contract and this may be a weapon that can be used to bargain with especially if the receiver wishes the contract to be completed.

As far as the contractor's accounts are concerned, SSAP9 requires that *the lower of cost and net realisable value* is included in the accounts. Where the employer is insolvent, the net realisable value will probably be the lower figure because the contractor may receive nothing, or a small percentage in the pound, in common with all other unsecured creditors and this will undoubtedly be less than cost. The contractor will need to be realistic as to the amount to be included in the accounts for the contract.

15.5.8 Contractor's insolvency

Where the contractor is insolvent, the employer is faced with a number of options, each of which will have to be considered in the light of the employer's legal duty under English law to mitigate his loss. This means being reasonable in the employer's claim for damages to complete the project albeit that, according to Davis (2011), the standard of 'reasonableness' expected by the courts will not be a high one as the insolvent contractor is the wrongdoer.

The impact of the main contractor's insolvency will be felt most by the employer, the subcontractors and trade suppliers but the most interested parties will be other contractors, which will see the insolvency as an opportunity to acquire the contract, or indeed the insolvent contractor's business or portfolio of contracts. This represents possible turnover and, if acquired on the right terms, profit. Here again, the quantity surveyor's skills will be called upon to value the contracts on offer and this must reflect the value of work in progress, any contingent liabilities for defective work and the projected cost to complete the contract.

What the employer must bear in mind, however, is that the contractor is insolvent and the employer is an unsecured creditor (unless there is a performance bond in place). Consequently, it is in the employer's best interests to complete the contract at the best price possible as it may well end up footing most if not all of the bill. The options open to the employer are:

1. decide not to have the works carried out and completed;
2. approach the original unsuccessful tenderers and invite tenders for the completion of the works;
3. re-tender the remaining work;
4. engage another contractor on a cost plus basis;
5. allow the contract(s) to be completed by insolvency practitioners;
6. novation.

Option 1

This may be an expensive option, as the employer may have bought the site, carried out enabling works and paid the insolvent contractor for work done prior to termination as well as having professional consultants on board. The employer may decide to 'mothball' the project until a later date and take time to consider the best solution.

Option 2

The original unsuccessful contractors probably did not win the contract because their prices were too high and, therefore, to ask them to tender for a partially completed contract is hardly likely to make them more competitive due to the reduced amount of work remaining and the effect of inflation since the date of tender. Conversely, the construction market may have hardened in the meantime and contractors may be looking for turnover and, therefore, may be more competitive. The tendering process will also take time and the project completion will thus be further delayed. In the meantime, the incomplete project and the materials on site may be suffering deterioration due to the weather, criminal damage and pilfering. In any event, a revised bill of quantities would need to be prepared in order to reflect the work done since the original tender competition resulting in additional professional fees.

Option 3

Re-tendering the contract suffers from the same disadvantages as Option 2 except that, in a more competitive market, the employer may be able to find other contractors more keen to win the work. A new tender list would suffer from the additional disadvantage of being completely unfamiliar with the project.

Option 4

Engaging a contractor on a cost plus basis would seem *prima facie* an expensive option. However, a contractor could be appointed quickly once a fee is agreed and work could re-commence without too much delay. This could benefit the employer commercially.

A target cost contract could be devised with a 'pain–gain' clause or a tight regime of cost control could be administered by the client's quantity surveyor, admittedly for a greater fee.

Option 5

It is likely that the receiver or administrator would only undertake completion of the contract if there was relatively little work to do and/or where a significant profit may be realised from finishing the job. This may, nevertheless, be an attractive option for creditors who could stand to regain more of their money. The insolvency practitioner would effectively be converting work in progress into a debt (due from the employer) by completing the contract. This may be preferable to creating another unsecured creditor (in the form of the employer) who would otherwise be claiming completion costs against the insolvent contractor's remaining assets.

The contract nevertheless remains in being even though the contractor's employment under the contract is terminated. However, the liquidator or administrator cannot re-write the contract and is bound by its terms. If the contract is a bad contract there may be no benefit accruing to this option.

Option 6

Novation is a legal arrangement where a third party steps into the shoes of one of the contracting parties with a view to completing the original contract. The idea is that the receiver or administrator finds another contractor acceptable to the employer to complete the works. The agreement of the parties to the existing contract and the new one is required.

In effect, the old contract terminates and a new one comes into being whereupon the insolvent company is released from its liability under the original contract. Subcontractors are not involved in the novation itself but it may be that the employer can arrange for subcontracts to be assigned to the new contractor provided that they have not already been terminated either automatically or by notice of the insolvent contractor.

Novations are common practise in the industry and it can be a quick and convenient means of continuing the contract without too much delay. The novated contract may be made on the same terms as the original contract or on varied or entirely new terms and much will depend upon the value remaining in the contract.

Novation works well for the employer and the receiver as the receiver sells the contract at the best price for creditors and the employer gets the job finished for a

Table 15.7 Contractor insolvency calculation.

Ref	Item	£	Notes
A	Cost to complete	590753	The total cost to the employer to complete the project following termination. Includes cost of engaging another contractor and any additional professional fees etc.
B	Payments made to the contractor	179449	The total amount paid to the original contractor prior to termination of his employment less retention monies held by the employer.
C	Total cost to employer to complete the work	770202	$C = A + B$
D	Total payable according to the contract	511672	The amount that the employer would have paid to the original contractor had it been able to finish the job. Takes into account any variations and extras carried out by the replacement contractor that would otherwise have been done by the original contractor.
E	Debt owed by the contractor to the employer	258 530	$E = C - D$ Represents the additional cost to the employer to complete the project as a result of the termination. This will normally be an unsecured debt and, therefore, the employer will be an unsecured creditor in the event of the contractor's insolvency. If (as is likely) the contractor is insolvent, the debt will be payable from the proceeds of the liquidation. If anything is left this is likely to be a few pence in the pound. The employer could carry insurance for this eventuality or there might be a performance bond which could be called in.

reasonable sum without the prospect of a challenge in court by the receiver for failure to mitigate its loss.

Table 15.7 illustrates a simple example of how the figures might work in the event of a contractor insolvency and shows the extent to which the employer might be 'at risk' in such circumstances.

References

Cooke, B. and Williams, P. (2009) *Construction Planning, Programming and Control*. John Wiley & Sons, Ltd., Chichester.

Cork, K (1982) Report of the Review Committee on Insolvency Law and Practice (Cmnd 8558) – Cork Report. The Stationery Office, London.

Davis, R. (2011) *Construction Insolvency: Security, Risk and Renewal in Construction Contracts*, 4th edn. Sweet & Maxwell Ltd, London.

Furst, S. and Ramsey, V. (2012) *Keating on Construction Contracts*, 9th edn. Sweet & Maxwell Ltd, London.

Keane, P.J. and Caletka, A.F. (2008) *Delay Analysis in Construction Contracts*. John Wiley & Sons, Ltd., Chichester.

Lowsley, S. and Linnett, C. (2006) *About Time-Delay Analysis in Construction*. RICS Books, London.

SCL (2002) The Society of Construction Law Delay and Disruption Protocol. The Society of Construction Law, Hinckley, UK.

16

Programme and progress

16.1 Contractor's obligations

Under the standard forms of contract employed in the industry, the contractor's obligation is to carry out and complete the works in the time allowed in the contract. This period is determined by the contract particulars, which may state a contract period or dates for possession and completion. This information may be found in the contract documentation:

Financial Management in Construction Contracting, First Edition. Andrew Ross and Peter Williams.
© 2013 Andrew Ross and Peter Williams. Published 2013 by John Wiley & Sons, Ltd.

- JCT SBC/Q 2011 – **Contract Particulars**
  - Date of possession of the site
  - Date for completion of the works.
- ICC – Measurement Version – **Appendix Part 1**
  - Works commencement date
  - Time for completion.
- NEC Engineering and Construction Contract – **Contract data**
  - Starting date
  - Possession date(s).

There may be sectional completion requirements written into the contract as well should the employer require specific parts of the project to be completed and handed over for occupation ahead of the overall completion date.

16.2   Programme

The financial consequence of time in a construction project is a major risk issue for the contractor. Despite the contractor's best efforts, and the sophisticated planning software available, there is no guarantee that the project will be completed in the time expected or in the time that the client wants. There are a great number of variables that can contribute to this situation some of which are:

- the client may have stipulated a completion date that is unrealistic;
- the contractor may have been over-optimistic in its assessment of the construction period;
- the weather;
- subsoil conditions and water table;
- ground contamination and the presence of hazardous materials;
- discovery of underground obstructions, underground services, underground water courses, historic artefacts – even unexploded bombs and dead bodies have been discovered on construction sites!
- late delivery of materials;
- delay on the part of subcontractors;
- insolvency of subcontractors;
- design changes and delay in receiving instructions from the architect/engineer.

It cannot be over emphasised that planning a construction project is NOT a precise science and that the contractor's programme – whether at the tender stage or when the contract has been awarded – can only represent the contractor's reasonable view of the project (Lowsley and Linnett, 2006). This is not to diminish the need for a programme or the importance of the programme in the management of the project but is merely a reflection on the quality of data used in the preparation of the programme. The programme may have been established on the basis of:

- the arbitrary construction period required by the employer;
- the contractor's empirical data (i.e. data of the time taken to complete previous similar projects in the past);
- the contractor's calculations for activity durations based on the quantity of work to be done and the expected output of the workforce (empirical or intuitive data);

- periods stated or indicated by subcontractors;
- intuitive data based on experience and a 'feel for the job'.

Notwithstanding this, best practice suggests that the contractor's planning of a project is closely linked to his financial planning and control procedures, and that failure to plan the time aspects of a project will inevitably lead to financial loss. Planning is necessary for effective control and the planning process should start at the tender stage. The plan may then be developed and refined through all stages of the project.

16.2.1 The tender stage

Not all contractors work in the same way and some pay little heed to the time issue at the tender stage. Some contractors, on the other hand, spend a good deal of time and effort in producing a pre-tender programme and use it for pricing the job and in their risk management of the bid (tender). Subcontractors, more often than not, price their tenders without any thought as to when the contract is likely to start, how long it will take or what resources are likely to be required to do the job if they are successful with their tender.

At the tender stage, the **pre-tender programme** is prepared as the basis of the contractor's bid. This programme is influential for a number of reasons:

- If the client has stipulated the contract period (or the site possession and works completion dates) the programme will indicate whether or not this is realistic (optimistic/pessimistic).
- If optimistic, the contractor can assess what is a reasonable construction period, can determine the likely overrun and then price any liquidated and ascertained damages (LADs) into its tender, for example:
  - Overrun 3 weeks
  - LADs=£2000 per day or part thereof
  - Tender risk allowance=3 weeks×7 days/week×£2000/day=**£42 000**
- If pessimistic, the contractor may be able to make a saving and thereby reduce its tender price, for example:
  - Time saving=4 weeks
  - Time-related costs=£15 000 per week
  - Saving=4 weeks×£15 000/week=**£60 000**
- The contractor's estimator will be better informed when pricing the resources needed to undertake the work and in the choice of the most appropriate plant and working methods.
- The contractor's time-related costs may be more accurately estimated:
  - Site supervision and transport
  - Site accommodation and security
  - Major items of plant such as mobile cranes, tower cranes, material handling equipment, access equipment
  - Hire period for scaffolding and other temporary works.
- Lead times for key materials and subcontractors may be assessed.
- The contractor's method of construction, safety and temporary works requirements may be indicated and realistically priced.

Some clients may ask for a programme to be submitted with the contractor's tender and this may be used as part of the tender assessment process. An example of a

PRE–TENDER PROGRAMME

Garmac Industries Project

ID	Name	Start	Duration	Finish	Feb	Mar	Apr	May	Jun	Jul	Aug	Sep
1	Site establishment	18/02/11	3w	10/03/11								
2	Groundworks	04/03/11	6w	14/04/11								
3	Drainage	04/03/11	3w	24/03/11								
4	Steel frame	15/04/11	3w	05/05/11								
5	Ground slab	06/05/11	4w	02/06/11								
6	Cladding	20/05/11	5w	23/06/11								
7	Brickwork	27/05/11	3w	16/06/11								
8	Internal fit out	24/06/11	6w	04/08/11								
9	Roller shutters	06/05/11	2w	19/05/11								
10	HVAC	24/06/11	5w	28/07/11								
11	Plumbing and electrics	24/06/11	4w	21/07/11								
12	Internal finishes	22/07/11	4w	18/08/11								
13	External works	05/08/11	5w	08/09/11								
14	Clear site	09/09/11	2w	22/09/11								
15	**PRELIMS ITEMS**											
16	Mobile crane	15/04/11	10w	23/06/11								
17	Telehandler	24/06/11	10w	01/09/11								
18	Cherry picker	15/04/11	3w	05/05/11								
19	Scaffolding	20/05/11	5w	23/06/11								
20	Safety nets	06/05/11	5w	09/06/11								

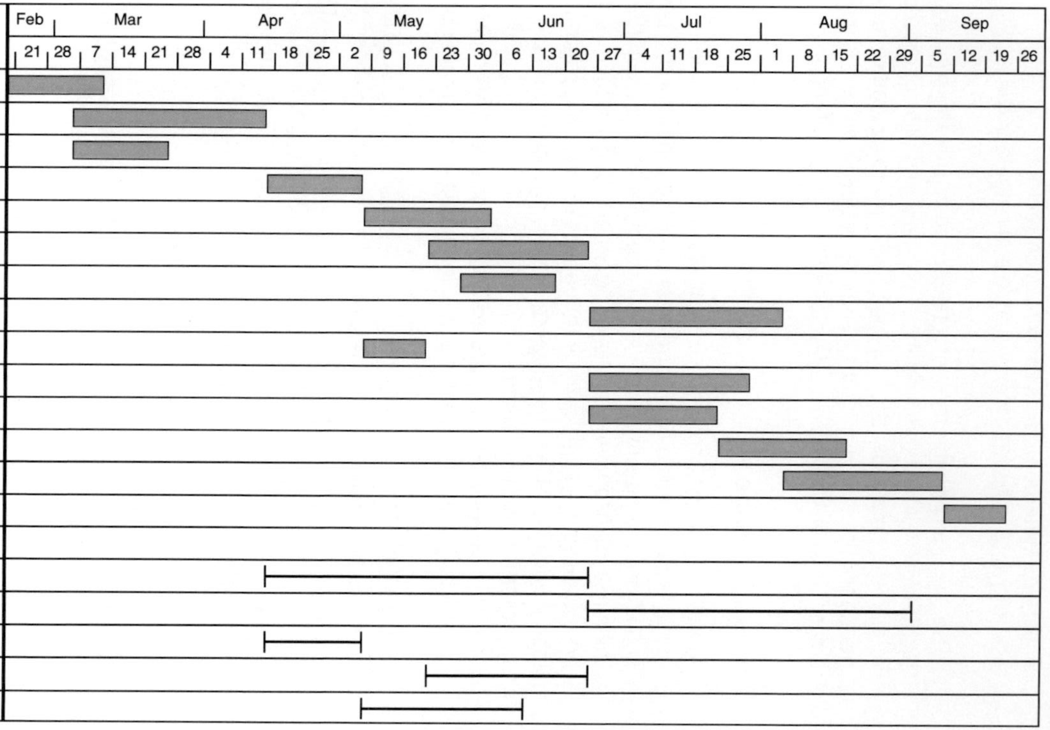

Figure 16.1 Pre-tender programme.

pre-tender programme is shown in Figure 16.1. It indicates the time period for key preliminaries items, which can then be priced into the preliminaries bill by the estimator.

16.2.2 The pre-contract stage

If the contractor has been successful in winning the contract, the pre-tender programme will be developed into the contract **master programme** at the pre-contract commencement stage. It may be that the master programme will differ from the pre-tender version because:

- site management staff have become involved in the planning process and may have completely different ideas to the estimator;
- the contractor may have had a 're-think' over proposed methods of construction and temporary works:
  - quicker or more economical methods may have been suggested
  - sequence studies may have been carried out on alternative methods along with cost to complete exercises for each method
  - subcontractors may have suggested different ways of working;
- subcontractors' availability may be a problem;
- the site possession date may have changed;
- potential problems may have been overlooked at tender stage.

It must be emphasised that the contractor's tender, if accepted, represents the contractor's budget for carrying out the contract works and any decision to change the method, order or sequence of the construction work must be made within this budget figure. Contractors, however, are always looking for ways to increase profitability and sometimes a trade-off can be made between cost and time which will increase profits overall:

- A more expensive method of construction may result in a shorter construction time, thereby reducing the overall construction period and the cost of site preliminaries, for example:
  - a single concrete pour for a suspended slab or bridge deck using a mobile truck mounted pump rather than several pours using skip and crane (Figure 16.2).
- A subcontractor may suggest a more expensive but quicker way of working, thereby saving time on the programme, for example:
  - trench fill foundations vs. traditional brickwork
  - dry lining vs. two coat plaster
  - spray painting vs. rollers.
- Greater concurrency in the programme (carrying out more items of work at the same time):
  - more resources on site and increased site supervision will cost more
  - the overall programme period will be shorter
  - savings in time-related costs could outweigh increased labour and supervision costs.

The master programme must be realistic bearing in mind the work to be carried out and it is important that sufficient time is allowed for work to be planned and carried out safely. Every contractor carrying out construction work to which the CDM Regulations (HSE, 2007) apply (i.e. most projects) has a statutory duty to *plan, manage and monitor construction work carried out by him ... so far as is reasonably practicable*. Whilst not necessitating a programme, it is difficult to imagine how the statutory duty can be complied with without one.

Prior to being given legal possession of the site, the contractor's programme will probably be presented at a **pre-contract meeting** attended by contractor's staff and the architect or engineer together with other representatives of the employer (e.g. quantity surveyor). Key issues to be discussed at this meeting will be:

- approval of subcontractors;
- communications including confirmation of verbal instructions, issue of variation instructions;
- dates of site meetings;
- dates of interim valuations.

16.2.3 The contract or master programme

The provision of a master programme is a specified condition under the ICC and NEC3 ECC contracts but the JCT the contract does not ask the contractor to provide a programme unless it has prepared one, in which case two copies are required. The preparation of a programme is left to the contractor's discretion and none of the standard contracts requires a specific programming format or the use of any particular software.

Under the ICC conditions, the engineer plays a significant role in the final production of the master programme and he may accept or reject the programme or ask for further information if he is not satisfied that it adequately describes the order of work or the contractor's intended arrangements and methods of construction. This programme is commonly referred to as the Clause 14 Programme. The ICC conditions also require the contractor to submit a written description of its proposed methods of construction.

The NEC3 ECC is very prescriptive as to what information the master programme shall contain. This includes key dates, method statements showing equipment and resources for each activity, provisions for float and time risk allowances, health and safety requirements and so on.

Using modern software programs, the master programme may easily be structured in such a way as to reflect the contractor's plant, temporary works and method of working as well as the order and sequence of the works. Figure 16.3 illustrates an example of this.

In the same way that the programme could be called a model of the timing and sequence of the works, so the bills of quantities is a model of the proposed building or structure in financial terms. Therefore, from a financial control point of view, it can be helpful to structure the master programme in such a way as to reflect the bills of quantities such that the programme can be used to control cost and value as well as time.

Problematically a programme contains operations or activities expressed against time whereas a bill of quantities is a list of quantities and prices expressed in accordance with the work classification system employed in one of the standard methods of measurement used in the industry (e.g. SMM7, CESMM4, Highways Method of Measurement). Table 16.1 shows some typical examples.

Table 16.1 illustrates the different phraseology used in measurement as opposed to the planning process. It should also be noted that the master programme necessarily contains far fewer items than the bills of quantities and, therefore, an activity on the programme is much more likely to be a 'composite' of a number of items that are separately measured and described in the bills of quantities. Table 16.2 illustrates this point using 'drainage' as an example.

Item	Crane and skip					Pumping				
	Quantity	Unit	Rate (£)	Totals £	£	Quantity	Unit	Rate (£)	Totals £	£
Materials										
Ready mixed concrete 25 N/mm²	200	m³	75.00	15 000.00		200	m³	75.00	15 000.00	
Premium for pump mix				—	15 000.00	200	m³	10.00	2000.00	17 000.00
Plant										
Mobile crane and skip	1	Day	300.00	300.00						—
Mobile concrete pump				—		1	Day	750.00	750.00	
Concrete vibrator	1	Day	30.00	30.00		2	Day	30.00	60.00	
Power beam tamp	1	Day	75.00	75.00		2	Day	75.00	150.00	
Duration	2	Day		405.00	810.00	1	Day		960.00	960.00
Labour										
Foreman	1	Day	120.00	120.00		1	Day	120.00	120.00	
Banksman	1	Day	96.00	96.00		1	Day	96.00	96.00	
Labourers	6	Man Day	90.00	540.00		8	Man Day	90.00	720.00	
Duration	2	Day		756.00	1512.00	1	Day		936.00	936.00
Total					17 322.00					18 896.00
Less										
Saving in Preliminaries at £12500/week					0	1 Day = 0.2 week	Week	12 500		2500.00
Total cost					17 322.00					16 396.00

Figure 16.2 Sequence study cost comparison (simplified).

Consequently, for effective financial control as well as for controlling the work on site, a **work breakdown structure (WBS)** is needed that reflects the practical work activities and subcontract packages and provides a vehicle for convenient analysis of the bills of quantities into suitable budgets for control purposes. Keane and Caletka (2008) suggest that 'the WBS defines every element of the completed project' and also that it 'cross refers these elements to their respective task'. The hierarchy (or levels) of the WBS facilitate the summary reporting of cost and the progress status of the programme and could be represented on the master programme as a numbered system referenced to a breakdown of the bills of quantities items.

Table 16.1 Programme versus measurement itemisation.

Ref	Typical programme activities	SMM7 Work Sections		CESMM4 work classification	
1	Bulk excavation	D20	Excavation and filling	E	Earthworks
2	Groundworks	D20	Excavation and	E	Eartworks
		E10	filling	F	In situ concrete
		F10	In situ concrete	U	Brickwork, blockwork
			Brick/block walling		and masonry
3	Piling	D30	Cast in place concrete piling	P	Piles
		D31	Preformed concrete piling	Q	Piling ancillaries
4	Steel frame	G10	Structural steel framing	M	Structural metalwork
5	Drainage	R12	Drainage below ground	I	Pipework – pipes
				J	Pipework – fittings and valves
6	Cladding	H20	Rigid sheet cladding	N	Miscellaneous metalwork
7	Painting and decorating	M52	Decorative papers/fabrics	V	Painting
		M51	Painting/clear finishing		

Table 16.2 Programme versus measurement phraseology.

Programme activity	SMM7	Work Sections	CESMM4 Work Classification	
Drainage	**R12**	**Drainage below ground:**	I	Pipework – pipes
Notes:	1	Excavating trenches	J	Pipework – fittings and valves
1. Drainage is a composite item including:	3	Disposal	K	Pipework – manholes and pipework ancillaries
• Excavation and pipework	4	Beds	L	Pipework – supports and protection, ancillaries to laying and excavation
• Disposal and backfilling	5	Beds and haunches	Notes:	
• Pipe bedding and surrounds	8	Pipes	1. Class I includes:	
• Pipe fittings (e.g. bends, junctions	10	Pipe accessories	• Excavation and backfilling	
• Manholes and gullies	11	Manholes	• Disposal	
• Testing			• Earthwork support	
			• Pipework, beds and surrounds	
	16	Connection to LA sewer	2. Manholes are not measured in detail	
	17	Testing and commissioning		
	Notes:			
2. Connection to the main sewer would be shown as a separate activity on the programme	1.	Manholes are measured in detail		

16.2.4 Contractor's method

In the ICC forms of contract, the contractor is required to submit a general description of his proposed arrangements and methods for the project and, if requested by the engineer, detailed information on construction methods, temporary works and contractor's equipment. Under the NEC3 ECC, the contractor's programme is required to include a method statement for each operation together with the equipment and resources he intends to use. JCT contracts are silent on these issues.

On occasion, the contractor may be asked to submit a programme and method statement with its tender, perhaps to show that it has recognised particular constructional problems or health and safety risks or phasing stipulations in the tender documents and allowed for them in its bid. The employer should be careful not to accidentally include the method statement as a contract document as any change to the method instructed by the architect/engineer, or forced by circumstances beyond the contractor's control, could lead to variations and claims. The case of *Yorkshire Water Authority v Sir Alfred McAlpine & Son (Northern) Ltd (1985)* refers.

A method statement is an expression of the intentions of the contractor for carrying out the works and will usually contain reference to all the main operations or activities in the project together with a description of the intended method of carrying each of them out. In addition to this, information can be included detailing the gang sizes and plant resources intended for each activity together with the quantity of work involved and the expected output of the gang. References may also be made to subcontracted work and to temporary works requirements.

Method statements may be presented in written prose or in a tabular format or may be included on the contractor's master programme. Cooke and Williams (2009) illustrate the various ways that method statements may be presented and provide a number of worked examples.

16.2.5 Shortened programmes

Standard construction contracts provide that the contractor shall complete the works within the prescribed contract completion time, which can be extended in certain circumstances in order to preserve the employer's right to deduct liquidated damages. However, some contractors submit programmes showing an earlier completion than that stipulated in the contract in order to capitalise on any opportunity to formulate a contractual claim for loss and expense. Contractors may also have an 'internal' programme showing a shorter construction period to that agreed with the employer. This is known as the 'target programme'.

Target programmes represent the construction period that the contractor would ideally like to achieve in order to save money on time-related costs. Target programmes put pressure on site staff and the greater intensity of work on site can result in additional expenditure on resources and supervision costs that need to be balanced against expected savings. Compressing the programme in this way may result in a lot more concurrent working (i.e. carrying out several activities at the same time) but the benefits will include significant savings in the contractor's time-related costs (preliminaries). On some projects, major roadworks for example, a shorter or compressed programme can help the contractor to avoid seasonal bad weather periods, which can be crucial for activities such as bulk excavation and filling. Shorter programmes can also be beneficial in reducing direct costs, reducing finance charges and increasing profitability.

Some employers do not accept shortened programmes because they are seen as predatory – in other words the contractor may be perceived to be pressurising the employer to release information earlier or quicker with any consequent delay resulting a contractual claim. Keane and Caletka (2008) argue, however, that whilst the employer must not hinder the contractor in his endeavours, there is no duty on the employer to produce information early. This view is based on the leading UK case of *Glenlion Construction v The Guinness Trust (1987)*, a JCT building contract, where it was held that the contractor will only have a valid claim if the contract is delayed beyond the contractual completion date and that, further, the architect is only obliged to furnish drawings and instructions *'within a reasonable time of the conclusion of the contract'*. Thus, it is only where the contract completion date is endangered in this way that the contractor will be entitled to a delay claim.

Keating (Furst and Ramsey, 2012) argues that, whilst the contractor is entitled to finish early, the employer is not under an implied obligation *'to enable the contractor to complete by the earlier date'*. Therefore, *'provided that the contractor can still complete within the contract period, he cannot recover prolongation expenses'* and *'the employer ...... is under no ...... obligation to ...... pay compensation if the contractor is unable to achieve an accelerated programme'*.

16.2.6   The contract stage

During the contract stage, Cooke and Williams (2009) report that some contractors prepare **short-term programmes** in order to plan the work ahead and keep progress on the master programme under close control. These programmes may be monthly or even weekly depending upon the size of the project and the complexity of the work. Short-term programmes enable problems and delays to be flagged up sooner rather than later, and thereby facilitate remedial action to prevent a delay to the master programme.

A further benefit of short-term programming is to contemplate alternative construction methods by carrying out further sequence studies in the light of prevailing circumstances on site. It frequently happens that better, easier, quicker or more economical working methods present themselves during work sequences on site. Ideas may be forthcoming from site management, subcontractors or from operatives during tool-box or task talks on site. Changes to planned methods need to be carefully considered, as different health and safety hazards may arise or there may be cost implications to weigh up. Time should therefore be allowed in order to re-assess risks and evaluate the associated cost implications of the proposed change.

Lowsley and Linnett (2006) suggest that it is not practical to plan too far ahead with any degree of detail and that the detailed planning of an activity may not take place until such time that the activity in question becomes more of a priority. They also suggest that short-term tactical programmes can be produced to demonstrate the timing and detailed sequencing of the activity in question.

16.3   Progress

16.3.1   Contractor's obligations

When the contractor is given possession of the site as prescribed by the contract he is under a duty to:

- commence and regularly and diligently proceed with the works;
- complete on or before the completion date.

The master programme is not a contract document and failure to observe the programme is not a breach of contract but failure to progress with the works in accordance with the intended programme may provide some evidence of the contractor's failure to proceed regularly and diligently. Persistent failure could lead to the drastic step of termination of the contract by the employer.

16.3.2 The baseline programme

The contractor's master programme is a useful tool for monitoring progress and, whilst it has no contractual status, the programme can be used for checking whether the contractor is likely to complete the contract on time. A simple Gantt (or bar) chart and colouring system is the traditional way of monitoring progress. Each week or month is given a colour and the appropriate percentage completion achieved for that period is coloured in on each bar (activity) of the programme. This will readily show which activities are late starting, which are on or behind programme and which are ahead of schedule.

More modern procedures are now generally employed by many contractors and this involves the use of project management software. These range from the relatively cheap and simple to use to the expensive and sophisticated programs used on large projects. Some packages use the **precedence** (activity on the node) format, some use the **arrow** (activity on the arrow) format and some employ the **linked bar chart** format. The linked bar chart format is the most popular because it resembles the traditional bar chart but also has the benefit of arrows which link each activity logically. This means that a critical path analysis may be carried out whilst retaining the visual advantages of a conventional bar chart layout.

Several of the project management software packages available include a **tracking** facility, which can be employed to compare planned progress with that actually achieved:

- the master programme is firstly 'frozen' at the beginning of the project and this creates an **as-planned** or **baseline** bar chart with a progress bar underneath the 'frozen' activities;
- when the programme is updated, the second bar moves according to the progress made;
- the position and length of the second bar relative to the baseline will depend upon when the activity was actually started, how much progress has been made at the review period and when the activity was completed.

Figure 16.4 illustrates this principle and shows:

- a baseline programme at the start of a project;
- the baseline programme updated for progress at week 14;
- a vertical progress line at week 14 which maps unfinished activities in order to indicate whether they are behind, ahead of or on programme.

16.3.3 Extensions of time

Under the standard forms of contract, the architect/engineer/project manager has extensive duties with regard to the control of time. If and when the contractor asks for more time for completion, it is the contract administrator's duty to:

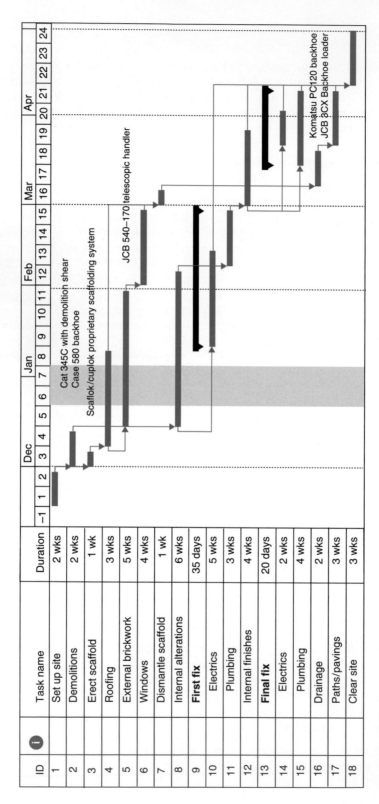

Figure 16.3   Master programme.

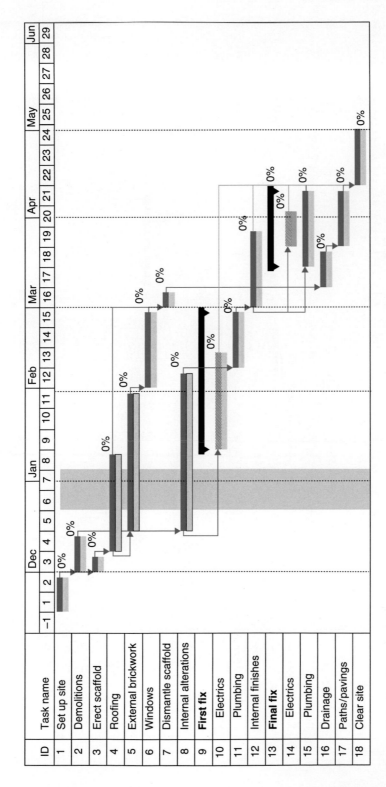

Figure 16.4 Baseline programme.

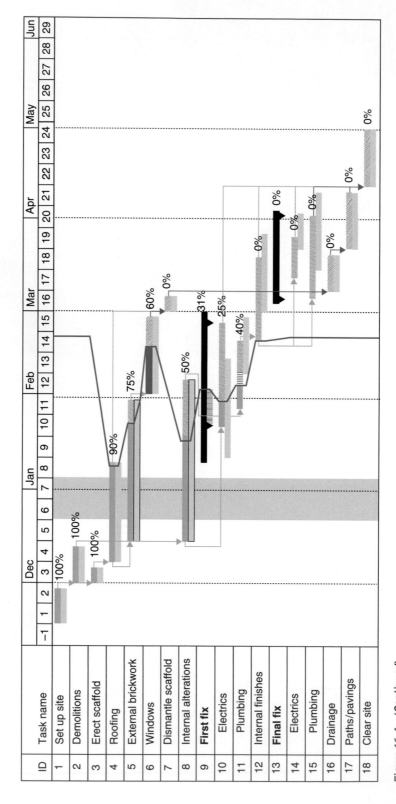

Figure 16.4 (*Continued*)

- assess the application and make a decision in accordance with the provisions of the contract;
- where the contractor is clearly getting behind with the work, to put the contractor 'on notice' and, if necessary, take firm action to ensure that the works are completed in accordance with the contract.

The master programme is an essential tool in this process and the contractor is required to furnish revised programmes from time to time when, for instance, progress is not as intended, or when there has been a change of method or if an extension of time has been granted.

From a financial perspective, extensions of time are of limited interest to a contractor except in circumstances where there is both culpable and non-culpable delay and liquidated and ascertained damages may be avoided with an additional time allowance. Contractors are very much more interested in extensions of time **with costs** which remunerate the contractor for time-related costs as well as the costs of dealing with the cause of delay and – hopefully – any resulting disruption to other site activities. The Society of Construction Law (SCL) provides detailed guidance – the Delay and Disruption Protocol (SCL, 2002) – for dealing with delay and disruption issues. This is a helpful non-mandatory guide to good practice in the preparation and submission of time and money claims.

16.3.4 Mitigation of loss

Where there is a delay to the progress of the works, the contractor is obliged to make its best endeavours to prevent delay and to mitigate its loss. In so doing it may be the case that the works completion date is not delayed but the contractor has, nevertheless, suffered disruption to his activities as a consequence. This might involve duplication of effort, out of sequence working or constructive acceleration of the works for which the contractor may seek compensation or loss and expense.

Keating (Furst and Ramsey, 2012) argues that the obligation to mitigate loss is an implied term of the contract and, therefore, any loss and/or expense claim for delay/disruption should take this into account. This is where the arguments start!

The contractor's quantity surveyor will need to be alert to this as far as the financial reporting of the contract is concerned and any potential losses and/or reduction in the value of work in progress and/or contractual entitlements will need to be recognised in the cost value reconciliation.

16.3.5 Measuring progress

Progress can be measured both physically and financially by comparing what was planned with what has actually happened. This can be done in several ways, including:

- comparing physical progress with the master programme;
- comparing planned and actual progress on critical activities;
- comparing forecast value (income) with actual interim payments;
- comparing actual spending against target expenditure.

A degree of financial control may also be achieved by comparing forecast value (income) with actual interim payments made to the contractor whilst also taking into account payment delay and retention. This topic is discussed below and earned value analysis was dealt with in detail in Chapter 13.

16.4 S-curves

16.4.1 Principles

Lowsely and Linnett (2006) remind us that the construction industry has a long history of using so-called 'S-curves' for forecasting and monitoring progress dating back to before the advent of the personal computer. They explain that the former Property Services Agency (PSA) developed a formalised approach to the use of S-curves which was called 'Planned Progress Monitoring'. This involved plotting the cumulative number of activity weeks resulting from the contractor's programme and comparing this with the actual performance to provide a useful if simple control mechanism on projects (it was called 'counting the squares').

Basically, S-curves are cumulative graphs which can be used for a variety of planned progress monitoring purposes, including:

- forecasting production output and comparing actual output with that planned;
- forecasting income from contracts and identifying any variance with the amounts received;
- forecasting costs incurred on contracts and identifying any variance with actual expenditure.

To create an S-curve depicting revenue from a project relative to the programme, money is plotted on the vertical axis and time along the horizontal axis. Similarly, units of production may be plotted on the vertical axis against time on the horizontal axis to show the actual output of a site activity against that expected (bricks laid or concrete poured for example).

The numbers (money or units of production) are recorded on a cumulative basis and an S-curve is easily achieved using a spreadsheet software package. Cooke and Williams (2008) suggest the following procedure to forecast 'value' relative to the contractor's programme:

- price each activity on the programme by allocating monies from the bills of quantities;
- divide the resulting sums by the duration of each activity (= £/week);
- allocate money (£/week) to each bar on the programme and total at the bottom;
- add together the totals cumulatively to enable the S-curve to be drawn.

Some standard contracts, including GC/Wks/1 (the Government contract) and the JCT Major Project Construction Contract 2011, include a provision for paying the contractor on the basis of S-curves, whereby the value to time graph indicates the cumulative valuation achieved at each payment period.

16.4.2 Measuring physical progress

Figure 16.5 illustrates the number of cubic metres of concrete planned against the volume of concrete actually poured over the 10-week cycle of a major concrete pour. It can be readily seen that actual concrete production is less than planned and

Concrete pours

Week No	Volume planned	Cumulative volume planned	Volume actual	Cumulative volume actual
1	100	100	80	80
2	120	220	90	170
3	150	370	120	290
4	150	520	150	440
5	180	700	190	630
6	220	920	210	840
7	250	1170		
8	250	1420		
9	200	1620		
10	150	1770		

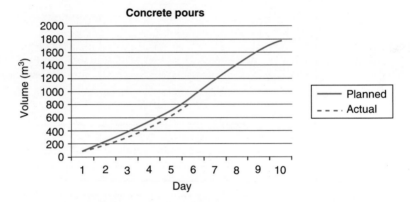

Figure 16.5 Measuring progress – units of production.

management needs to take action to increase output in order to avoid the concreting activity falling behind programme and losing money. Management action is particularly important if the concreting activity is on the critical path.

An alternative use of S-curves is to plot activity weeks against time. A programme activity such as 'plumbing and electrical' with a planned duration of three weeks is allocated three activity weeks. All activity weeks appearing in individual weeks of the programme are added together and cumulative totals are derived. The cumulative totals are plotted on a line graph. The revised programme showing actual progress at the review period is similarly allocated with activity weeks and the actual cumulative totals plotted on the line graph.

This process is illustrated in Figure 16.6, which depicts a simple refurbishment project, planned over a 12-week construction period, with the activities shown on a bar chart programme. Progress is reviewed at week 7, where it can be seen that progress is ahead of schedule after a slow start.

16.4.3 Measuring financial progress

Forecasts of the contractor's monthly income and expenditure can be readily produced using S-curves, which can be based on formulae, the ¼:⅓ rule (both

Refurbishment programme

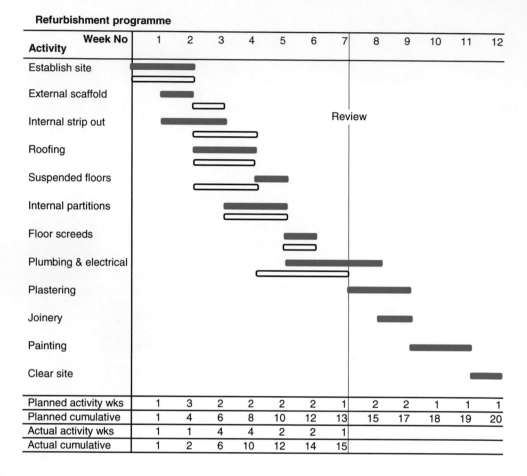

Week No	1	2	3	4	5	6	7	8	9	10	11	12
Activity												
Planned activity wks	1	3	2	2	2	2	1	2	2	1	1	1
Planned cumulative	1	4	6	8	10	12	13	15	17	18	19	20
Actual activity wks	1	1	4	4	2	2	1					
Actual cumulative	1	2	6	10	12	14	15					

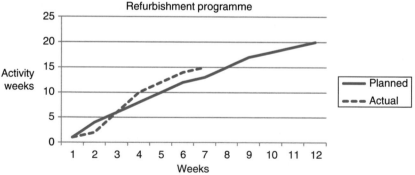

Figure 16.6 Measuring progress – activity weeks.

explained in Chapter 11) or empirical data and also by allocating money to the contractor's programme.

Figure 16.7 shows the contractor's Clause 14 Programme for a 10-month sewer project. The normal durations have been supplemented by the tender allowances for each activity. Using this information an approximation of the expected revenue from

SEWER project

Activity	Month Duration (weeks)	£	1	2	3	4	5	6	7	8	9	10
Site establishment	4	100 000	10	10	10	10	10	10	10	10	10	10
Access road	8	10 000		5	5							
Divert services	16	16 000		4	4	4	4					
Outfall headwall	8	6000			3	3						
Sewer lines 1–4	16	100 000				25	25	25	25			
Manholes	16	28 000					7	7	7	7		
Road crossing	8	10 000					5	5				
Sewer lines 4–5	8	50 000						25	25			
Pumping station	28	70 000				10	10	10	10	10	10	10
Test & commission	8	6000								3	3	
Clear site	8	10 000									5	5
Totals		**406 000**	10	19	32	59	61	82	77	23	28	15
Cumulative			10	29	61	120	181	263	340	363	391	406
Actual			12	35	48	130	198					

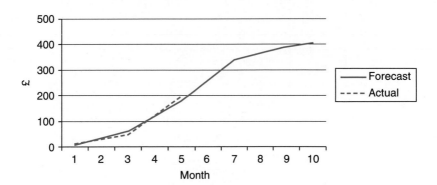

Figure 16.7 Measuring progress – earned value.

the project can be calculated and shown as an S-curve for easier reading. In Figure 16.7 the forecast cumulative income is indicated together with the actual valuations (no allowance has been made for payment delay). This indicates graphically whether the contractor is on or ahead of programme or whether remedial action is needed. It can be seen that, initially, the contractor was doing well but then fell behind programme in month 3. Progress then picked up in months 4 and 5, by which time the contractor was ahead of programme.

The graph could equally be used to compare actual expenditure with the forecast budget but Winch (2010) considers this to be of limited use for the proactive

SEWER project

Month / Activity	Duration (weeks)	£	1	2	3	4	5	6	7	8	9	10
Site establishment	4	100 000	10	10	10	10	10	10	10	10	10	10
Access road	8	10 000	5	5								
					5	5						
Divert services	16	16 000	4	4	4	4						
				4	4	4	4					
Outfall headwall	8	6000		3	3							
						3	3					
Sewer lines 1–4	16	100 000			25	25	25	25				
							25	25	25	25		
Manholes	16	28 000			7	7	7	7				
								7	7	7	7	
Road crossing	8	10 000				5	5					
								5	5			
Sewer line 4–5	8	50 000					25	25				
								25	25			
Pumping station	28	70 000		10	10	10	10	10	10	10		
					10	10	10	10	10	10	10	
Test & commission	8	6000								3	3	
											3	3
Clear site	8	10 000									3	3
											5	5
											5	5
Totals ES		**406 000**	19	32	59	61	82	77	23	23	15	15
Totals LS			10	14	29	32	52	82	82	52	35	18
Cumulative ES			19	51	110	171	253	330	353	376	391	406
Cumulative LS			10	24	53	85	137	219	301	353	388	406
Actual			12	35	48	130	198					

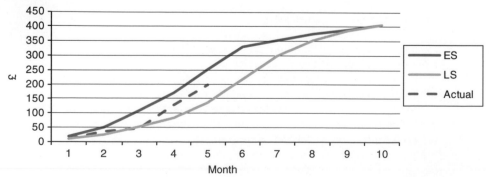

Figure 16.8   Measuring progress – value envelope.

management of projects. Some would argue that any effort, however simple, to exercise control of a project is better than nothing but Winch (2010) prefers the more sophisticated **earned value analysis** approach. This topic is covered in detail in Chapter 13.

Further sophistication could be added to the display by recognising that payments are actually delayed in practice under the normal contract conditions and that retention monies are kept back by the employer to allow for potential defects in the contractors work. The payment delay may be up to one month following the valuation date and the retention may be 3% or 5% depending on the specific details of the contract.

Cooke and Williams (2009) suggest the use of a 'value envelope' as a means of recording progress by earned value. This is shown in Figure 16.8. The 'envelope' consists of two S-curves – produced exactly as illustrated in Figure 16.7 – but based on a critical path analysis of the programme. One S-curve is produced with the programme activities commencing at their earliest start (ES) times and the other commencing at their latest start (LS) times (e.g. Access road, planned for week 2, could start in week 1 or in week 3). This is done by using the 'float' or free time of non-critical activities on the programme. Actual progress is plotted in the usual way and the boundaries of the 'envelope' delineate progress within programme. If the boundary of the envelope is breached above, the line then the contractor is ahead of programme; if breached below the line, the contractor is behind schedule. Figure 16.8 shows actual value plotted against the ES/LS envelope where it can be seen that month 3 is behind programme but the delay has been recovered by month 5.

The problem with the control envelope idea is that two programmes have to be produced in order to plot the curves. As far as financial S-curves generally are concerned, comparing actual value with planned can be misleading because the influence of variations, extra work and unforeseen events on site can impact actual value to the extent that comparisons become distorted. It is important to compare 'eggs with eggs'! If the contractor's programme is regularly updated then this becomes less of a problem but it must be borne in mind that S-curves are more indicative than accurate but, nevertheless, provide a useful, if under-used, control technique.

16.5 Project acceleration

Projects may be delayed for many reasons but it may be in the interests of the parties to complete the works by the original completion date or as close to it as possible. The contractor may wish to bring forward the completion date for its own reasons, such as to avoid paying liquidated and ascertained damages or to release resources for another project. Alternatively, the employer may ask the contractor to complete earlier to ensure the original handover and occupation dates are achieved where the employer has been responsible for the delay (e.g. delayed possession, late supply of information, too many design changes etc.).

Some standard conditions of contract (ICC for example) provide a clause for accelerated completion in order to contractually require the contractor to speed up its operations if behind programme or, where there has been an employer fault delay, to enable the employer to complete the project by the intended date. Under the NEC3 ECC, acceleration is intended as a means of completing the works *prior* to the contract completion date and the contractor may be asked to submit a quotation for this.

Where the contractor is at fault it will be for the contractor to bear the cost of acceleration but, if not, the contractor is entitled to be paid for his efforts and the precise details and cost implications should be agreed with the employer beforehand. Where there is no provision in the contract, an acceleration agreement should be drafted separately to the main contract agreement.

Cooke and Williams (2009) explain that 'acceleration' is a widely used word with several legal meanings:

- **Pure** acceleration – where the contractor speeds up work on site so as to finish earlier than scheduled at the request of the employer. This is usually paid for under a separate agreement.
- **Constructive** acceleration – where the contractor is effectively 'forced' to work at a faster rate because the contract administrator has delayed or refused consideration of a legitimate application for an extension of time.
- **Expedite** – when the contractor is in culpable delay and is asked to increase the pace of work on site so as to get back on programme.

They further explain that acceleration, or speeding up the work, can be achieved by reorganising the work more efficiently or by increasing the resources on site or both. Keane and Caletka (2008) advise that acceleration is not synonymous with efficiency and that an accelerated unit of production can be less efficient and more expensive than the equivalent work carried out normally. Cooke and Williams (2009) suggest that acceleration may be demonstrated by using the management technique of **time-cost optimisation**, whereby the direct cost of **doing the work** (labour, plant, equipment) is balanced with the indirect costs of **managing the process** (time-related preliminaries). The resultant analysis provides an **optimum time and cost** solution.

Time–cost optimisation is a theoretical technique with a practical application but it does not take into account all the issues and practicalities of achieving acceleration on site. For instance, increasing the concurrency of site operations may create inefficiencies or accidents or there may simply be too much congestion on site for safe and efficient working. Quality of work may be adversely affected or there may be interface problems with subcontractors working too close together. The practicalities of speeding up the work therefore need to be balanced with the financial benefits so as to achieve the break-even point or **optimum time and cost solution**.

Figure 16.9 illustrates how this might be demonstrated in a simple time–cost optimisation graph, which identifies the ideal balance between direct costs and indirect costs with respect to time.

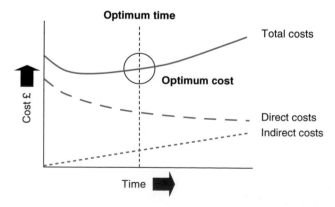

Figure 16.9   Time–cost optimisation.

References

Cooke, B. and Williams, P. (2009) *Construction Planning, Programming and Control*. John Wiley & Sons, Ltd., Chichester.

Furst, S. and Ramsey, V. (2012) *Keating on Construction Contracts*, 9th edn. Sweet & Maxwell Ltd, London.

HSE (2007) Construction (Design and Management) Regulations 2007. Regulation **13**(2), UK Health and Safety Executive (HSE), London.

Keane, P.K. and Caletka, A.F. (2008) *Delay Analysis in Construction Contracts*. John Wiley & Sons, Ltd., Chichester.

Lowsley, S. and Linnett, C. (2006) *About Time-Delay Analysis in Construction*. RICS Books, London.

SCL (2002) Delay and Disruption Protocol. The Society of Construction Law (SCL), Hinckley, UK (www.scl.org.uk).

Winch, G.M. (2010) *Managing Construction Projects*, 2nd edn. John Wiley & Sons, Ltd, Chichester.

17

Valuations and payments

Financial Management in Construction Contracting, First Edition. Andrew Ross
and Peter Williams.
© 2013 Andrew Ross and Peter Williams. Published 2013 by John Wiley & Sons, Ltd.

17.1 Valuations and interim certificates

17.1.1 The purpose of valuations

On the majority of construction projects it is normal for the contractor to be paid for work carried out at regular intervals. This might happen at discrete stages of completion (e.g. work up to DPC, work up to eaves etc.) or at certain time intervals (e.g. fortnightly or monthly). The amount due for payment is stated in an **interim certificate** and this is usually, but not always, supported by a **valuation** of the work completed. Interim valuations are carried out either:

- by the employer's quantity surveyor (QS), usually in consultation with the contractor, or
- by the contractor, whose valuation will normally be checked by the employer's quantity surveyor or the engineer's representative (normally a quantity surveyor or measurement engineer).

The purpose of the valuation is to value the work done so as to ascertain the amount to be recommended for payment by the client (employer). This is done by taking account of:

- the physical work done on site;
- any extra work ordered by the client;
- the work of any specialist nominated subcontractors;
- the cost of running the site (the contractor's preliminaries);
- the value of any materials delivered to site.

The issue of the interim certificate determines the **payment due date** under the Construction Act and the **final date for payment**, as provided for in the Act, is stated in the civil contract between the employer and contractor. It should be noted that it is only the contract administrator (e.g. architect, project manager or engineer) who has the power to *certify* payment and, thereby, trigger the statutory notices required under the Act (Chapter 5).

17.1.2 The timing of valuations

Interim valuations occur at intervals as defined in the applicable conditions of contract. For example:

- In accordance with the JCT SBC/Q 2011, *Interim Valuations shall be made by the Quantity Surveyor whenever the Architect/Contract Administrator considers them necessary for the purpose of ascertaining the amount to be stated as due in an Interim Certificate*. The quantity surveyor in this case is engaged by the 'employer' under the contract who is normally referred to as the 'client'.
- Under the ICC – Measurement Version, the methodology is different because it is the Contractor who *shall submit to the Engineer … a statement showing the estimated contract value … carried out up to the end of that month*. It is the opinion of the Engineer that decides the amount to be certified for payment.

In both cases someone working for the employer (client) has to decide how much the contractor is to be paid. This needs to be done even-handedly. If the contractor is overpaid then the client is put at risk in the event of the contractor's default or insolvency and, if he is underpaid, the contractor faces the same risk and also the possibility of cash flow problems.

The timing of subcontract valuations or payment applications will be defined in the conditions of subcontract. The subcontract may also define the form that payment applications shall take and sometimes the main contractor will provide the subcontractor with a standard form or spreadsheet which must be used as the basis for the valuation or payment application.

17.1.3 The timing of interim certificates

Most contracts for construction work provide for interim payments at regular intervals in order to help the contractor's cash flow. Alternatively, the contract may provide for payments to the contractor at various stages of completion. Interim certificates are usually issued monthly during a contract but this can vary in practice. Whatever the case, interim certificates are a prerequisite for making payment to the contractor.

The period for certificates is stated in:

- the Contract Particulars (JCT Contracts);
- the Contract Appendix (ICC Contracts);
- the Contract Data (NEC Contracts).

In the JCT 2011 Standard and Intermediate Contracts, provision is made for monthly payment linked to the architect's or contract administrator's interim certificate. In simple terms the procedure is that a certificate stating the amount due to the contractor is issued at monthly intervals and this is then due for payment by the client (the employer under the contract) within 14 days.

Other forms of contract have different payment intervals but in this respect the Housing Grants, Construction and Regeneration Act 1996 has to be noted, as there are now statutory controls regarding contractual payments.

Interim certificates are not applicable in the case of subcontract payments but the main contractor may issue a formal notice of the amount agreed for payment prior to the actual payment being processed by the accounts department.

17.1.4 The status of interim certificates

The issue of an interim certificate by the architect (or contract administrator) is important as such a certificate is a condition precedent to the contractor's entitlement to be paid that sum.

Should there be a mistake in the amount certified, then the Construction Act provides for the issue of a payment notice and a withholding notice where adjustments may be made. Should these notices not be issued, or if the contractor issues a default notice, then the amount stated in the interim certificate must be paid. The case of *Lubenham Fidelities & Investment Co Ltd v South Pembrokeshire DC* (1985) is worthy of note with regard to mistakes in certificates.

17.2 Interim payment

17.2.1 Contractual provisions

Different forms of contract have different provisions for making contractual payments and these vary even within the JCT family of contracts. For instance, under the JCT 2011 Standard and Intermediate forms, there are two ways in which to arrive at the contractor's payment:

1. by the employer's QS valuing the work;
2. by acceptance of the contractor's application for payment.

Under the Minor Works Building Contract 2011, however, the architect or contract administrator must certify the value of work properly executed, including any agreed variations and so on, but the contractor has no right to make a payment application as such. Under the JCT Design and Build Contract 2011, it is for the contractor to apply for interim payments according to one of two alternatives allowed under the contract:

1. *Alternative A* – predetermined stage payments as set out in Appendix 2 of the contract based on the cumulative value at each relevant stage,
2. *Alternative B* – regular (e.g. monthly) payment applications from the contractor where the contractor determines the value of work completed.

Interim payments are subject to the deduction of retention money at the rate prescribed in the contract.

17.2.2 Methods of payment

There are a number of payment methods used in the construction industry, the choice of which depends upon the contractual arrangement adopted for the particular project in question.

For small projects of short duration where a lump sum price has been agreed, payment may be made either on completion (an 'entire' contract) or on completion of particular stages of the work (e.g. work to damp-proof course, work to eaves etc.). The stages in question may be dictated by the lender to the property owner (e.g. bank or building society), which will only release money to pay the contractor when work has been certified as complete and satisfactory.

For longer-term projects, interim payments may be made (usually monthly) where the contractor receives a regular income from the project related to the value of the work completed to date. Alternatively, a stage payment approach may be adopted; this will either be linked to a schedule or **contract sum analysis** or to the work activities shown on the contractor's programme.

Occasionally, interim contractual payments may be made by reference to an agreed value profile or **'S-curve'**, where work completed to date is paid for by reading off a graphical display. Problems with this approach arise if and when the contractor is behind or ahead of programme or where variations have been ordered by the client. In such cases, the 'S-curve' may have to be adjusted to reflect the different circumstances to make sure that the contractor is not over or under paid for what he has done. This methodology is employed under the GC/Works/1 Government contract and is also operated under the NEC3 Engineering and Construction Contract as well as in bespoke contracts by private clients.

For re-measurement contracts, payment will be made on a **measure and value** basis. This approach requires site measurement of work done, which is then valued according to the priced schedule submitted by the contractor at tender stage.

17.2.3 Payment notification

The contractor is notified of the amount due through the architect, engineer or contract administrator, who normally sends a copy of the certificate to the contractor and to the employer. The Construction Act also requires the issue of a payment notice with details of the payment and how it has been arrived at. Failure to issue this notice may lead to a default notice from the contractor and the sum then payable will be that which is stated in the interim certificate.

The amount stated as due in the payment notice may well differ from that stated in the interim certificate. The employer may disagree with the valuation, for instance, or there may be other reasons for withholding monies, such as the deduction of liquidated and ascertained damages or set-off as provided for in the contract. Whatever the eventual amount, payment will usually be made by way of a cheque or inter-bank transfer subject to the issue of a withholding notice prior to the final date for payment.

17.2.4 Retention

Once the employer's quantity surveyor has determined the gross interim valuation, a percentage is deducted from it – usually between 3% and 5% depending upon the prevailing contract conditions. The deduction of retention has been a traditional feature of construction contracts for many years, the reason being to create a fund, held by the employer, to pay for the rectification of faulty work should the contractor be unable or unwilling to carry it out.

Under some contracts (e.g. JCT SBC/Q 2011), the contractor can ask the employer to keep the money in a separate bank account (this does not apply to local authority employers). Whilst the employer is entitled to any interest earned, the money legally

Table 17.1 Application of retention limit.

			£700 000		
Tender sum			£700 000		
Retention	5%				
Limit of retention	3% ×		£700 000	= £21 000	
			£	**£**	**£**
Valuation No 1					
Gross valuation to date					170 000
Retention	5% ×		170 000	8500	8500
Retention limit	5% ×		420 000	21 000	
Net valuation					**161 500**
Valuation No 2					
Gross valuation to date					310 000
Retention	5% ×		310 000	15 500	15 500
Retention limit	5% ×		420 000	21 000	
Net valuation					**294 500**
Valuation No 3					
Gross valuation to date					490 000
Retention	5% ×		490 000	24 500	
Retention limit	5% ×		420 000	21 000	
	0% ×		70 000	0	21 000
Net valuation					**469 000**

belongs to the contractor. However, the employer has recourse to the money and in certain circumstances may use the retention fund to 'set-off' monies owed by the contractor to the employer. Unless a dispute arises, it is normal practice for the retention fund to be released to the contractor in two parts:

1. 50% is released when practical completion of the contract has been certified (or substantial completion in civil engineering contracts);
2. the other 50% is released when the certificate of final completion is issued which takes place when all defects have been corrected by the contractor.

Under ICC conditions of contract, there is usually a retention limit which, when reached, signals that further payments to the contractor shall be made without deduction of retention. Table 17.1 illustrates that 5% retention is deducted until the valuation reaches £420 000 (5% of £420 000 = £21 000) where after retention is 0%.

Retention release is important for the contractor as this frees up money 'locked-up' in the contract and improves the contractor's cash flow and working capital position. The second part of the retention release takes some time as the defects correction period is usually a minimum of six months duration and commonly lasts for 12 months from practical (or substantial) completion.

17.2.5 Alternatives to retention

In the famous Banwell Report of 1964, Sir Harold Banwell urged that the practice of deducting retention money should be gradually phased out. Despite much discussion in the construction press over the years, there is little sign of this happening albeit that some construction contracts deal with the issue to some extent.

Under some forms of contract the deduction of retention is an option (e.g. Option P NEC3 Engineering and Construction Contract) and in other cases the contractor may be asked to provide a 'retention bond' in consideration of there being no retention deducted from interim payments (JCT SBC/Q 2011).

17.3 Principles and procedures

17.3.1 Valuation principles

Over the years, quantity surveyors have established a few simple principles which underpin the valuation process. These principles have not been formalised in any way but have simply developed by 'custom and practice' as 'the best way to go about things':

Principle 1 – Accumulation

Valuations are carried out on a cumulative basis. This means that at any point in time the valuation is the total value of work carried out to date. For example, the valuation at month 3 of a contract includes the work done in months 1 and 2.

The cumulative method of valuation is much simpler to operate in practice than might be imagined. This method is also less prone to error as mistakes made in the previous month are likely to be picked up in the current month's valuation. Valuing work carried out in a particular period (e.g. a month) in isolation from that which has been carried out previously is very tricky. To do this, previous work has to be effectively 'freeze-framed' and work done subsequently highlighted. This requires very accurate records to be kept. It is far easier to value the total of work done to date and then deduct previous payments made.

Principle 2 – Previous payments

The principle of deducting previous payments works in conjunction with Principle 1 in order to arrive at the value of work done in a particular period (e.g. a month). For example, the total valuation of work done up to and including month 3 is ascertained and the total of payments made in months 1 and 2 is deducted to give the value of work done in month 3.

Principle 3 – Pro rata

Where an item of work has to be valued and there is no applicable rate for it in the bills of quantities, a 'pro rata' rate may be established pending further negotiation between the parties. For example, the rate for a 150-mm thick block wall priced at £36.95/m² may be used to establish a rate for a 100-mm thick block wall of the same specification:

$$£36.95 \times \frac{100}{150} = \textbf{£24.63} \, / \, \textbf{m}^2$$

Table 17.2 illustrates the principle of pro rata using unit rates from a price book, such as Spon's (Davis Langdon, 2012) or BCIS (formerly Wessex). The price book

Table 17.2   Principle of pro rata using price books.

Variation No 8:
Omit half brick thick facings in gauged mortar 1:1:3 **pointed one side** as the work proceeds and *add* half brick thick facings in gauged mortar 1:1:3 **pointed both sides** as the work proceeds.
Half brick thick facings in GM 1:1:3 **pointed one side** as the work proceeds
Bills of quantities rate   =   £53.72 m²
Half brick thick facings in GM 1:1:3 **pointed both sides** as the work proceeds
Price Book rates: • Facings pointed one side   =   £46.48 m² • Facings pointed both sides   =   £48.60 m² **New Rate** = $\frac{£48.60 \text{ m}^2}{£46.48 \text{ m}^2}$ = 1.045611 × £53.72 = **£56.17 m²**

rates are used to create a ratio that establishes the difference in cost of brickwork pointed one side and both sides and this ratio is then applied to the billed rate in order to arrive at a rate for the new item of work.

17.3.2   Valuation procedures

The usual valuation procedure is to establish the total value to date and then to deduct retention at the rate stated in the contract. The total of previous payments made are then deducted which leaves the net payment due to the contractor.

To determine what is to be included in an interim valuation, the relevant conditions of contract must be scrutinised because there are some subtle and some distinct differences between the various standard forms. However, there is no prescribed format as to *how* this should be done. Even where the contract states that the contractor is to be paid particular sums of money for reaching specific stages of completion, the valuation may be 'blurred' by variations, claims or defects in the work.

Different quantity surveyors have their own individual approaches to the valuation process and there are no set ways to do this. Care needs to be taken to be as accurate as possible because the professional quantity surveyor, or the quantity surveyor acting as the engineer's representative, has a duty of care to the employer. Keating (Furst and Ramsey, 2012) suggests that such a duty does not extend to the contractor following *Pacific Associates v Baxter [1990]* Q.B. 993(CA). In any event, however, most participants consider that the professional quantity surveyor will act in a fair and even handed manner.

In all but the simplest of cases, a formal valuation of work in progress is necessary in order to facilitate the issue of an interim certificate. This is because continuing construction projects are mostly in a partial state of completion and even completed parts or sections cannot be considered complete because of the possibility of patent defects. Consequently, there are complexities in determining actual 'value' and this cannot be done simply by guesswork.

Assessing value involves making informed judgements and arriving at carefully considered decisions such as:

- how much of the permanent works has been completed to a satisfactory extent so as to warrant payment;
- has satisfactory progress been made to date with regard to time-related payments to the contractor;
- have materials been brought to site at an appropriate stage of construction and have they been stored or protected suitably so as to preserve their value.

17.3.3 Types of valuation

There are three types of valuation:

1. external valuation
2. internal valuation
3. subcontractor valuation.

The distinguishing features of the three types of valuation are identified in Sections 17.3.4-17.3.6 and the procedures involved in their preparation are illustrated in Figure 17.1, which also identifies their respective relationships to the three environments - project, internal and external.

External valuation

- This is carried out by the employer's quantity surveyor, usually with the contractor's quantity surveyor in attendance.
- Alternatively, the contractor may prepare the valuation and submit an application for payment. This will be checked by the employer's quantity surveyor or by the engineer's representative.
- If the main contractor prepares the external valuation, subcontract payment applications will inform the quantum of work in progress for the payment period (Figure 17.1).
- The external valuation does not state the 'true' value of the work done but merely measures and values the work in progress according to the contract documents.
- The external valuation does not give a 'true' value because:
  - the contract rates may not represent 'true' values due to the contractor's pricing strategy and/or
  - the work on site may have been over or under measured or may contain defects which will need to be corrected later.

Internal valuation

- This is a control measure and is carried out by the contractor's quantity surveyor.
- The internal valuation is not seen by the employer's quantity surveyor.
- Its purpose is to reconcile the external valuation with the contractor's costs.
- This requires adjustments to be made to the external valuation to reflect the 'true' value and to bring value and costs to a common date.
- This process enables cost and value to be compared at a given date (e.g. the 30th of the month) which will then show whether or not:
  - the contract is making a profit
  - the contract is on programme.

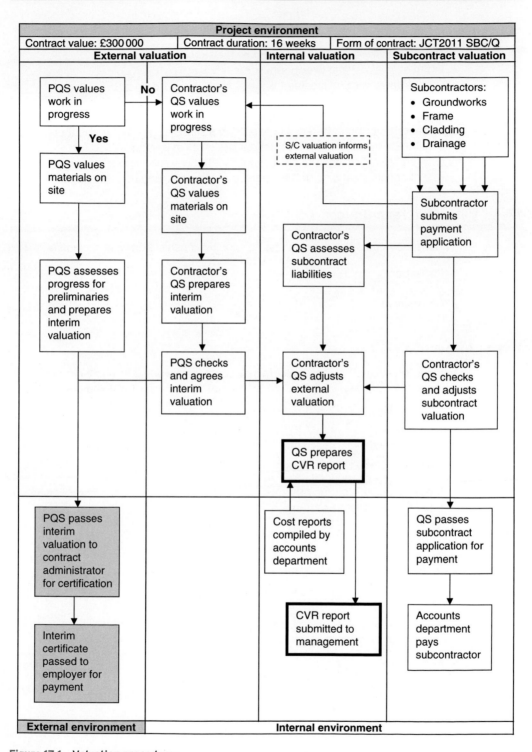

Figure 17.1 Valuation procedures.

Subcontractor valuation

- Normally carried out independently by the subcontractor.
- Often overvalued and over-optimistic.
- Often littered with errors due to:
  - lack of understanding of the conditions of contract;
  - lack of knowledge of measurement rules;
  - ignorance of the difference between lump sum and re-measurement contracts;
  - failure to insist that site instructions are given in writing;
  - failure to confirm oral site instructions by letter or e-mail before commencing work;
  - lack of understanding of the meaning of 'daywork' and how it operates.

17.4 Valuation techniques

Various methodologies are available for valuing construction work including:

- Inspection
- Measurement
- Ogive curve
- Gantt chart

17.4.1 Inspection

Value is determined by assessing the percentage of work that has been physically completed or by assessing the proportion of work carried out judged against a specific amount of money included in the bills of quantities.

- Advantages:
  - popular and widely used;
  - quick and simple;
  - avoids getting 'bogged down' in fine detail;
  - easy to agree with the other party;
  - practice improves accuracy.
- Disadvantages:
  - rough and ready method;
  - work can easily be over or undervalued;
  - requires experience to achieve a reasonable degree of accuracy.

17.4.2 Measurement

Value is established by physical site measurement, measuring from revised drawings or a combination of both.

- Advantages:
  - useful for measuring variations and extra work;
  - can be accurate provided the method of measurement is respected;
  - agreement is assured when carried out collaboratively (e.g. two QSs).
- Disadvantages:
  - extensive re-measurement is very time consuming;
  - site measurement is prone to overmeasurement, especially where excavations are concerned;
  - difficult to isolate variations from other work already carried out.

17.4.3 Ogive curve

Value is ascertained by reference to a model of planned expenditure based on the contract sum and an appropriate formula which predicts the expected cumulative value to be achieved each month.

- Advantages:
  - fair and reasonable;
  - simple to operate;
  - forecasts can be adjusted according to progress.
- Disadvantages:
  - not accurate;
  - results may be skewed depending upon the formula used;
  - slow progress may result in over-payment.

17.4.4 Gantt chart

Actual progress is compared with the contractor's programme in order to arrive at the proportion of work completed at a specific point in time.

- Advantages:
  - easy to operate;
  - simple to agree;
  - can be related to a stage payment arrangement.
- Disadvantages:
  - inaccurate as programmes are only 'guesstimates';
  - progress difficult to assess accurately;
  - open to abuse due to front loading of prices;
  - time consuming as contract price must be allocated amongst programme activities;
  - programme activities will differ from bills of quantities work sections and monies will need to be reallocated accordingly.

17.4.5 Adjustment and judgement

The valuation process is not an exact science and it is likely that 10 quantity surveyors valuing the same work will arrive at 10 different answers! In fact, ask a quantity surveyor to resolve the equation $2+3$ and the answer will be '5 – but I have a way of making it $5\frac{1}{2}$'!!

Whilst a fairly accurate valuation may be established by careful measurement, this is not as foolproof or easy as it sounds – it is also very time consuming and, therefore, costly. Consequently, a valuation based on judgement is the next best thing and, with experience, this can be quite accurate within reasonable confidence limits. The trick for the contractor's quantity surveyor – and for the subcontractor, too – is to ensure that the work is overvalued (for cash flow reasons) without being greedy. Asking for 35% completion when the truth is 25% may result in agreement at 30%, which is not too bad.

Notwithstanding the on-site valuation of work done, it may be necessary to reflect on the outcome and make appropriate adjustments prior to certification. Discussion with the architect, for example, may reveal that some of the work carried out is not up to the required standard or work may be covered up without verification that it complies with the specification.

Subcontract valuations may need to be adjusted for contra-charges where the main contractor has supplied materials, plant, equipment or facilities that were written into the contract as being the subcontractor's responsibility or it may be that a subcontractor has damaged the work of other trades which will ultimately be making a claim to the main contractor.

In any event, the Construction Act allows changes to be made to certified payments via the mechanism of notices introduced in 2006.

17.5 Materials on site

Contractors clearly need to have materials on site in order to work efficiently and avoid delays to the programme waiting for deliveries. Materials commonly found on site include stocks of drainage goods, bricks and blocks, bags of plaster, door and window frames and so on. Most standard forms of contract provide for the contractor to be paid for these materials provided that they have not been brought to site prematurely (some contractors do this so as to boost the valuation and, thereby, their positive cash flow on the project).

Off-site materials, such as fabricated structural steelwork, may also be included in the interim valuation. The JCT SBC/Q 2011 and the ICC – Measurement Version, for example, permit this subject to strict controls.

17.5.1 Valuing materials on site

For the external valuation the employer's QS will normally value the materials on site. This is an important part of his job so as to protect the employer from overpaying the contractor and to make sure that, when the employer pays for the materials, they belong to him. This requires knowledge of the law relating to ownership of unfixed materials on site and the exercise of some judgement and decision making. Among the considerations to be made will be:

- What materials are there on site?
- Have they been brought to site prematurely?
- What materials were there last month?
- Are the materials well protected from damage or theft?
- Has the contractor supplied a copy of the goods received sheet?
- Does the contractor have evidence that retention of title has been relinquished (Section 17.5.2).

The QS's approach to the valuation of materials on site is very much down to individual preference:

- Some quantity surveyors will simply accept the contractor's valuation.
- Others will make a tour of the site, compile their own list of materials and make a judgement as to the quantities. They may use their own knowledge of materials prices or ask the contractor for a list of basic materials costs.

- Another approach is to ask the contractor for a list of materials delivered to site (the 'goods received' sheet) and once the materials have been identified on site they can be quantified and priced.

The QS will have to be alert to the possibility that the contractor has brought materials to the site simply to boost the valuation and reduce negative cash flow. If he feels this is the case, then the value of such materials should not be included in the valuation.

Materials on site should be valued at net cost (i.e. after deducting trade discounts and with no mark-up for overheads and profit). Strictly speaking the prices used to value the materials on site should be the materials' cost allowances used in the contractor's tender. This might be to the contractor's advantage cash flow-wise especially where significant buying gains have been made.

When the materials on site have been incorporated in finished items of work, the completed work will be valued as normal and the value of the materials on site will be adjusted in the subsequent valuation.

17.5.2 Retention of title

As mentioned above, the question arises as to exactly who has title (ownership) of the materials on site. The answer is that it depends!

The employer should not be paying for materials that, ultimately, do not belong to him as this puts then employer at risk in the event of the contractor's insolvency. Whilst both the JCT SBC/Q 2011 and the ICC Infrastructure Conditions stipulate that materials on site paid for by the employer belong to him, the existence of retention of title clauses overrides the contractual provision. This is because most contracts for the supply of materials include a of **retention of title** clause which has statutory status under the Sale of Goods Act 1979.

This means that the goods delivered to site remain the legal property of the supplier until paid for by the contractor. Frequently, however, the contractor will be paid by the employer (client) under the contract before the materials are due to be paid for under the contract of supply and, therefore, the employer's quantity surveyor must ensure that legal title in the goods is passed to the employer before certifying payment to the contractor.

Consider the following common scenarios:

Scenario 1
The main contractor orders bricks from a builders' merchant which are delivered to site but not fixed and with no conditions attaching to the sale.

Sale of Goods Act 1979 applies where ownership passes to the contractor upon delivery at the latest.

Scenario 2
The main contractor orders bricks from a builders' merchant which are delivered to site but not fixed where a retention of title clause applies to the conditions of sale.

Materials belong to the builders' merchant until such time as they are fixed or paid for by the contractor.

Scenario 3
As scenario 2 but the value of the bricks is included in an interim valuation and paid for by the employer.

Materials belong to the builders' merchant until such time as they are fixed or paid for by the contractor.

Scenario 4

As scenario 3 but the main contract contains a clause which states that the value of materials included in an interim valuation and paid for by the employer belong to the employer.

Where a retention of title clause applies to the conditions of sale, the materials belong to the builders' merchant until such time as they are fixed or paid for by the contractor.

Scenario 5

A subcontractor delivers materials (which he owns) to site ready for his work and the main contractor is paid for materials on site but does not pay the subcontractor.

1. *Materials belong to the subcontractor until such time as they are fixed or paid for by the contractor.*
2. *Materials belong to the employer where there is a clause in the subcontract to that effect.*
3. *Materials belong to employer when they are fixed.*

Scenario 5 happened in the case of *Dawber Williamson Roofing Ltd v Humberside County Council* (1979) where the subcontractor was entitled to his materials even though the employer had paid the insolvent contractor for them. The JCT SBC/Q 2011 provides for this possibility whereby materials that are included in an interim certificate, and paid for, become the property of the employer and a like provision is included in the JCT Standard Form of Subcontract.

In all cases, however, the QS should make sure that title in the materials is relinquished by the supplier before including such sums in the interim valuation. This is because most suppliers have a retention of title clause included in their standard conditions of sale. These are called 'Romalpa' clauses following the case of *Aluminium Industrie v Romalpa* (1976) when a retention of title clause was deemed by the court to be effective.

The situation is different, however, when the materials have been incorporated in the building. In English law, whenever anything becomes attached to the realty of any property (e.g. land or buildings), then title vests in the property owner and retention of title clauses become ineffective.

17.6   Basic valuation procedure

The basic steps when preparing a valuation are:

- ascertain the *cumulative* total value of work carried out to date;
- ascertain the total value of materials on site;
- deduct retention monies;
- deduct the total of previous payments;
- the remaining sum is the amount to be paid in the current payment period;
- **NB**: In the case of subcontract valuations there may be an additional step where there is a cash discount to be deducted.

This is best illustrated by a practical example. Table 17.3 shows how monthly valuations are developed for a 16-week contract over four monthly periods. There is a six-month defects correction period. The valuations are shown in summary form for simplicity – that is the various work sections making up the project are given as

Table 17.3 Developing monthly valuations.

	Contract sum £	Month 1 % Complete	Month 1 Value £	Month 2 % Complete	Month 2 Value £	Month 3 % Complete	Month 3 Value £	Month 4 % Complete	Month 4 Value £	Month 10 % Complete	Month 10 Value £
Preliminaries	30000	25	7500	44	13200	75	22500	100	30000	100	30000
Groundworks	30000	60	18000	100	30000	100	30000	100	30000	100	30000
Frame	60000		0	90	54000	100	60000	100	60000	100	60000
Cladding	40000		0	60	24000	100	40000	100	40000	100	40000
Services	100000		0		0	65	65000	100	100000	100	100000
Finishes	20000		0		0	25	5000	100	20000	100	20000
Drainage and external works	20000	20	4000	30	6000	50	10000	100	20000	100	20000
		Subtotal	29500	Subtotal	127200	Subtotal	232500	Subtotal	300000	Subtotal	300000
		Materials on site	10000	Materials on site	30000	Materials on site	15000	Materials on site	0	Materials on site	0
		Subtotal	39500		157200		247500		300000		300000
		Retention at 5%	1975	Retention at 5%	7860	Retention at 5%	12375	Retention at 2.5%	7500	Retention at 0%	0
			37525		149340		235125		292500		300000
		Previous payments	0	Previous payments	37525	Previous payments	149340	Previous payments	235125	Previous payments	292500
Total	300000	Sum due	37525	Sum due	111815	Sum due	85785	Sum due	57375	Sum due	7500

Interim valuation

totals but not subdivided into their component parts. For example 'Groundworks' is shown as a total figure of £30 000 without breaking this down into excavation, filling, concrete work and so on. For simplicity it is assumed that there are no variations on this contract.

The form of contract is the JCT SBC/Q 2011 and the contract works is made up of six work sections and preliminaries totalling £300 000.

The four monthly valuations are calculated as follows:

Month 1

- One month having elapsed 'Preliminaries' is valued at 25% ($^4/_{16}$) of the total contract allowance on the basis of four week's elapsed time from the 16-week total. This valuation assumes that the contract is on programme.
- Work in progress concerns Groundworks and Drainage, which are valued at 60% and 20% complete, respectively.
- Materials on site is valued at £10 000, which comprise items such as filling materials, bricks and blocks, drainage products and probably structural steelwork brought to site ready to commence this work in Month 2.
- There are no previous payments to be deducted as this is the first valuation.

Month 2

- At the end of Month 2 the contract is one week behind programme.
- Preliminaries is therefore valued at 44% ($^7/_{16}$) of the contract total.
- Groundworks is complete.
- The Frame, Cladding and Drainage work sections are underway being 90, 60 and 30% complete respectively
- Materials on site is £30 000 reflecting that higher value items are probably now on site. Such items will include roof and wall cladding materials, roof lights and heating and air conditioning materials ready to start this work in the next valuation period.
- It should be noted that the value of materials on site is the value of materials actually on the site at the time that the valuation is made.
- The total cumulative value of work carried out to date is £157 200, which represents the total value of work done in Months 1 and 2 and materials on site. This is subject to the deduction of 5% retention.
- It should be noted that the amount of retention held by the employer is the cumulative total to date, that is £7860.
- As this is the second valuation a previous payment will have been paid during Month 2. The sum of £37 525 is therefore deducted from the cumulative valuation total.
- The payment that the contractor should receive during Month 3 is £111 815, unless a payment notice or withholding notice states a lesser sum.

Month 3

- By the end of Month 3 the contractor has recovered the one week delay.
- Preliminaries are consequently valued at 75% ($^{12}/_{16}$) of the contract total.
- The contract works have now reached the peak of activity with Groundworks, Frame and Cladding already complete.
- Services and Drainage are now 65 and 50% complete, respectively, and Finishing trades have reached 25% completion.

- The value of materials on site is now £15 000, which indicates that major items of expenditure have now been made.
- The cumulative value of work to date is £235 125 after 5% retention is deducted.
- The amount of money that the contractor will receive during Month 4 will be £85 785, subject to notification of any legitimate adjustments by the employer.

Month 4

- Valuation 4 indicates a cumulative value to date of £300 000, which is equal to the contract sum.
- In practice there would normally be variations to take into account which should be measured and valued and added to the total valuation.
- There is no work in progress and, consequently, the contract is now complete subject to the rectification of defects during the defects correction period.
- The QS will only certify 100% completion if the architect is satisfied that all 'snagging' items have been completed and that the building work has reached practical completion ready for occupation.
- Valuation 4 is subject to only half of the usual 5% retention. This is because the contract states that 50% of the retention sum will be released (paid) to the contractor when the works are complete.
- Previous payments totalling £235 125 are deducted from the valuation leaving a total of £57 375 to be paid to the contractor during Month 5.
- The retention monies held by the employer legally belong to the contractor but the employer has legal recourse to this money under the contract.

Month 10

- At the end of the defects correction period the contractor is entitled to receive outstanding retention monies provided that the contractor's obligation to rectify outstanding defects is satisfied.
- A payment to the contractor of £7500 will be made when the architect has certified the end of the defects correction period. This should happen during Month 11.

17.7 External valuation

The JCT SBC/Q 2011 requires, where deemed necessary by the architect, the contract works to be **valued** by the quantity surveyor. Such valuations are usually done at regular intervals and are called 'interim valuations'. A contractor would distinguish such valuations as 'external' compared with the company's 'internal' procedure of valuation linked to the financial control system.

The quantity surveyor's duty under the contract is to 'value' the work carried out by the contractor so that the contractor can be paid for what he has done. This is done monthly (usually) and, whilst relatively straightforward to do, there can be complexities requiring professional judgement. In this context it is important that the QS does not over or undervalue the work. The contractor is entitled to be paid for the work done under the contract but, at the same time, the client will not want to overpay the contractor in case of insolvency or other problems in the future.

Table 17.4   Comparative example of contract phraseology.

JCT SBC/Q 2011	ICC – Measurement Version
Contract provision: • The total value of the work properly executed by the Contractor • The total value of the materials and goods delivered to the Works • The total value of any Listed Items (off-site materials and goods) • Any amounts in respect of adjustments to the Contract Sum (e.g. statutory fees and charges, insurance payments etc)* • Any amounts ascertained for loss and expense* • Any amount allowable under Fluctuations Option A or B*	**Contract provision:** • The estimated value of the Permanent Works carried out • A list of any goods and materials delivered to the site • A list of any goods and materials not yet delivered to the site • The estimated amounts to which the Contractor considers himself entitled under the contract • Amounts payable in respect of Nominated Sub-contracts

* Not subject to retention.

17.7.1   Contract provisions

The contract terms applicable to a particular project will determine the elements to be included in the gross valuation of work carried out. Different contracts use different phraseology as indicated in the comparative example in Table 17.4.

The amount stated as due in the interim certificates under both contracts is *exclusive* of VAT. VAT is normally payable on contract work, of course, and the usual practice is for the contractor to invoice for the certified amount plus VAT, which is then paid by the employer. The employer will thereby have a VAT invoice for its own accounting purposes.

By contrast, the NEC3 ECC is much more concise and states that the '*amount due is the Price for Work Done to Date plus other amounts to be paid to the Contractor less amounts to be paid by or retained from the Contractor*'. In this contract, VAT is to be *included* in the amount due, which is sensible and practical bearing in mind that certified sums may change before payment if a withholding (or pay-less) notice is issued.

In any event, what all the contracts mean in practice is that a valuation will include:

- measured work;
- variations to the contract;
- work carried out pursuant to prime cost sums;
- work carried out pursuant to provisional sums;
- preliminaries.

17.7.2   Components of an interim valuation

The following non-exhaustive list illustrates the sort of items that may be included in an external valuation of work in progress:

- measured work completed;
- incomplete measured work;
- work which may not be built to specification (but overlooked);
- work completed or partially completed which may be a variation to the contract;
- extra work carried out to the verbal instructions of the architect/engineer;
- work done by the contractor which may have to be valued on a prime cost basis (i.e. daywork);
- preliminaries, that is the contractor's fixed and time-related costs of running the site;
- materials brought to site which the contractor is entitled to be paid for;
- materials or components being prepared or fabricated off-site;
- the impact of progress which might be ahead of or behind the master programme;
- the contractor's entitlement to loss and expense for non-culpable delay and/or disruption;
- circumstances where the contractor may be entitled to extra preliminaries;
- the impact of adjustments to the time for completion of the contract;
- deductions from the contractor's payment due to set-off by the employer;
- deductions for liquidated and ascertained damages where the contractor has exceeded the time for completion of the contract.

The external valuation is determined in accordance with the specific provisions of the appropriate contract. For instance, the JCT SBC/Q 2011 form prescribes how the 'gross valuation' is to be calculated and states what is to be included and what is not included in the calculation. Other forms of contract have similar provisions.

The valuation is struck at a particular point in time and, under the JCT SBC/Q 2011 form, this is specified as being *up to and including a date not more than seven days before the date of the Interim Certificate*. Other contracts vary.

17.8 Preparing the external valuation

Conventional practice is that valuations are cumulative. In other words, the total value of the contract works to date is determined in order to produce the valuation and then the cumulative amount paid to the contractor to date is deducted. This calculation gives the net amount payable to the contractor for a particular month's work and enables errors or omissions in the valuations to be more easily rectified as work proceeds. An example is shown in Table 17.5.

Provided that the architect/contract administrator is in agreement with the quantity surveyor's valuation, £69 783 is the value of the payment that the contractor will receive.

The gross value of £240 494 to date in the example (Table 17.5) would be made up of a number of different items, as shown in Table 17.6.

Table 17.5   Interim Valuation No 4 – Summary (£).

Gross value to date (valuations 1–4)		240 494
less Retention	at 3%	7215
Net value to date (valuations 1–4)		233 279
less Previous payments (valuations 1–3)		163 496
Net Valuation No 4		69 783

Table 17.6   Gross value of project (£).

Measured Work				
Site clearance		23 932		
Groundworks		75 887		
Brickwork		24 984		
Roofing		12 348		
Drainage		17 916	155 067	
Preliminaries				
Fixed charges		18 000		
Time-related	3 months			
charges	at 10 000	30 000	48 000	203 067
Variations to date				12 848
Materials on site				24 579
Gross value to date				240 494

17.8.1   Measured work

The value of measured work carried out by the contractor to date is essentially made up from the total quantity of work done multiplied by the contractor's prices or rates as shown in the bills of quantities. Suppose that the total quantity of a measured item of facing brickwork in the bills of quantities is 1900 m^2 and that a rate of £94.70 per m^2 has been priced by the contractor. If the quantity of facing brickwork completed to date is 40% of the total, then the measured item included in the interim valuation would be valued as shown in Table 17.7.

Table 17.7   A measured work example.

	Quantity	Unit	Rate (£)	Total (£)
Total value of facing brickwork as per bills of quantities	1900	m^2	94.70	179 930
Quantity of facing brickwork completed to date	40	%	×£179 930	
Total value of facing brickwork to date				**£71 972**

17.8.2   Variations

Bearing in mind the complexity and uncertainty of construction work, it is almost inevitable during a contract that there will be changes to the design in order to account for unforeseen events or changes of mind by the employer or architect. Where a design change occurs, this is called a 'variation to the contract' or, under NEC3 conditions, a 'compensation event'. In order to pay the contractor for such work it has to be valued and included in the appropriate interim certificate.

JCT SBC/Q 2011, for instance, defines 'variation' as *the alteration or modification of the design, quality or quantity of the Works*' and sets out the rules by which variations to measurable work are to be valued. This form of contract requires that the gross valuation shall include variations provided that their value has been established under the contract.

The principles of valuating variations are well established in most standard conditions of contract:

- For similar items of work carried out under similar conditions with no significant difference in quantity – use bills of quantities rates.
- For similar items of work carried out under dissimilar conditions and/or with significant differences in quantity – use bills of quantities rates as a basis for the valuation.
- For dissimilar items of work carried out under dissimilar conditions and/or with significant differences in quantity – value the work at fair rates.
- Where an approximate quantity proves to be a reasonably accurate forecast of the quantity of work required – use the rate or price in the bills of quantities.
- Where an approximate quantity is not a reasonably accurate forecast of the quantity of work required – use the rate or price in the bills of quantities as a basis for valuation.
- Where the work involved in a variation cannot be properly valued by measurement – use daywork as the means of valuation.

Notwithstanding the valuation 'rules' for variations, contractors often complain that the true cost of variations is not fully recovered and this is an issue that should be considered when reporting the value of work in progress. Under some conditions of contract, the contractor may be asked to provide a quotation ahead of being instructed to carry out a variation and this gives the contractor more 'comfort' that the associated costs will be fully recovered.

In practice, variations are frequently not valued and agreed promptly. The usual problem is waiting for the formal variation order from the contract administrator. This is frustrating for contractors and subcontractors alike because potential income is locked-up in the contract which has a detrimental impact on cash flow and, hence, on working capital availability.

17.8.3 Daywork

Daywork is a method of valuation and payment for work that cannot be measured and valued normally. Sometimes, variation instructions are given, or work is ordered by the architect or engineer, and the only solution to arriving at a fair valuation of the cost is to record the time spent doing the work and the materials used (e.g excavating trial holes to find existing services). This is **daywork**. The basis for payment on a daywork basis is the **prime cost of daywork** to which an addition is made to allow for on-costs, overheads and profit.

The prime cost of labour, for example, is the basic cost of employing a labourer or tradesman to include:

- guaranteed minimum earnings at nationally agreed rates;
- employer's national insurance contributions;
- training levy;
- holidays with pay;
- welfare benefits.

The percentage addition allows for all the other costs of employment, such as additional wages paid over the 'basic' minimum, any bonus payments or incentives, travel

Table 17.8 Calculation of daywork rates.

Daywork rates			
Prime costs		£	£
Basic pay p/a	46.2 weeks	401.70	18558.54
Employers' NIC	as per HMRC NI contributions calculator		1679.81
Hoildays with pay	46.2 weeks	32.07	1481.63
Public holidays	64 hours	10.30	659.20
Training levy	0.5%	on basic pay + holiday pay	103.50
			22482.68
Basic hours	1802 hours		
Prime cost per hour			**12.48**
On-costs			
Cost p/a	calculations omitted for clarity		26132.72
Supervision	10%		2613.27
			28745.99
Overheads & profit	25%		7186.50
			35932.48
Basic hours	1802 hours		
On-cost per hour			**19.94**
Hourly daywork rate	12.48+19.94 =		**32.42**
Uplift on prime cost	$\dfrac{£32.42 - 12.48}{£12.48} \times 100 =$		**160%**

money, payments for skills, the cost of supervision, employment and public liability insurances and so on, and the contractor's required margin for overheads and profit.

For consistency, the prime cost of labour is set out in the **Definition of Prime Cost of Daywork for Building Work** published by the RICS (BCIS, 2007) and the **Schedules of Dayworks Carried Out Incidental to Contract Work** issued by the Civil Engineering Contractors Association (CECA, 2011). These definitions include schedules of the cost of various types of construction plant and equipment to which the contractor also adds a percentage to cover his overheads and profit.

The reason for defining a prime cost of daywork is to establish a universal hourly rate that will be the same figure for all contractors. This figure is effectively a 'known-known' to all parties to the contract including the employer's quantity surveyor. The percentage addition is a variable and each contractor will have its own idea as to what the percentage should be. Consequently, the percentage addition introduces a competitive element (at tender stage) which varies from contractor to contractor.

Table 17.8 illustrates how daywork rates are calculated:

- Firstly, the **prime cost** per hour is calculated by dividing the annual cost by the number of hours worked each year = £12.48/h.
- Secondly, the **on-costs** are calculated which are the additional costs of employment per year *over and above* the prime cost plus an addition for overheads and profit. This is then converted into an hourly rate as well = £19.94/h.

- The total daywork rate is therefore £12.48 + £19.94 = £32.42/h.
- The daywork rate may be given as a rate per hour or may be expressed as a percentage over and above the prime cost of labour:

$$\frac{£32.42 - £12.48}{£12.48} \times 100 = \mathbf{160\%}$$

Therefore, the cost of labour for, say, six hours of daywork can be calculated two ways:

1. 6 h labour at £32.42 = **£194.52**
2. 6 h labour at £12.48 =   £74.88
              +160%   £119.81
   Total   **£194.69**

It can be seen that both methods arrive at more or less the same answer (allowing for rounding errors).

The pre-requisite for payment on a daywork basis is a completed daywork sheet signed by someone in authority as a true record of the time spent by various classes of labour and the materials expended doing the job. Main contractors usually submit daywork sheets for signature by the architect or engineer at the end of each week and subcontractors similarly submit theirs to the main contractor. Figure 17.2 shows a simple example.

A signed daywork sheet does not necessarily mean that the work concerned will be paid for on a daywork basis and it is up to the employer's quantity surveyor to decide what is the appropriate method of valuation in the circumstances. This is why daywork sheets are often signed 'for record purposes only' (RPO). Many subcontractors do not understand this.

17.8.4 Prime cost sums

Prime cost (PC) sums are included in bills of quantities to provide for specialist services that the main contractor cannot provide and where the architect/engineer wishes the specialist work to be carried out by a particular company. This will often involve the provision of a design service as well as installation and the specialist company may well have been involved in the design development prior to inviting main contract tenders.

Prime cost sums are expended on instructions from the architect or engineer whereupon the main contractor will enter into a subcontract with the specialist company. This happens at some stage during the contract period. The creation of a nominated subcontractor means that, to all intents and purposes, the subcontractor will be managed and be accountable to the main contractor, albeit that the architect/engineer will also have some involvement especially on the design side of the subcontract works.

Prime cost sums are usually based on a budget price or quotation from the specialist contractor and are simply a means of including a 'provision' for the specialist work in the main contractor's tender figure. It is very likely that the eventual price for the specialist work will differ from the value of the PC sum and this will be adjusted in the final account.

Prime cost sums in the bills of quantities are normally accompanied by items for the contractor's profit and for attendance. This enables the main contractor to include a margin on the specialist work as a reward for managing the specialist

| DAYWORK | Contract: *University Building B Refurb* | | | | | | | Variation: | Date: *w/e 6ᵗʰ Sept* | | | | |

Description and location of work:

Repaint completed plastered walls damaged by other trades as per SI 764.

Name	Trade	M	T	W	Th	F	S	Su	Total Hrs	Rate		£	P	£	p
Hughes J	*P*			*8*					*8*	*12*	*48*	*99*	*84*		
Ashworth M	*P*				*8*	*6*			*14*	*12*	*48*	*174*	*72*		
°												*274*	*56*		
								Add *160* %				*439*	*30*	*713*	*86*

Machine	Ref	M	T	W	Th	F	S	Su	Total Hrs	Rate		£	p		
							Add		%						

Materials							Quant	Unit	Rate		£	p			
5L Emulsion							*1*	*No*	*25*	*00*	*25*	*00*			
Filler							*2*	*No*	*2*	*50*	*5*	*00*			
											30	*00*			
							Add *75* %				*22*	*50*	*52*	*50*	

| Signed **AJB** | for **Maincontractor RPO** | | | | | | | | | | **Total £** | | *766* | *36* | |

Figure 17.2 Completed daywork sheet.

Table 17.9 Typical prime cost sum layout in the bills of quantities.

Ref	Item	Quant	Unit	Rate	£	p
A	Allow the following prime cost sum for roll-formed roof sheeting:				160 000	00
B	Allow for main contractor's profit	3	%		4 800	00
C	Allow for general attendance		Sum		2 000	00
D	Allow for special attendance		Sum		5 000	00
	Total				171 800	00

subcontractor and for 'attending' to the needs of the subcontractor as regards the provision of services such as offloading of materials, power and lighting, space for cabins and so on.

A typical PC sum layout in the bills of quantities is shown in Table 17.9.

The PC sum of £160 000 shown in Table 17.9 is inserted by the architect/engineer in the tender bills of quantities and the contractor prices the profit and attendance to bring the specialist work to a total. When the order for the nominated subcontract work is placed, this might be for say £150 000. Work in progress is then valued on the basis of the proportion of work carried out relative to the value of the order with the addition of the profit percentage and the proportion of attendances carried out by the main contractor.

It should be pointed out that the nomination of subcontractors is somewhat 'out of fashion' these days and, under the JCT SBC/Q 2011, it is more likely that specialist work will be carried out by 'named' or 'listed' subcontractors suggested by the architect/engineer but organised by the main contractor entirely.

17.8.5 Provisional sums

Provisional sums are allowances in the tender bills of quantities for work that is envisaged but not precisely determined at tender stage. A provisional sum is also normally included for work that is 'unexpected' – this type of provisional sum is called a 'contingency'. The money included in the contract bills for provisional sums and general contingencies is expended as and when required during the contract and may be spent in its entirety or partially or, rarely, not at all. The work done under a provisional sum will be valued according to the rates in the bills of quantities where appropriate or, alternatively, the 'bill rates' will be used as a basis for agreeing a 'fair rate'.

In SMM7 the distinction is drawn between 'defined' and 'undefined' provisional work, whereby the main contractor is deemed to allow in his programme for the inclusion of defined work but not for undefined work.

17.8.6 Preliminaries

Preliminaries consist of fixed and time-related charges and represent the contractor's costs for organising and managing the site. The valuation of preliminaries is a question of assessing:

- which work related to fixed charges has been done;
- what time has expired to date as a proportion of the contract period for the valuation of time-related items.

Complications in the valuation of preliminaries can arise should the contractor be ahead of or behind programme, whereupon judgement will be needed to determine the correct value to date.

The following simplified example in Table 17.10 illustrates how part of the 'prelims' might be valued at week 12 on a 26 week contract.

The valuation of preliminaries to be included in the contractor's interim payment will depend on the progress that the contractor is making on site. If the contractor is behind programme then due consideration has to be taken so as not to overpay the contractor. For instance, with a contract period of 26 weeks and an elapsed time of 12 weeks, time-related preliminaries would be paid at $^{12}/_{26}$ of the total included in the contractor's tender (Table 17.10). However, if the physical work on site has only progressed as far as week 10, then this signifies that the contractor is two weeks behind programme. Consequently, the assessment of preliminaries would be made

Table 17.10   Simplified illustration of Preliminaries valuation.

Preliminaries	Item	Tender allowance (£)	Progress to date	Value of Prelims to date (£)
Fixed charges	Transport cabins to and from site	1000 × 50%		500
	Connect temporary services	5000 × 100%		5000
Time-related charges	Site supervision	52000 × $^{12}/_{26}$		24000
	Hire of cabins	13000 × $^{12}/_{26}$		6000
	Site security	19500 × $^{12}/_{26}$		9000
Total				44500

in relation to progress, and not the time elapsed to date, as a proportion of the over-all contract period (i.e. $^{10}/_{26}$ of the total).

17.8.7   Progress

The external valuation is commonly used as a useful benchmark with which to measure progress on a contract. The basic idea is to compare the cumulative value attained at the review period with the planned value. If the actual value (or earned value) exceeds the planned value, indications are that the project is ahead of programme. Where the actual value is less than that planned, the project is behind programme.

In practical terms this is somewhat over-simplistic as there are many factors that can increase actual value, such as variations and loss and expense claims, and therefore comparing this with planned value is not comparing like with like.

The planned versus actual value method of assessing progress is one of several ways that the contractor has to measure physical progress. These issues were explored in greater detail in Chapter 16.

17.9   Internal valuation

17.9.1   Purpose

To determine whether a contract is making a profit or loss at any particular point in time, it is necessary to compare the contractor's actual costs with the true value of work carried out to date. For a number of reasons the external valuation is not reliable enough to do this, so the contractor needs an accurate **internal valuation** in order to make the cost/value comparison accurate.

There are many reasons why the external valuation cannot be relied upon, including:

- Work in progress may have been overmeasured or undermeasured by mistake or by faulty assessment.
- The contractor may have persuaded the client's quantity surveyor that more work had been done than was actually the case so as to increase income on the contract and, thereby, reduce negative cash flow.

- The contractor's tender prices may have been manipulated so as to create positive cash flow early in the contract and, consequently, do not represent the true value of work carried out.
- Materials on site may have been overvalued in the external valuation and, therefore, do not bear true comparison with actual costs to date.
- For cash flow reasons, the contractor's preliminaries may include the profit and overheads from measured work items that have yet to be carried out.
- Work may have been included in the external valuation which the contractor nevertheless knows is defective. This work will have to be rectified – at a cost – and, thereby, represents a contingent liability that the contractor cannot truly take into value.

To compare actual cost with true earned value, adjustments to the external valuation need to be made so as not to give a false impression of profit or loss on a contract. This is the purpose of the internal valuation, which is carried out every month by the contractor's own quantity surveyor in firms that recognise the importance of reporting the true financial position on their contracts. On any decent sized project, preparation of the internal valuation will occupy up to three weeks of the quantity surveyor's time each month.

Whilst the internal valuation is indispensable in the cost value reconciliation (CVR) process, not all contractors and subcontractors have a CVR procedure and therefore have to rely upon the external valuation to reflect monthly earned value. It may well be a fair argument that such a 'rough and ready' approach to financial management is justifiable in terms of the time and cost of carrying out a monthly CVR and, for many subcontractors in particular, it is probable that the payment application is an acceptably rough approximation to the true value of the work carried out, especially as they are unlikely to be in a position to reconcile costs correctly either.

Unless tender prices have been manipulated to improve cash flow, and provided that the work has not been unreasonably over or undervalued, the external valuation or payment application may well be 'near enough' to the true value of the work for effective financial control provided that work in progress is added into the calculation. In such circumstances, the real problem is aligning costs to value such that the profit/loss situation may be accurately reported and this is where some contractors and many subcontractors 'lose the plot'! There is no substitute for a CVR in all honesty, and Chapter 18 covers such issues in more detail.

17.9.2 Link to the external valuation

The contractor's internal valuation is closely linked to the external valuation because the external valuation is the starting point for preparing the internal valuation. As part of the cost value reconciliation procedure, the contractor's quantity surveyor has, firstly, to adjust the external valuation to account for the inconsistencies referred to earlier in order to arrive at a 'true value' and, secondly, to account for any work in progress undertaken during the reporting period that has not been included in the external valuation.

The contractor's CVR procedure is geared to regular reporting periods which are delineated by **cut-off dates** determined by management each year. The cut-off dates may be the last day of each month, for example. The CVR report has to demonstrate value and attributable cost at these dates. With this in mind, Figures 17.2–17.4 demonstrate the relationship between the external and internal valuations for the first three months of a contract based on the JCT SBC/Q 2011.

Month 1 (Figure 17.3)
- The contract starts on the 8th February.
- The cut-off date precedes the external valuation date. Therefore there is no external valuation in the reporting period and no payment will be due.
- There is work in progress during the period and this is the work done between the start of the contract and the cut-off date.
- The contractor's quantity surveyor values the work in progress and compiles costs and accruals for the period.
- The cost information is generated by the accounts department and will be available around one week after the cut-off date.
- A CVR meeting takes place where the CVR is discussed, adjusted if necessary, and finalised.
- The CVR report is then presented to top management some 12–14 days after the cut-off date.

Month 2 (Figure 17.4)
- The first interim certificate is due (one month after the contract start date).
- The first external valuation precedes the certificate (usually one week before).
- The valuation is the value of the contractor's work done between the contract start date and the first valuation date.
- Payment follows 14 days after the certificate is issued.
- Work in progress accumulates after the valuation date and up to the second cut-off date.
- The contractor's quantity surveyor adjusts the external valuation for inconsistencies and values the work in progress.
- Costs and accruals are compiled for the period.
- The second CVR report is submitted to management.

Month 3 (Figure 17.5)
- The second interim certificate is due preceded by the second external valuation.
- The valuation is the cumulative value of work done between the contract start date and the second valuation date.
- Payment follows 14 days after the certificate is issued.
- Work in progress accumulates after the valuation date and up to the third cut-off date.
- The external valuation is adjusted for inconsistencies and the work in progress is valued to produce the internal valuation.
- Costs and accruals are compiled for the period.
- The third CVR report is submitted to management.

This example demonstrates how the external valuation forms the basis for the internal valuation, how work in progress is absorbed by the external valuation month by month and how the contractor's quantity surveyor converts this information, along with attributable costs for the review period, into a cost-value reconciliation report for management. Chapter 18 further develops this process.

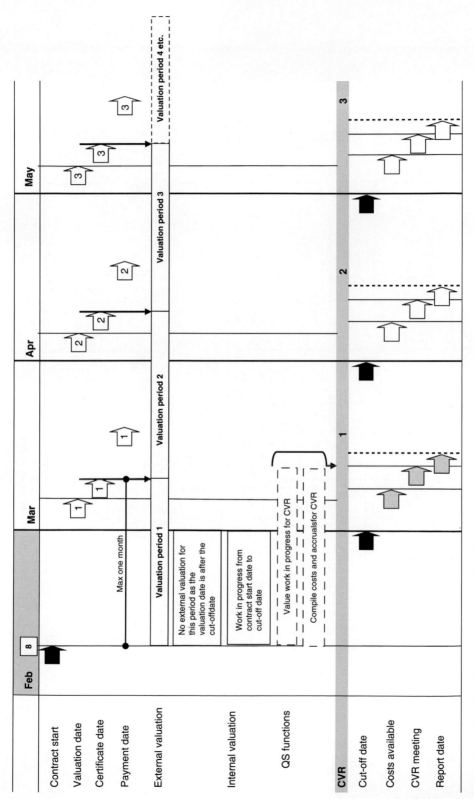

Figure 17.3 Relationship of external and internal valuations – month 1.

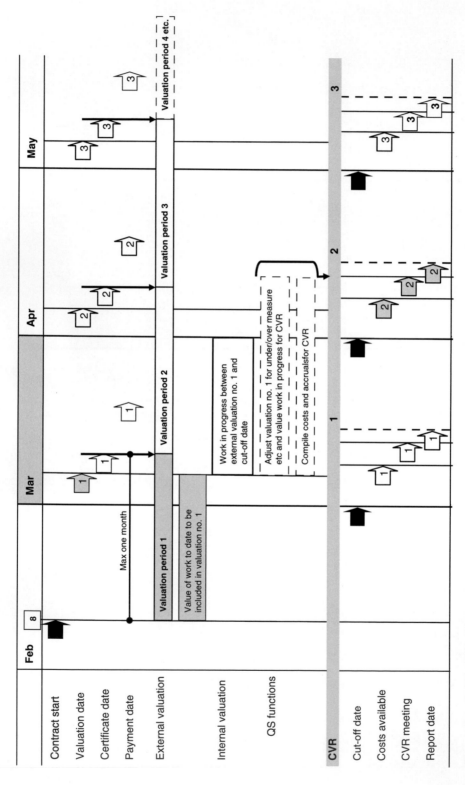

Figure 17.4 Relationship of external and internal valuations – month 2.

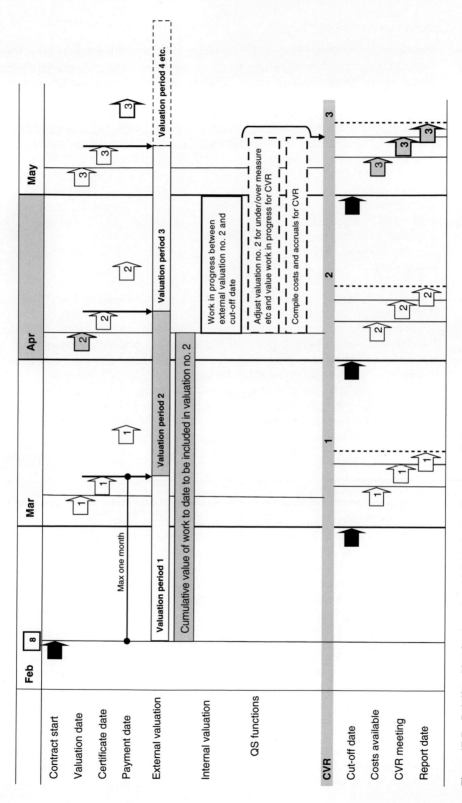

Figure 17.5 Relationship of external and internal valuations – month 3.

17.9.3 Link to the estimate

The contractor's tender stage estimate of the cost of construction is inextricably linked to the internal valuation and forms the basis for making appropriate adjustments to the external valuation in order to determine 'true value' during the construction phase. The estimate is effectively the budget for the project and, once a contract has been awarded, the estimator (usually) will provide an analysis of the estimate broken down into labour, materials, plant and on-costs for each billed item. This creates a 'bill of allowances' or 'net bill' that the quantity surveyor can refer to when adjusting for:

- overmeasure
- undermeasure
- on-costs in advance
- buying gains/losses
- production gains/losses
- and so on.

Realistically speaking, the estimate can never aspire to represent anything more than an approximation of the cost of carrying out the work on site, as the costs of construction can fluctuate significantly between tender and contract award. Additionally, the contracts management staff may decide to employ completely different construction methods compared to those envisaged by the estimator with quite different cost implications to those anticipated at tender stage.

Consequently, the bill of allowances will frequently display wild 'swings and roundabouts' when compared with actual costs and it is often part of the quantity surveyor's job to smooth out these fluctuations with astute buying and sub-contract procurement strategies (depending on the company structure).

17.10 Subcontract valuation

Most subcontracts require the subcontractor to prepare and submit a valuation of work in progress at predetermined times during the currency of their work on site. Late submissions will invariably result in late payment which is a 'no-no' as far as cash flow is concerned.

Additionally, in circumstances where the main contractor is responsible for preparing the external valuation, subcontract payment applications will form the basis of the valuation of the quantum of work in progress (i.e. how much work has been done in the period). This input to the external valuation process is illustrated in Figure 17.1. Notwithstanding, the external value of work in progress will be calculated using the main contractor's rates because (a) they are the contract rates and (b) subcontractors' rates will usually be discounted as part of the main contractor's margin.

The construction industry has a wide variety of types and sizes of subcontractors. Some are very large businesses, some are substantial but the vast majority are small or relatively small companies. Consequently, the quality and sophistication of the payment applications received by main contractors varies enormously. Whilst the principles and common practices that apply to the preparation of the external valuation referred to earlier apply equally well to the preparation of a subcontractor application they are not always universally adopted. This can cause problems.

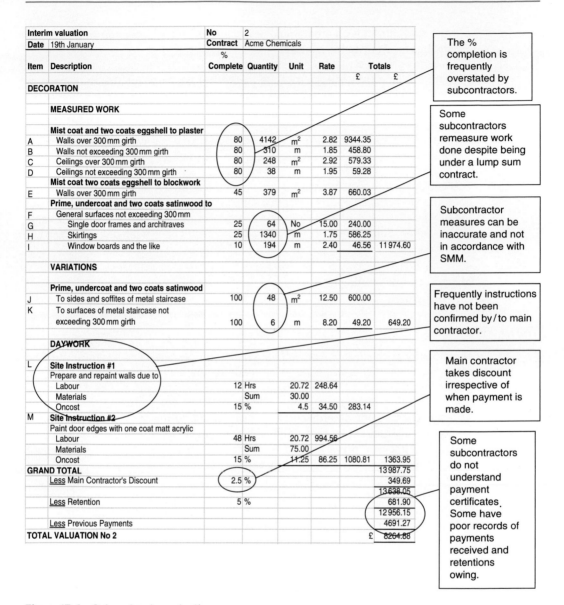

Item	Description	% Complete	Quantity	Unit	Rate	Totals	
	Interim valuation		No	2			
	Date 19th January		Contract	Acme Chemicals			
						£	£
	DECORATION						
	MEASURED WORK						
	Mist coat and two coats eggshell to plaster						
A	Walls over 300 mm girth	80	4142	m²	2.82	9344.35	
B	Walls not exceeding 300 mm girth	80	310	m	1.85	458.80	
C	Ceilings over 300 mm girth	80	248	m²	2.92	579.33	
D	Ceilings not exceeding 300 mm girth	80	38	m	1.95	59.28	
	Mist coat two coats eggshell to blockwork						
E	Walls over 300 mm girth	45	379	m²	3.87	660.03	
	Prime, undercoat and two coats satinwood to						
F	General surfaces not exceeding 300 mm						
G	Single door frames and architraves	25	64	No	15.00	240.00	
H	Skirtings	25	1340	m	1.75	586.25	
I	Window boards and the like	10	194	m	2.40	46.56	11974.60
	VARIATIONS						
	Prime, undercoat and two coats satinwood						
J	To sides and soffites of metal staircase	100	48	m²	12.50	600.00	
K	To surfaces of metal staircase not exceeding 300 mm girth	100	6	m	8.20	49.20	649.20
	DAYWORK						
L	**Site Instruction #1**						
	Prepare and repaint walls due to						
	Labour		12	Hrs	20.72	248.64	
	Materials			Sum	30.00		
	Oncost	15		%	4.5	34.50	283.14
M	**Site Instruction #2**						
	Paint door edges with one coat matt acrylic						
	Labour		48	Hrs	20.72	994.56	
	Materials			Sum	75.00		
	Oncost	15		%	11.25	86.25	1080.81	1363.95
	GRAND TOTAL						13987.75
	Less Main Contractor's Discount	2.5		%			349.69
							13638.05
	Less Retention	5		%			681.90
							12956.15
	Less Previous Payments						4691.27
	TOTAL VALUATION No 2					£	8264.88

Annotations:

The % completion is frequently overstated by subcontractors.

Some subcontractors remeasure work done despite being under a lump sum contract.

Subcontractor measures can be inaccurate and not in accordance with SMM.

Frequently instructions have not been confirmed by/to main contractor.

Main contractor takes discount irrespective of when payment is made.

Some subcontractors do not understand payment certificates. Some have poor records of payments received and retentions owing.

Figure 17.6 Subcontractor valuation.

A typical subcontractor's payment application is illustrated in Figure 17.6, which identifies potential issues that might cause disagreement or dispute. It is the job of the contractor's quantity surveyor to check and adjust subcontract payment applications and it is important to establish:

● The arithmetic correctness or otherwise of the application.
● Whether the subcontract work claimed for payment has actually been done.
● Whether any materials on site claimed are actually on site.
● Whether the site manager knows of any defects in the subcontractor's work that could influence the amount paid.

- If the subcontract is on programme or if there are any possible future contingent liabilities for delay to the contract.
- That daywork sheets have been signed and the work has been authorised.
- Whether daywork is the appropriate method of payment for the work done or whether there are agreed rates that could be used instead.
- Whether any contra-charges are outstanding for set-off against the payment applied for.
- Whether the subcontractor has delayed or damaged the work of other subcontractors that could lead to claims and future subcontract liabilities.

The checking and adjustment of subcontract applications and the recognition of potential future liabilities arising from a subcontractor's work are important parts of the main contractor's internal reporting procedures that culminate in the cost value reconciliation. Subcontract applications and payments have been discussed further in Chapter 6.

17.11   Final accounts

17.11.1   Purpose

The majority of construction projects are not straightforward and it is invariably the case that adjustments will have to be made to the agreed price for a variety of reasons including:

- Design changes and other eventualities that occur during construction leading to variation instructions from the architect or engineer.
- The adjustment of prime cost sums (if permitted under the contract) to account for the actual cost of specialist work carried out.
- Adjustment to allowances made in the contract bills in respect of work carried out under approximate quantities and provisional sums where such work was uncertain at tender stage.
- To take account of contractual entitlement, for example:
  ○ loss and expense for delay and/or disruption
  ○ liquidated damages where the contractor is late completing the contract.

Such items will require changes to the original contract sum in order to determine the amount of the final payment on the contract. Where there is no contract sum, on a re-measurement contract for example, the final sum payable by the employer will, nevertheless, have to be determined taking into account the actual quantities of work carried out and any variations and other eventualities, such as contractual claims and other entitlements.

It is, therefore, usual to prepare a **final account** at the end of a project, the purpose of which is to ascertain the total amount payable to the contractor less previous payments. The term 'final account' does not actually appear in standard JCT contracts and it is more accurate to refer to the **final certificate**, which states the total value of the contract at the end of the project. The ICC standard contracts do refer to a final account, to be submitted by the contractor, whilst the NEC3 contracts simply refer to 'assessment of the amount due after completion of the whole of the works'. For all practical purposes, however, the term 'final account' is widely understood by employers, main contractors and subcontractors alike irrespective of the terminology of the contract.

Main contractors will normally prepare a final account for subcontractors irrespective of whether or not the subcontractors submit their own for the main contractor's verification. This is where the fun starts!

17.11.2 Timing

Under most contracts there is a timetable for preparing the 'final account', for issuing the final certificate and for making the final payment (if any) to the contractor.

The JCT SBC/Q 2011, for example, requires that:

- the contractor provides all necessary documentation to enable a final account to be prepared not later than six months after practical completion;
- the PQS prepares the final account no more than three months later;
- the architect issues the final certificate not later than two months after the defects correction period or after the certificate of completion of making good defects;
- the employer pays the contractor 28 days after issue of the final certificate.

The final account process is usually linked to the defects correction period in that the time for issuing the final certificate follows on from the issue of the certificate of completion of making good defects or the end of the defects correction period, whichever is the later event. Under JCT standard contracts, this is all triggered by the architect's final defects list, which is issued no later than 14 days before the end of the defects correction period.

Individual forms of contract need to be referred to in order to appreciate how the contract sum is to be adjusted.

17.11.3 Preparation

As most building contracts are 'lump sum' contracts, final account adjustments are made on an 'add and omit' basis. This is because the contract sum may not be adjusted other than as permitted under the contract. The JCT SBC/Q 2011 contract, for example, specifies how the contract sum shall be adjusted and what can be added to or deducted from it. The value of variations shall be added to the contract sum when agreed and variations of omission shall be deducted from it. Provisional sums and quantities marked 'approximate' shall be deducted and shall be replaced by the value of the work actually done, quantified in accordance with the method of measurement. Under lump sum contracts, provisional sums for daywork are omitted from the contract sum in the final account and a 'daywork account' is added.

Where there is a re-measurement contract, there is no contract sum until the final account has been settled. Consequently, the actual quantities of work carried out must be ascertained and any variations, contractual entitlements, prime cost or provisional sums must be appropriately dealt with. Thus, provisional sums will be omitted and the value of variations added back. Prime cost sums (if any) will be omitted and the value of the actual order placed by the contractor substituted. Contractor's profit would be added to the actual invoice as a percentage. Tender allowances for attendances on nominated subcontractors would be added without adjustment provided that no material change in the circumstances or value of the work was encountered.

FINAL ACCOUNT SUMMARY

Contract	Goss Street Redevelopment	JCT SBC/Q 2011 Conditions
Main contractor	Relton Construction Ltd	
Date	6th May	

	Omissions (£)	Additions (£)	Notes Commentary
Contract Sum	—	**583 496**	The contract sum shall not be altered other than by reference to a specific contract clause
Adjustment of Contract Sum			
1. Provisional Sums	165 000	136 579	Provisional sums for defined or un-defined work are expended in whole or in part by architect's instruction
2. Approximate Quantities	1000	1765	Approximate quantities are identified in the contract bills and the actual work carried out is re-measured
Variations Account	49 070	83 651	Variations are omitted or added work which is measured and valued according to rules set out in the contract
Daywork Account	5000	8349	Variations that cannot be measured are valued by Daywork
Fluctuations	—	—	Allowances for price fluctuations may be allowable under the contract
Loss and Expense	—	—	Payable where relevant matters materially affect regular progress of the works
Liquidated & Ascertained Damages	—	—	The employer may deduct LADs where the contractor is in culpable delay
Totals	220 070	813 840 220 070	Total omissions are deducted from the adjusted contract sum
		593 770	This represents the contract sum adjusted in accordance with the contract and stated in the Final Certificate
Previous Payments		506 367	Previous payments to the contractor are deducted from the adjusted contract sum
BALANCE DUE TO CONTRACTOR		**£87 403**	The difference between the adjusted contract sum and amounts previously certified is payable to the contractor

Figure 17.7   Final account summary.

Traditionally, quantity surveyors use special paper for the final account process, called 'double cash' paper, which has two cash columns on the right-hand side rather than the usual one – one for omissions and one for additions. Spreadsheets are now commonly used instead.

Figure 17.7 illustrates a simple example of how a final account may be set out in summary form for a JCT SBC/Q 2011 lump sum contract. Notes are included by way of brief description for each item. It should be noted that the final account would contain extensive justification of the final value of the contract and that separate detailed accounts would be prepared for variations, daywork and so on. All additional work carried out should be referenced to architect's instructions and the work should be valued according to contract rates and prices or fair rates if applicable.

17.11.4 The final certificate

The final certificate is issued some time after the completion of the defects correction period or issue of the certificate of completion of making good defects (JCT SBC/Q 2011 - two months). However, the process will hinge on the contractor's obligation to supply information regarding the contractor's entitlement under loss and expense provisions, for example, which would be required to complete the final account calculations.

The final certificate states the difference between the contract sum, as adjusted by the final account process, and the interim payments previously made to the contractor. The balance becomes a debt due to one party by the other. If the PQS has done his/her job properly this money will be owed by the employer to the contractor.

17.11.5 Contractual significance of final certificate

The issue of the final certificate is an important step in any contract. Under the JCT SBC/Q 2011, for example, the final certificate has the effect, in any adjudication, arbitration or legal proceedings arising out of the contract, of being:

- Conclusive evidence that where the standards of materials and workmanship were to be approved by the architect, the quality or standard was to his/her reasonable satisfaction.
- Save for errors, conclusive evidence that any adjustments to the contract sum have been done correctly.
- Conclusive evidence that all extensions of time have been properly given.
- Conclusive evidence that any direct loss and expense payments made to the contractor are in full and final settlement of all such claims.
- Where adjudication, arbitration or legal proceedings have been commenced before issue of the final certificate, the final certificate shall be subject to the terms of any decision, award or settlement made.
- Where adjudication, arbitration or legal proceedings have been commenced within 28 days of the final certificate being issued, the final certificate may only be challenged with respect to the specific issues concerned in the proceedings.
- Where an adjudicator gives a decision after the final certificate has been issued, arbitration or litigation proceedings must be commenced within 28 days of the adjudicator's decision.

The legal issues referred to above can be complex and there is a body of case law on such matters. These are beyond the scope of this book.

References

Banwell, H. (1964) *The Placing and Management of Contracts for Building and Civil Engineering Work*. HMSO, London.

BCIS (2007) *Definition of Prime Cost of Daywork Carried Out Under A Building Contract*. Building Cost Information Service of the RICS (Royal Institution of Chartered Surveyors), London.

BCIS (2012) *BCIS Comprehensive Building Price Book 2013*, 30th edn. Building Cost Information Service of the RICS (Royal Institution of Chartered Surveyors), London.

CECA (2011) *Schedules of Dayworks Carried out Incidental to Contract Works*. Civil Engineering Contractors Association (CECA), London.

Davis Langdon (2012) *Spon's Architects' and Builders' Price Book 2013*, 138th edn. Spon Press, Abingdon, UK.

Furst, S. and Ramsey, V. (2012) *Keating on Construction Contracts*, 9th edn. Sweet & Maxwell Ltd, London.

18

Cost value reconciliation

Financial Management in Construction Contracting, First Edition. Andrew Ross and Peter Williams.

18.1 Introduction

Accounting tools generally fulfil three functions: stewardship, decision accounting and control accounting (Sizer, 1986:18).This chapter seeks to summarise the previous chapters and consider how contractors interpret and apply company level accounting policies to their projects. It reviews how a contractor brings together the costs and allowances to form a management report which is the focus of control accounting.[1] Chapter 2 considered the stakeholder and regulatory environment which *inter alia* described some of the accounting standards that are applied while Chapter 8 considered the issue of governance and stewardship of shareholders investment. This chapter will refer back to these chapters as they form an important backdrop to the cost value reconciliation (CVR) process.

Financial accounting is concerned with stewardship, the preparation of periodic statements in the form of profit and loss accounts and balance sheets that provide a true and fair view to shareholders and meet the requirements of the law. Whilst this chapter considers *management* accounting techniques there is a connection to *financial* accounting. Management accounting combines decision and control accounting. The former includes estimates of costs and revenues associated with particular alternatives; the latter, information to assist management to plan and control effectively.

The reader is referred to Barrett's (1992) seminal text which provided the first detailed consideration of the cost value reconciliation process and highlighted the link to financial management and accounting policies. This chapter refers to Barrett's approach in illustrating how policies are interpreted and applied to project reporting.

18.1.1 Accounting standards

Chapter 2 considered how accounting standards inform the production of a company's published accounts and stressed the importance of the basic principles of accounting:

- prudence,
- true and fair representation,
- the need for recorded information to support this representation.

The company accounts are a representation of the aggregate position of all of its 'live' projects. Live projects can be defined as:

- projects that have been completed and awaiting final account settlement;
- projects that have commenced but for which payment has yet to be made;
- projects that are continuing.

These live projects can be considered as work in progress (WIP) and which will, when summarised with others, form the basis of the published accounts. The

---

[1] Sizer identifies the following features of each:
- '*Stewardship* (incl. tax accounting), e.g. periodic financial statements for external users.
- *Decision accounting*, e.g. estimates of costs and revenues associated with particular alternatives.
- *Control accounting*, e.g. information to assist management to plan and control effectively.'
(Sizer, 1986:18, emphasis in original)

Statement of Standard Accounting Practice No. 9 (SSAP9) for stocks and long-term contracts is intended to standardise the variety of practices that can be adopted by organisations and to provide confidence to stakeholders who are reliant on the financial report that WIP has been calculated consistently (as between companies, and years) and, appropriately, having regard to the detailed requirements of the statement (ICAEW, 1989).

The valuation of a construction company's WIP is difficult as its projects are subject to a wide range of unique project related factors, such as project value, length of contract, contract type, contract payment provisions, valuation procedures, resource costs, types of estimate, credit payment terms, overhead allocation and so on.

The company has to design and implement a system of governance that:

- can apply uniformly across all its projects to provide a true and fair representation between projects and between accounting periods;
- takes account of the external and internal contexts of projects;
- will produce records that can be audited if required;
- demonstrate the application of the concept of prudence at a project level.

The organisation's accounting policies consequently influence the reporting of aggregated business units within organisations and also the reporting of the financial position at a project level. Construction organisations are faced with unique difficulties when compared with organisations in other sectors in reporting on their financial position; these difficulties relate to the idiosyncratic nature of the financial management of projects. The main external factor that influences the reporting processes adopted by organisations is the client procurement arrangement that generally prescribes explicit processes for the valuation of completed work and work in progress. The internal factors that influence the processes are discussed later within this chapter. The key precepts SSAP9 identifies that need to be considered by organisations in their reporting processes relate to cost definition, timing, overheads, net realisable value, losses, claims and variations, profit release and reconciliation.

Figure 18.1 illustrates how the inputs and outputs of the company's accounting system informs the cost value reconciliation process and provides a useful illustration of some of the concepts considered by this final chapter.

18.2 Guiding principles

18.2.1 Costs

Chapter 10 considered the systems used by companies to capture, store and report cost information. The cost report which is generated at the end of the accounting period forms one part of the CVR process and the reader should ensure they have a good understanding of the terms:

- cost cut off;
- direct and accrued costs;
- period and cumulative costs;
- cost provisions.

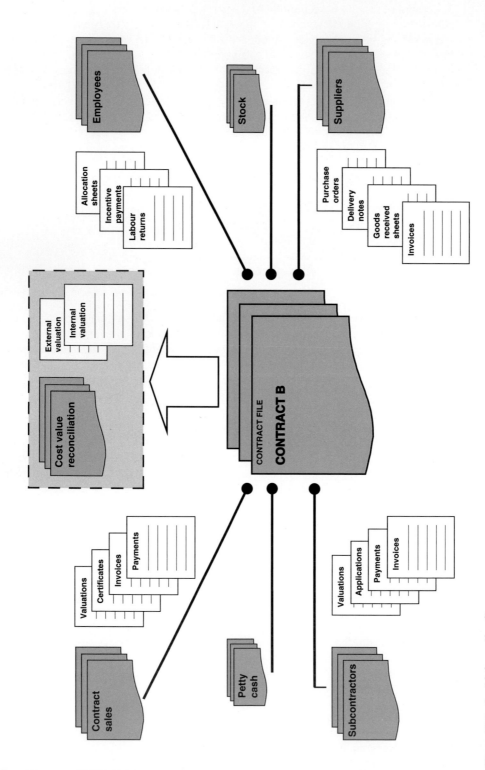

Figure 18.1 Cost information flows.

18.2.2 Site cost information

At a recent three-day course on claim preparation and negotiation, the main speaker, a very experienced claims consultant and barrister, was musing on the frailties of the human memory and suggested that within a period of three weeks most of the delegates would have forgotten 95% of the course content. When asked what was the most important thing to remember, he identified three aspects – records, records and records!

Cost information has a number of attributes. It has a value to the organisation in that it reflects resource usage, it is specific to one activity and site and it has a currency that relates to time and its place within the reporting process. It should be recorded and stored in a systematic manner as it can be used for:

- identifying variations from the original estimate or plan;
- as a basis for the valuation of work;
- a method of resolving disputes;
- supporting applications for payment;
- its storage is an audit requirement;
- it can also be used as a basis for analysis which can improve a company's approach to estimation and planning of future works.

The site office is a key component in any system. It can be considered as a hub for information collection and transmission to the accounts department. The type of information that comes to the site office is often closely associated with the resource, these include:

- labour – this can be company employed or subcontractor;
- materials;
- plant;
- subcontractor;
- temporary work;
- site overheads, for example site staff, cabins, welfare facilities;
- company overheads.

Each of these resources relies upon slightly different systems that reflect the different nature of their deployment on site. A project cost summary report will be produced at the end of each monthly accounting period; an example is shown in Table 18.1.

18.2.3 Labour

There are often two systems in place for the recording of labour costs, one that relates to the payment of the labour that is usually developed from an accounting perspective and one that relates to capturing information about what the labour is doing. The accounting system will require the site to confirm the hours worked and any bonus due, the accounts department will take this information and arrange for payment.

Labour usage is usually recorded weekly on a labour allocation form. The form would include the name of the person, his trade, the activity that was being undertaken and an indication of the time spent. An example of a form is given in Table 18.2.

Table 18.1 Typical cost report format.

Description	Recorded Cost	Accrual	Accrual Adjustment	MOS	S/C Adjustment	Total Cost	Previous Cost	Period Cost
Other G.E.'s								
G.E. Subcontractors								
Staff (and cars)								
Professional Fees / Planning / Building Regs								
Materials								
Subcontract								
Plant								
Bid Costs								
Labour								
Provisional Sums								
Fixed Price–Final price adjustment								
Contractors Contingency / Risk								
Adjudication Bond								
PFI Levy								
Generated Additional Margin								
LADs								
Maintenance Defects								
Tender Margin								
Claims Paid								
Totals								

Table 18.2 Labour allocation sheet.

Contract	XYZ	Labour allocation sheet			
Week commencing					
Day	**Name**		**Activity**	**hours**	**notes**
Mon	F Smith	Labour	Cleaning fmwk	3	
			Moving plasterboad	2	Contra S/c?
			Protecting paintwork	3	

The form is often completed retrospectively by the person carrying out the work, usually at the request of site management and can be the subject of a lot of error. The labour often see little value in reporting what they have been doing, have little knowledge of how the information they report will be used and can bias the data to support a view of their activity that is advantageous to their personal position. For example, a labour-only bricklaying subcontractor may have a contractual agreement to be paid a rate based on a week site measure, the measure may relate to the square metres or number of bricks laid. Often the contract will have provision for payment for non-productive time at an agreed rate per hour. The consequence of this can be that the subcontractor may use the non-productive provision to compensate for lower than expected outputs from the bricklayer. The site management has to be careful about the verification of any hours that relate to non-productive time.

The inaccurate recording of non-productive time can also be a difficulty when capturing data on the resource usage of the company's own labour. To provide an incentive to achieve optimum production targets, many companies use bonus schemes. For example, a joiner may have an agreed bonus rate per square metre of shuttering formwork if completed over an established period of time. (The benefits to the company are that the time-related costs of any other resource related to the formwork such as support are minimised.) The implementation of the bonus scheme needs to be carefully considered as misalignment of benefits can cause one party to over-record the time spent doing non-productive activities and, consequently, increase the bonus paid.

The benefits of records of the deployment of site resources cannot be overstressed. They provide a detailed picture of the day to day activities on site that can be used at a later date to assemble the 'as built' picture. This is often required if a contractor is required to substantiate a claim for additional monies and used to contrast the actual position and compare it with that envisaged at tender. Quantity surveyors acknowledge the lack of accuracy of labour site records, particularly in the number of hours recorded, as they can provide a useful insight into how the project is constructed and provide a basis for further analysis to identify any variations to contract.

18.2.4 Materials

The basis for material cost recording differs from labour as there is usually a process of sending out for quotations and then the placement of an order with a supplier. Consequently, a company has an agreed basis for payment to the supplier and the

Table 18.3 Materials received sheet.

Contract	XYZ					
Week commencing		Materials received sheet				
Date	Supplier	Order number	Description	Quantity	Units	Signature
12-Feb	RMC	123ac	C35 Concrete	30	m³	

site's involvement is usually limited to providing information on materials that are received. A materials received sheet (MRS) (Table 18.3) is usually used and is completed at regular intervals. The MRS are forwarded to the accounts department which uses them as a basis for the establishment of accruals. The site management needs to establish robust systems for input of data as the resultant costs reports produced by the accounts department can be inaccurate if deliveries are misreported or forgotten.

The processes of reporting of costs are typically as follows.

A delivery of material arrives on site, the driver produces a delivery ticket for signature by one of the site staff, the site staff sign to confirm that the materials have been received. A copy of the delivery ticket is held by the contractor and the driver's copy is used by the supplier as a basis for invoice; invoicing is usually monthly. The materials received sheet records the date and details of the delivery, such as quantity, type and the supplier's details.

The MRS will be sent to the accounts department which will use this information along with the suppliers order information to calculate an accrual value for the delivery. The cost will be allocated to a cost heading and aggregated with other suppliers costs for the same heading and then reported back to site. The calculation of accrual values can be a source of error – if an incorrect order value is used or an incorrect quantity allowed for, the resultant accrual could be inaccurate, which would make the summary cost too high or low. The consequences of this are that a project could be significantly over or under-reporting its profitability. Most cost reports therefore include a column which indicates movement within an accounting period, if a cost movement appears to be unreasonably high (or low) it would prompt further investigation of the detailed cost records.

The supplier's usually invoice at the end of each month. The delivery tickets are used as a basis for establishing the amount delivered to site and substantiate the invoiced total. Inaccuracies in reporting costs from a contractor's perspective can relate to lost delivery tickets. This is easily done as often the materials are delivered directly to the area of operation, for example concrete, and the tickets can be signed by an operative who is focused upon the construction activity.

18.2.5 Plant costs

Plant costs are generated by either owning or hiring plant. In both cases a system of recording what plant is on site, who owns it, when it was delivered and when off hired is required. Plant can include anything from site cabins to rubbish skips. The system of recording and establishing a cost is similar to that of materials. If the plant is hired

Table 18.4   Plant hire record.

Contract XYZ		Plant hire record				
Week commencing						
date	Supplier	Plant	Delivered	Collected	Days hired	Consumables

it is usually delivered to site and a hire ticket is signed to confirm delivery and commencement of the hire period. This ticket forms the basis for the plant hire company's monthly invoice.

A plant hire form is usually used; an example is given in Table 18.4.

18.2.6   Subcontract costs

Subcontract costs are usually estimated by the quantity surveyor on a monthly basis and are often referred to as subcontract liabilities. They are called liabilities as they are what the contractor estimates the subcontractor is due if all claims are pursued. In this way they can differ from what has been paid as by the subcontractor are often included. The subcontract liabilities are calculated on a cumulative and gross basis and any discount offered will be shown on the form. The quantity surveyor has to demonstrate to the senior commercial management that a prudent approach has been taken to the calculation of the liability and that any future losses/profits against the subcontractors work package can be estimated. Contracting companies have developed forms that capture the essential information; an example is given in the following.

Muckshift Contracting Ltd has a subcontract order based on a priced bill of quantities for the groundwork element of contract 123a. It has completed its second valuation of the work and has submitted an application for with an gross order value of £550 000; their first application was £60 000. During negotiations cumulative to agree a contract sum, it agreed to gross offer the contractor 8% discount. This gives a net application of £133 400.

The main contractor's quantity surveyor has calculated an allowance for the works completed using the rates in the original estimate; this allowance is usually net of profit or overheads. In the case of Muckshift's work package the allowance has been calculated at £150 000. The quantity surveyor has identified that Muckshift could potentially pursue a claim for out of hours working as it had been carrying out works on the instruction of the site agent during the weekend. The quantity surveyor calculates this to be worth a gross total of £4000 and consequently calculates the cumulative gross liability to be £137 080. A cumulative of £133 400 (145 000 less discount) has been made for application 2. This is illustrated in Table 18.5.

To provide further explanation of the form, the expanded details of each column are:

● Subcontractor order number – some companies use an alpha numeric system to indicate the type of subcontractor, for example the prefix LO could represent labour only.
● The subcontractors name is self-explanatory.
● The subcontractor application (S/C appl.) is inserted here as a cumulative net figure, it will not include any claims and variations, it will also exclude materials on

Table 18.5 Subcontract liability form.

S/C Order No	Name	S/C appl.	Var.	Allow.	Valn	Paymt	S/C adj.	MOS	Total Liability	Variance (Allowance/ Total Liability)
					Contract To Date					
12	Muckshift	133	0	150	137	133	4		137	13

Table 18.6 Subcontract liability form indicating monthly variance.

Contract To Date									Period		
S/C appl.	Var.	Allow.	Valn	Paymt	S/C adj.	MOS	Total Liability	Variance (Allowance / Total Liability)	Allowance	Liability	Variance (Allowance / Liability)
133	0	150	137	137	4		137	13	64	55	9

site as they will be dealt with in a separate column. The reason for this is explained in Chapter 8.

- The variance (Var.) column - this figure is included to provide senior management with a reference to how much the subcontractor considers its work package to be worth including all claims and variations and when comparing with the total liability can give an indication of potential conflicting views of valuation between site staff and subcontractor. In the above example a figure of zero is included as the subcontractor has not recognized its entitlement for payment for out of hours working.
- The allowance (Allow.) column - this figure is the 'external' amount included within the estimate, that is that which was originally allowed by the estimator to carry out the works. It differs from the subcontract valuation as it is usually included in the estimate at an earlier stage than when the subcontract package was let. It is usually calculated net of overheads and profit. This is £150 000 which is the contractor's cumulative budget for carrying out the works undertaken by the subcontractor to date.
- The valuation (Valn) column - this figure is the contractor's quantity surveyor's valuation of the subcontract work package, it will include for any variations and, claims. It can differ significantly from the subcontractors application as there can be differences of opinion on measurement, status of claims, variation evaluation and so on\in Table 18.5 this is the sum of the net valuation (£133 400) and the net valuation of the out of hours working (£3680).
- The payment column indicates the cumulative net payment due to the subcontractor.
- The subcontract adjustment column indicates any adjustments made to the subcontract valuation; in the case of Muckshift's valuation an inclusion of £4000 has been made for out of hours working.

The form could be expanded to include the movement in the accounting period under consideration, as illustrated in Table 18.6.

This information is passed to the accounts department for the production of the cost reports. It provides information that allows judgements to be made on the progress of the works and the profitability of the subcontract element of the works (as well as the origin of the profit, i.e. whether it has been made as a result of a 'buying' gain).

The calculation of subcontract accruals is usually based upon when the subcontractor is paid and/or the status of the paid figure on the subcontract liability sheet. Some companies consider that the paid figure can be considered as 'direct cost' and that the difference between the valuation and the subcontract payment could be defined as an accrual. In this example this would £4000.

18.3 Cost reporting

The costs are classified under defined headings and in some cases the classification system is very comprehensive. Table 18.7 shows an example of a classification scheme for staff costs and site cabins. In some cases the classification system can encompass over 230 headings.

It would be unwieldy to report upon this number of headings and most companies

Table 18.7   Overhead classifications.

Code	Description
G0110	SALARIES & EXPENSES
G0115	AGENCY STAFF
G0168	MISCELLANEOUS STAFF
G0180	ENTERTN/DISALLOW EXP
G0210	S/KEEPER & WELFARE

reduce the cost report to a summary that is then used in reconciliation. An example of a summary report is given in Table 18.8.

A robust system of capture and classification of resource data from site is essential to provide the basis for accurate and reliable cost reports. The accountant or commercial manager will allocate head office overheads to projects in order to ensure that non-project costs are covered. The classification system will allow for information to be aggregated into a summary report that can be 'drilled down' into if further analysis is required. Companies will have monthly cost cut-off dates which will be used to calculate accrual values, period end adjustments and period movements. The cost report is in a standard format and most companies have training schemes for their staff to ensure that human error in classification or calculation is minimised.

18.3.1   Timing

SAPP9 stresses the importance of establishment of profit in a particular accounting year and the need to match costs with revenue. This can be sometimes problematic in its interpretation at contract level. Often a contract will incur a cost for which there will be an expectation of future revenue. For example, a contractor may have incurred some costs which, due to an oversight, have not been included within the application for payment. There are several ways the contractor could deal with this situation. A negative accrual could be put into the costs to adjust them for the

Table 18.8 Summary cost report.

Description	Recorded Cost	Accrual	Accrual Adjustment	MOS	S/C Adjustment	Total Cost	Previous Cost	Period Cost
Other G.E.'s								
G.E. Subcontractors								
Staff (and cars)								
Professional Fees / Planning / Building Regs								
Materials								
Subcontract								
Plant								
Bid Costs								
Labour								
Provisional Sums								
Fixed Price - Final price adjustment								
Contractors Contingency / Risk								
Adjudication Bond								
PFI Levy								
Generated Additional Margin								
LAD's								
Maintenance Defects								

non-payment in the period under consideration, an undermeasure could be included within the net valuation to adjust for the expected under-recovery of monies or the costs could be carried forward into the next accounting year. Obviously, a rule is required for consistency and SAPP9 states clearly this rule, it is to match cost and revenue in the year in which the revenue arises rather in the year the cost is incurred. This would mean the cost would be carried forward into the next accounting period.

This approach is straightforward. If costs are incurred and there is no chance of future revenue to which they could be apportioned the costs should be taken into account in the current accounting period. This is where the prudence concept comes to the fore but, it is possible to see how a bullish contractor could interpret this concept loosely. In many cases during a contract, claims are made for additional monies for which costs have been incurred. Some of these claims are likely to be extremely speculative-, it is not unknown for contractors 'to fly kites' during the contract to see whether a potential claim may be accepted. The contractor, therefore, must be prudent when assessing future revenue and it should be certain and demonstrable before it is taken to value.

18.3.2 Overheads

A company has to allocate its overhead costs, which can be categorised as those relating to site and those relating to head office. This is overly simplistic; some head office overheads, such as planning, often provide a service to sites. Senior contracts managers will spend proportions of their time on individual contracts but will also spend time supporting the estimating function. Some definitions and rules for allocation and apportionment are required. SAPP 9 states that production overheads:

> 'should be classified according to function (e.g, production, selling or administration) so as to include the inclsion, in cost of conversion, of those overheads which relate to production, notwithstanding that these may accre wholly or partly on a time basis.'

This definition gives rise to some interesting questions regarding overheads associated with the various different categories. The head office departments such as estimating, buying and planning do not directly relate to the production function indentified in the above SAPP9 definition. They may be cost centres in their own right and have to report on their cost recovery. Most companies will have a policy that is followed for staff head office overhead costs; they may be either allocated as a percentage of project costs on a month by month basis, charged to projects based on an daily allocation or a combination of the two. The only advantage of daily allocation to projects is that it can be perceived as keeping overheads low but can run into difficulties in the time spent recording the data.

The company will maintain a monthly turnover budget and overhead budget to monitor the recovery of the head office overheads. All projects will project their turnover on a monthly basis until completion and this will form the basis for allocation. If the predicted turnover is lower than expected the contractor has only two options, either increase the overhead percentage addition to the ongoing projects or reduce the costs.

Project overhead costs are reported in a similar way to other resources. Headings such as staff, plant, utilities and subcontractors are often used to categorise the overhead costs. It is important to report upon these separately as any variances in revenue to cost provide useful information to the site team about the performance of the site. Often 'unfair' allocations of cost are made by head office; for example under-deployed site-based staff will be allocated to continuing contracts, irrespective of whether they have been used by the site, or an over-allocation of a contract manager's time may be made to a contract being completed. Unfair?, only in the sense that the site team cannot recover these costs and, consequently, the project has to carry the unplanned costs; in such a case, it would be unfair to judge the site teams' performance on the overall project profitability which will have been reduced as a consequence.

Some occasions arise where the revenue for a site overhead is received at a different time to the costs being incurred:

- A design and build project may have an arrangement whereby the design team is paid a high proportion of its fee on winning the project. The recovery of this cost may be spread over the length of the contract. In this case, using the principle of matching costs to revenues, the design fees would be paid but the cost allocated to the contract would be in proportion to the fees recovery from the client over the duration of the contract.
- A scaffolding company may require payment for the supply and erection of a brickwork access system and include it within its application for payment. The estimator may have built the costs of the system into the unit rates which will be recovered on a basis of production. In this case, the subcontract costs would appear to be too high at the start of the brickwork operation and a rule is required. The rule for calculation of project overheads in this case is based upon the principle of matching cost with revenue which can again be applied. The scaffolding subcontract liability should be based upon the brickwork production achieved rather than the subcontract valuation. This means that the subcontractors (correct) valuation would be higher than the accrual value (assuming no buying gain). This may cause difficulty when the commercial manager identifies a payment (i.e. that includes the access system) exceeding the accrual value; however, an explanation can be given.

The allocation of head office and site overhead costs can be problematic. However, it is important for companies to have clear rules for apportionment and allocation and to have systems which allow for the auditability of the costs should variances from what was expected occur.

As discussed in Chapter 10, the cost report is produced by the accounts department, which relies upon the company's accounting information system. The reports are completed monthly and can be either downloaded from the company's intranet or emailed to the site. The report should contain no surprises for the quantity surveyors; the subcontract costs originate from the liabilities, which, in turn, relate to the subcontractors valuation of the work. The quantity surveyor has completed the liabilities and the payments, therefore this element does not take too long to check. The period costs for each of the resource headings is the area that is checked first; any unusual costs within the accounting period will highlight potential under/over accruals and be a prompt for further investigation that may require inspection of site and accounting records.

Almost all cost information needed for analysis is stored within linked databases in the company system; it is relatively easy for run a query to identify the records that relate to a particular variance. The next task is to explain the variance from the allowance; this requires an understanding of the value component of the CVR.

18.4 Net sales value (NSV)

Net value, net realisable value, net sales value are all interchangeable terms that are used to describe the value of a project. It is important for the reader to understand that the net value is **not** any of the following:

- contractor's value of the works as stated within the application for payment;
- client's representative's value of works as stated in the interim certificate;
- costs of the work as reported in the cost statement.

The net realisable value is defined by SAPP 9 as:

'The actual or estimated selling price (net of trade but before settlement discounts) less:

(a) *All further costs to completion; and*
(b) *All costs to be incurred in marketing, selling and distributing'.*

This is better defined by Barrett (1992) as *'the contractor's valuation of the works in progress in net terms after allowing for future losses and provisions for the defects liability period, remedial works or any other costs that can be foreseen'.*

As discussed in Chapter 17, the convention within the industry is to value projects using gross cumulative figures. However, whilst this is useful to give a macro picture of how a project is performing, most companies will undertake monthly calculations of net value and report upon both cumulative and period positions. A comparison of a project's net value with costs will provide an assessment of whether the project is progressing to the financial plan developed by the estimator at tender stage. If the NSV is below the cost, the project is in trouble and the site team needs to know where the problem originates from; if this is the case the problem is one of cost, however the underlying reason for the cost increase may be:

- Inability to procure the resources within the budget established at tender stage (buying loss). This could be subcontract, materials, plant-, for example subcontract post-tender quotations higher than budget.
- Inability to manage the resources within the outpts established at tender stage (prodction loss). This could also relate to any resource, for example labour unable to meet production outputs used for shuttering formwork.
- Poor control of resources on site, for example materials wastage, theft.
- Inability to meet programme and the incurring of additional overhead costs.

It is important for the governance of the company that the system of reporting NSV and project performance captures the site team's assessment of the variances from that of the original position. It is also critical that the variances are extrapolated, where appropriate, to the completion of the project and a final NSV/cost position be established.

As is clear from the above, there can be differences in definition which can have an influence on how a company reports the performance of its projects. Often notes are attached to the annual company accounts explaining how the accounting policies relating NSV, cost , attributable profit and losses are applied.

18.4.1 Calculation of net value

Cost acts as a common denominator by representing the deployment of resources in a way which permits their accounting. Cost *sits* on one side of the simplest equation whose object is to identify the profit from the activities of the enterprise. On the other side of the equation value has to be calculated. Accounting conventions require the identification of 'attributable profit', and it is therefore necessary to present a sufficiently detailed account of the project's accomplishments for profit to be identified in such a way that there is confidence it will satisfy the accountancy tests. As illustrated in Chapter 17 the approach that most organisations adopt is to commence with the application for payment and make a number of adjustments for over and undervaluation, claims and work in progress. The next section of this chapter will use a simple example to illustrate the concepts.

The application for payment (the processes for this are outlined in Chapter 17) forms the starting point for the calculation of the net valuation of the project. The main precept for the evaluation of the works is to assess the net value of the works complete at the same date as the costs are reported. As identified earlier, organisations apply a common date for the production of a cost statement; this date is referred to as the cost cut-off date.

The other time-related process that is relevant to the assessment of net value relates to the timing of receipt of income. To minimise an organisation's cash requirement it will identify a date by which income is expected to be received, a **cash cut-off** date. The cash cut-off date can influence when the contractor agrees the date of the external valuation of the works with time allowed for negotiation, certificate issue and client payment of certificates. Consequently, the date of the external valuation and cost cut-off date are often misaligned and there is a need for adjustment for work in progress (WIP) to ensure an equitable comparison is made between net sales value (NSV) and costs. This was considered in Chapter 8. The two other adjustments to the external valuation in order to arrive at the NSV are for overvaluation and undervaluation of the works; this is considered later in this chapter.

In Table 18.9 the cost report is produced at the end of week 4, the valuation for the project is undertaken in week 6. The company has to decide whether to report

Table 18.9   Adjustments for work in progress.

Period	Week				
1	1				
	2				
	3				
	4	Cost report		Cost cut off p1	CVR 1
2	5			Work in progress	
	6	Valuation			
	7				
	8	Cost report		Cost cut off p2	CVR 1a
	9			Cash received cut off	

the project position at week 4 or week 8. If it reports at week 4 (CVR 1), the WIP can be considered as an overvaluation; if in week 8 (CVR 1a), the WIP will be treated as an undervaluation. This is to ensure that the contract NSV is adjusted to the corresponding cost cut of date.

18.4.2 Calculation of profit

The statement of accounting practice defines 'attributable profits' as:

> '*That part of the total profit currently estimated to arise over the duration of the contract (after allowing for likely increases in costs so far as not recoverable under the terms of the contract), which fairly reflects the profit attributable to that part of the work performed at the accounting date.*' (SSAP9, Part 2, paragraph 23)

The release of the profits generated by projects is an area where most companies diverge from the recommendation of the SSAP9 guidelines, which categorise projects into short and long term. A long-term contract '*is entered into for the ... manufacture or provision of a service for a period exceeding a year*'. There can be no attributable profit until the profitable outcome of the contract can be assessed with reasonable certainty, that is profit can be taken at interim stages only if the outcome at final account can be assessed with some confidence.

SSAP9 also suggests that if the contract is short term no profit can be taken until completion. Due to the scale of projects within construction, projects completed in a year can be valued at £50m, or a company may have a number of projects that will not be completed within an accounting year. Thus, most companies will release profit from their projects at an interim stage. They do, however, comply with the guidelines in preparing a statement of assessment of the outcome of a project.

As identified previously, a CVR that will allow for the accurate assessment of cost, net value and attributable profits (or losses) will be undertaken every month. These will be stated both cumulatively (year to date) and monthly (accounting period) and be forecast to contract completion.

A **cost and value to complete exercise** is often undertaken as part of the process of producing the reconciliation. The relative certainty of outcome obviously changes as the project progresses and this is often reflected in the documentation supporting the cost and value to complete exercise. In the early stages the estimate and allowances are used. In the later stages resource programmes and subcontract procurement schedules can used; this is a lot more accurate. In order to assess the extent of profit release a regular assessment of the financial progress of a project is required.

18.5 Losses

18.5.1 Foreseeable losses

The next component (the second adjustment to 'cost') is '*foreseeable* losses', which are defined in SSAP9 as '*Where a loss on a contract as a whole is foreseen, a proportion of the overall loss, calculated either by reference to time .... or to the expenditure incurred may normally be taken into account ...*' which must be incorporated as a deduction. This self-evidently requires an assessment of events in the future, both their occurrence and their consequences, expressed in economic terms.

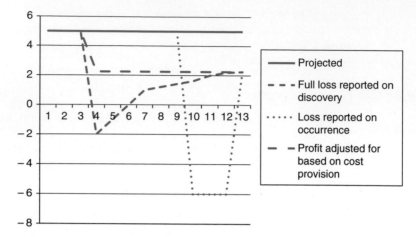

Figure 18.2 Impact of the timing of the reporting of losses.

This assessment therefore entails the formulation of probable events, the resource implications and the translation of the resources into some economic measure. All of this is problematic, clearly and manifestly. If an organisation is likely to make a procurement loss, which has been identified during the project, a cost provision is normally made within the cost statement; the extent of the provision would usually relate to expenditure. The application of judgement as to the scale of the loss and when to declare the loss within the accounts is a further area of uncertainty in cost accounting.

An example of this in practice may be where a loss is made on 'asphalting' due to rises in the oil price. Say, the asphalting on a contract is to be undertaken in month 10 but the main contract is based on a fixed price and, due to pressures to remain competitive, the estimator may have based the tender price upon a subcontract quotation, which was only fixed for three months. In month 4, the site quantity surveyor seeks prices for the asphalt only to find that the contract allowances (based on the subcontract quotation) cannot be achieved and consequently estimates that the buying loss will be £40 000. If the buying loss is reported when first identified it would unnaturally skew the profitability of the project; if it is ignored until the work is carried out the profit would be shown as being too high in the preceding periods (Figure 18.2).

The ideal scenario would be to take a cost provision identified against time for each month of the contract and report the adjusted profit at a constant amount from month 4, as indicated by the last data line in Figure 18.2.

18.6 Claims and variations

Due to the uncertainty of the construction process, variations and claims arise on virtually every project. The evaluation of a variation or a claim is often prescribed by the contract and can vary from being based on the contract estimate to quantum meruit. The difficulty in accounting for the costs and value for variations relates primarily to the likelihood of recovering costs. SSAP9 identifies that if the variation is approved (there is a judgement to be made regarding the definition of 'approved') and that an amount is to be received, but settlement has yet to be reached, then a conservative estimate should be made.

Some organisations require the commercial manager to identify the extent of the NRV that is derived from conservative estimates of claims and variations. Others require the amount to be identified separately and senior management takes the decision as to the 'safety' of taking the claim or variation to value.

SSAP9 identifies that, in order for a value allowance to made for income from claim, *'Provision can only be made when negotiations have reached an advanced stage and there is evidence in writing of the acceptance of the claim and an indications to the sum entitled to'*.

However, the reality of construction contract administration means that often claims and variations are aggregated in order to settle accounts at the end of a project and parties tend to be reluctant to agree to the acceptance let alone the sum entitled to. Consequently, senior management must take account of the relationships that exist between organisations before making a judgement concerning the value of claims and variations

Almost all conditions of contract have explicit procedures as to how to manage uncertainty and variations that arise on contracts. The definition of 'claim' and 'variation' is usually clear, notices or instructions usually precede the events and valuation is done using a set of explicit processes, which refer to either the original priced documentation, market rates or quantum meruit.

The evaluation of variations and claims within a project CVR is problematic, as identified earlier; the internal allowances associated with a variation or claim are allocated to the appropriate resource (labour, plant, materials, subcontractor or on cost) and the senior management has to decide when and how much of a claim and the associated allowances to take to value. Some organisations net the internal valuation of all variations and claims that are in dispute and rank them as likely, probable or unlikely.

18.7 Valuation: application and internal valuation

The internal valuation of a project can be considered as an evaluation of the resource allowances that the estimator made for the project at tender stage. The allowances are usually 'internal' to the reporting organisation and typically are for labour, plant, materials, subcontractors and on-cost. It is rare for the allowances to be disaggregated to a finer detail despite this information being produced during the estimating process. Typically, the documentation that the quantity surveyor will have to develop the internal valuation will be a 'bill' or schedule of allowances under the above headings. The external valuation is the basis for the calculation of the project resource allowances that will be adjusted and then reconciled with the costs Table 18.10. A variance from this practice would be when a valuation had not taken place and there was a need to reconcile work in progress to cost cut off. An assessment of WIP would be made based on site records.

The internal valuation commences almost immediately after the external valuation has been submitted, the time taken for its compilation is dependent of the complexity and scope of the project and two weeks for its calculation is not unusual. Often allowances will be reallocated as subcontract workpackages may not be as originally intended at tender stage. If variations or claims have been included within the external valuation the commercial manager will use his/her judgement as to where to allocate the appropriate allowances. The principle of matching expenditure to income is generally followed; however, there are few explicit rules on this. The completed internal valuation should correspond to the external application for

Table 18.10 External and internal bill.

External BQ ⟵⟶ Internal BQ

Ref	Description	Quantity	Unit	Rate	£	Materials Rate	Total	Labour Rate	Total	Plant Rate	Total	Subcontract Rate	Total	Temp Materials Rate	Total
A	Excavating To reduce levels 1 m maximum depth	7290	m³	4.16	30 326.40		0		0		0	3.62	26 389.80		0
B	For pile caps and ground beams; between piles 1m maximum depth; commencing from reduced level	1043	m³	12.95	13 506.85		0	3.09	3222.87	8.17	8521.31		0		0
C	In situ concrete Ground beams Reinforced ≤15%	269	m³	105.77	28 452.13	76.86	20 675.34	15.11	4064.59		0		0		0
D	Formwork Sides of ground beams Plain vertical: height 0.5–1.00m	186	m	32.92	6123.12		0	17.27	3212.22		0		0	11.36	2112.96

Notes:

1 The shaded area is an extract from the contract bills ie the **External bill of quantities.**

2 To the right of this is the **Internal bill of quantities.** This is a breakdown of the contract bill rates into materials, labour, plant and temporary materials less the contractor's margin for overheads and profit (in this case = 15%).

3 Temporary materials include items such as timber for formwork and earthwork support. They are not part of the permanent works and therefore cannot be included in the valuation of materials on site.

4 The total of the Internal BQ represents the contractor's overall budget for the contract less overheads and profit.

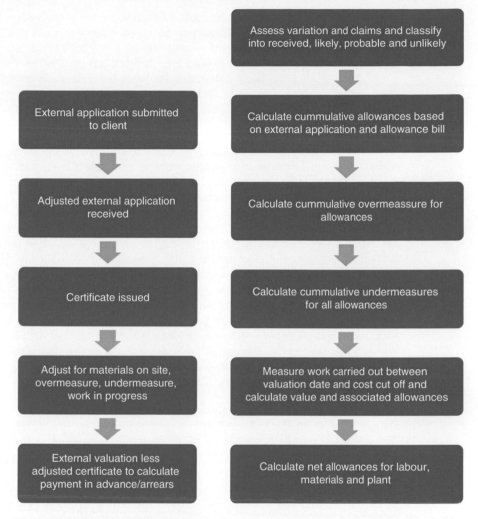

Figure 18.3   Adjustments to valuation.

payment. The next process to be undertaken is the adjustments in order to arrive at the net valuation of the works (Figure 18.3).

18.7.1   Adjustments: overmeasure

The commercial benefit of overvaluing the works is primarily an enhancement of cash flow and the information asymmetry that exists between the parties in most procurement arrangements makes it a common practice. The client's representatives are aware of the practice and often apply hidden retentions i.e. purposively undervalue elements of the work to balance against the overvaluation practice. Overvaluation can take place due to unbalancing of tenders, recovery of enhanced margin due to variations, claiming for works that have been completed prior to their actual completion, and claiming for entitlements such as claims and variations whose entitlement is questionable. The overvaluation of the works can be defined as 'application for monies in advance of their eligibility'. The allowances for these items should be deducted from the gross internal valuation of the works in order to arrive at the net sales value. These can be categorised into on-cost in advance, measured work and disputed items. In order for a contractor's senior management to judge whether the profits indicated on a project are secure, the site quantity surveyor is often required to categorise the disputed valuations to indicate the likelihood and extent of recovery at the end of the project.

18.7.2 Adjustments: undermeasure

In a similar fashion to overmeasure, an undermeasure schedule will be produced which will identify those items where work is completed that costs have been incurred for but have not been included in the application for payment. The inclusion of a large element of undermeasure is a signal of failure of the commercial management systems in identifying work completed or entitlement. The allowances that are derived from the undermeasure calculation process are added to the internal valuation in order to arrive at the net internal valuation of the works. Work in progress, - defined as 'work that has been carried out in the period between the cost cut-off date and the valuation date'-, is treated as an undermeasure if the cost cut-off date is after the date of the external valuation. This is the usual position on the reporting of projects and can cause the commercial manager a lot of administrative efforts as it entails two evaluations of the work, one at the date of external valuation and a further one at the date of the cost cut off. Normally, the WIP is only adjusted for internal labour and materials, as the subcontractor liabilities are taken as costs and are aligned with the external valuation.

18.8 Development of the internal valuation: an example

An example of an allowance bill for three items has been used as the basis for a valuation. The allowances are for totals for labour, materials, plant and subcontract (Table 18.11). These originate from a more detailed breakdown, which includes production outputs for labour and plant, wastage allowances for materials, subcontract rates and comparisons and discounts for subcontracting. These breakdowns will be referred to as the example develops. The estimator will allow for an on-cost which may include general expense items, commercial opportunity and /or margin.

The rate per unit can be considered as an external rate as it includes an 'allowance' for the components that make it up and it is this rate that appears in the priced bills of quantities.

A site measure to support the application for payment for the works is completed up to a certain time (Table 18.12). The contractor's quantity surveyor may have inadvertently over or under applied for certain items in the application and one of the first processes in arriving at the net valuation is to adjust for this.

Table 18.11 An example from an estimate indicating allowances for each resource heading.

Item	Quant	Rate	Labour	Materials	Plant	Subcontract	On-cost	Total
1	3400	12.45	10582.50	21165.00	4233.00	4233.00	2116.50	42330.00
2	2500	34.78				82602.50	4347.50	86950.00
3	150	55.34				7885.95	415.05	8301.00

As can be seen from Table 18.12, item 1 has been undermeasured by 10%, items 2 and 3 both overmeasured by 10%. In order to arrive at the net allowances for these items an adjustment is required. Firstly, the external application for payment is broken down into resource headings (Table 18.13).

The allowances before adjustment for labour (6879), materials (13 757), plant (2751), subcontract (43 866) and on-cost (3540) shown in Table 18.13 are calculated by multiplying the application percentages to the total allowances in Table 18.11. In order to adjust or "net" these allowances to represent work complete, an overmeasure and undermeasure calculation is undertaken, as indicated in Tables 18.14 and 18.15.

The net allowances are calculated by deducting the total overmeasure allowances from the applied for allowances and then adding the undermeasure allowances. This

Table 18.12 Contractors application for payment.

Valuation No. 2			
Item	Application %	Site measure %	Ext. Valuation
1	65	75	27 514
2	45	35	39 128
3	50	40	4151
Total			70 793

Table 18.13 Application for payment allocated to resource allowances.

Valn 2								
				Allowances				
Item	Application %	Site msr. %	Ext. Valn	Labour	Materials	Plant	Subcontract	On-cost
1	0.65	0.75	27 515	6879	13 757	2751	2751	1376
2	0.45	0.35	39 128	0	0	0	37 171	1956
3	0.5	0.4	4151	0	0	0	3943	208
		Total	70 793	6879	13 757	2751	43 866	3540

Table 18.14 Overmeasured items and associated allowances.

Valn 2	Overmeasure schedule				Allowances		
Item	Overmeasure	Labour	Materials	Plant	Subcontract	On-cost	Total
2	0.1	0	0	0	8260	435	8695
3	0.1	0	0	0	789	42	830
Total					9049	476	9525

Table 18.15   Undermeasured items and associated allowances.

Vaⁱn 2	Undermeasure schedule						
Item	Undermeasure	Labour	Materials	Plant	Subcontract	On-cost	Total
1	0.1	1058	2117	423	423	212	4233
Total		1058	2117	423	423	212	4233

Table 18.16   Total net allowances.

		Net allowances					
Valuation 2	Labour	Materials	Plant	Subcontract	On-cost	Total
Applied for	6879	13757	2751	43866	3540	70793
Overmeasure	0	0	0	9049	476	9525
Undermeasure	1058	2117	423	423	212	4233
Total net allowance	7937	15874	3175	35240	3275	65500

calculation arrives at the allowances that would be derived from applying the site measure; however, the external application for payment is usually used as it provides an external representation of the valuation of the works. Table 18.16 indicates the calculation of the net allowances.

18.9   Reconciliation

The costs against each resource heading are input from the cost report and the calculation of the cumulative and period variances is completed. The variances give the senior management an indication of the performance of the project in the last accounting period and also the contract to date. This allows strategies for remedial action to be discussed at the monthly reconciliation meeting.

The above simple example has used just three items to illustrate the general principle of arriving at net allowances from a contractor's application for payment. In this example, the estimate was broken down under separate headings of labour, plant, materials, subcontract and on-cost. In some cases, the estimate is significantly more complex than that illustrated, preliminaries are apportioned to measured items or to general expense items within the preliminaries section of the documentation sent to the client, rates may be loaded for commercial reasons and professional fees and provisional sums may be included (particularly if the project is design and build).

18.9.1   On-costs in advance

On-cost was used in the previous example as a separate heading. It can be defined as project contribution and can include profit as well as overhead contribution. All evaluation of work includes an element of on-cost. Consequently, if works have been

Table 18.17   Amended estimate with on-cost loaded onto item 2 and unloaded from items 1 and 3.

Item	Quant	Rate	Labour	Materials	Plant	Subcontract	On-cost	Total
1	3400	11.97	10583	21165	4233	4233	500	40714
2	2500	35.59				82603	6379	88982
3	150	52.57				7886	0	7886

Table 18.18   Revised gross allowances with on-cost loadings.

Item	Application %	Site msr. %	Ext. Valn	Labour	Materials	Plant	Subcontract	On-cost
1	0.65	0.75	26464	6879	13757	2751	2751	325
2	0.45	0.35	40042	0	0	0	37171	2871
3	0.5	0.4	3943	0	0	0	3943	0
Total			70448	6879	13757	2751	43866	3196

claimed for that has not yet been carried out there is an element of on-cost generated in advance. If the valuation of the works is not adjusted for on-cost generated in advance an over-reporting of the allowances for overheads and profit will occur, giving an overoptimistic assessment of the project's position. There is a range of techniques for the evaluation of on-cost in advance. Usually, the on-cost associated with the net valuation of the works is deducted from the gross on cost, which will include on-cost associated with variations, claims, overmeasure and preliminaries; the residual amount will then be the on-cost in advance of entitlement. This would be considered as an overmeasure and shown as such.

To illustrate how such rate loadings are treated in practice, the example used above is changed slightly; the on-cost allowed within the estimate was 5% and was distributed evenly between the items. If, for commercial reasons, the contractor loaded item 2 with additional on-cost and unloaded items 1 and 3, the total percentage on-cost would remain the same. However, the allocation between items would now be uneven. The bill of quantities rates for the loaded items would increase slightly. The client's representative should be alert to such amendments, as they could lead to overvaluation of the works or overpriced variations should the rates be used for valuation purposes. The amended estimate is shown in Table 18.17.

When the application for payment percentages are applied to calculate the gross allowances (Table 18.18), it can be seen that the on-cost allowance for item 2 has now increased to 2871 compared with 1956, which was indicated in Table 18.13.

When the overmeasure/undermeasure calculations are undertaken the summary is as shown in Table 18.19.

The generated allowance for on-cost has now fallen to 4% from the 5% that was calculated before the loading of item 2 and unloading of items 1 and 3. Although in this example this is not a large amount, if this was magnified on a large contract it could suggest that there was a difficulty that requires rectifying. In this case an additional undermeasure of on-cost generated in arrears can be calculated; this is simply the difference between that generated if no loading had been included and that which is now generated. In the example above this would be 3275 (from Table 18.16) less 2608, that is 667.

Table 18.19   Summary of revised on-cost allowance after rate loading.

| Valn 2 | Labour | Materials | Plant | Net allowances | | |
				Subcontract	On-cost	Total
Applied for	6879	13757	2751	43866	3196	70448
Overmeasure	0	0	0	9049	638	9525
Undermeasure	1058	2117	423	423	50	4233
Total net allowance	7937	15874	3175	35240	2608	64833

18.9.2   CVR in practice

The previous sections of this chapter have considered how the costs are defined, recorded and reported. The accounting definitions of value and attributable profit and the approach to dealing with losses as applied at a project level have been dis-cussed. This last section deals with the reconciliation of the project's costs and value and the reporting of the project's performance to senior management. Good man-agement practice suggests that frequent reporting is required in order to underpin effective interventions if the project is not going to plan. This applies to all aspects of the project. In the case of the CVR process, the reconciliations usually occur monthly. A CVR will be produced for all 'live' projects, where live is defined as 'having some activity or likely future activity'. Often it can take many months to complete the snagging (which incurs costs) or settle final accounts (which generate value) and it is normal for a summary CVR to be produced for these contracts – even if it is only to remind senior management to chase for the release of retention or final account agreement.

The documentation that would be provided in the CVR reports for the live con-tinuing contracts is:

- a single A4 summary of the position of the contract, an example of which is given in Figure 18.4;
- a cost report indicating cumulative to-date costs and period costs;
- a buying report indicating buying gains/losses;
- list of subcontract liabilities and payments;
- architects certificate;
- explanations for variance between allowances and costs;
- cost to complete exercise showing final account value, cost and margin;
- projected turnover until completion;
- any items for management action.

A great deal of a quantity surveyor's time is taken up with the compilation of these reports and they often complain that more time is spent looking backwards and analysing what has happened in the past, over which they can have no influence, rather than looking forwards and anticipating future opportunities, which they may have some influence over.

A simple example of a project CVR is shown in Table 18.20.

Cost value reconciliation				Contract performance			
Contract No	XYZ			Contract period elapsed			9 weeks
Contract Title	Byrom Street			Contract period remaining			45 weeks
Accounts period	3			EOT			0
				% complete by value			0
Application	450 768			% complete by time			0
Adjust	25 645						
Cum. Net	425 123						

Cost			Reconciliation				
Cum. to last period	Current period	Total		Cum. net Val.	Cum. cost	Variance	Profit
325 675	78 654	404 329		391 113	404 329	−13 216	−3%
				Period net val.	Period cost	Variance	Profit
				69 564	78 654	−9090	−13%

Cost to complete		Tender	Final account	Cash flow			
Valuation		2 567 800	2 952 970	Projected value			
Cost		2 362 376	2 787 604	Next month	month 2	month 3	
Profit		205 424	165 366	754 500	85 500	94 000	
	%	8%	6%				

Figure 18.4   A simplified CVR pro forma.

Table 18.20   Accounts period and NSV.

Contract No.	XYZ
Contract title	Byrom Street
Accounts period	3
Application	450 768
Adjust	25 645
Cum. net	425 123

As shown in Table 18.20, all CVRs will have a statement of the accounts period, the contractors application for payment and the appropriate adjustments to demonstrate a cumulative net position. The reference to the external application is important as it provides some externality to the CVR process that an auditor could track from the NSV via the adjustments to the architects certificate.

A good snapshot of the progress of a project can be gathered by comparing its value against time elapsed, as shown in Table 18.21. A summary of the time elapsed, time remaining and the percentage complete against value or time can be made.

As shown in Table 18.22, a cost statement that is taken from the cost report will give the cumulative costs to date and those of the current period under consideration. This will allow for calculation of margin to date as well as the period movements.

Table 18.21   Time versus value performance.

Contract period elapsed	9 weeks
Contract period remaining	45 weeks
EOT	0
% complete by value	20
% complete by time	14

Table 18.22   Cumulative and period costs.

Cost		
Cum. to last period	Current period	Total
325 675	78 654	404 329

Table 18.23   Reconciliation of cost and value.

Reconciliation			
Cum. net val.	Cum. cost	Variance	Profit
391 113	404 329	−13 216	−3%
Period net val.	Period cost	Variance	Profit
69 564	78 654	−9090	−13%

Table 18.24   Cost to complete.

Cost to complete		Tender	Final account
Valuation		2 567 800	2 952 970
Cost		2 362 376	2 787 604
Profit		205 424	165 366
	%	8%	6%

The reconciliation shown in Table 18.23 will consider the to-date position as well as the performance in the month. The data shown in Table 18.23 indicate that the cumulative net valuation is £391 000 and the cumulative costs are £404 000, which indicates a negative margin of 3%. An examination of the period value and cost shows that a 13% loss has been made on £69 500 of turnover. This would prompt an examination of the period records.

The cost to complete exercise shown in Table 18.24 indicates the forecast final account and the outturn final margin. This is required before interim profits are taken.

To complete the record, a projection of cash flow is provided for the remaining three months of the contract (Table 18.25).

A more detailed CVR format is shown in Figure 18.5; both cumulative and period allowance figures for each of the resource headings are given to indicate the variances to date and also in the period. A more detailed breakdown of the overmeasure is also given into different categories to reflect the likelihood of its recovery to value.

Table 18.25 Projected turnover.

Cash flow		
Projected value		
Next month	month 2	month 3
754 500	85 500	94 000

18.10 Explaining variances

The quantity surveyor and contracts manager will consider their explanations of the variances within the CVR prior to the formal monthly meeting. If there are variances which relate to production, the site manager will have to explain why they have occurred; they may be due to resource levels, plant hire or materials control. If the variances relate to subcontract procurement or payment for what is considered to be entitlement, the quantity surveyor will have to provide an explanation.

The monthly CVR meeting will be chaired by a senior member of staff and attended by the commercial manager, contracts manager and sometimes the accountant. The site team is represented by the site quantity surveyor and construction manager. The meetings are to establish the true position on the contract and to identify management action.

The predicted and current margin is the most important consideration and the site quantity surveyor is careful to report a prudent and constant margin throughout the project. The reporting of widely fluctuating margins from one month to another is to be avoided, as it would indicate a poor predictive ability.

It is also through experience that the margin figures are not reported too bullishly, as an overoptimistic assessment can lead to unrealistic final account projections, which can lead to protracted negotiations with clients and subcontractors in trying to achieve this margin. Most contracting firms will assign the surveyor to the next project without allowing sufficient time for the settlement of subcontract final accounts. The quantity surveyor usually builds in some autonomy to quickly settle the accounts by building into the CVR a hidden reserve and obscuring the true financial performance.

The commercial manager has a choice of two options to obscure the financial performance of the project: either to enhance the project's costs or to adjust the project's valuation. To enhance a project's costs, the site reporting team generally chooses a cost area that it has the opportunity to adjust without monies flowing to another party; in many cases this tends to be subcontractor liabilities: the subcontract accruals which form the basis for the subcontract cost headings can be artificially increased, thus hiding a contingency that can be used at a later date; a corresponding increase in subcontract allowance will ensure that there will be no unexplainable subcontract variances. Another manipulation of the cost statement may be to include a cost provision for future losses; this, however, is more observable and more likely to raise senior management's attention to potentially available additional on-cost. The adjustment of value may be another tactic used by the site team; it may increase the extent of overmeasure, consequently reducing the allowances that are taken to value and, consequently, under-reporting the profitability of the project. The most comprehensive cost reports reviewed when researching this book required the date and amount of the last subcontract certificate to be included. This suggested

Contract:			Contract period elapsed		9	weeks
Accounting period 2			Contract period remaining		45	weeks
Architect Cert. No. 2			EOT		0	
Claimed		750	% complete by value		0	
MOS		10	% complete by time		0	
Omsr		147	Overmsr	On-cost	20	
Umsr		4		Likely	40	
WIP		0		Prob	75	
NSV		589		Unlikely	12	
Arch Cert 1		730	Undermeasure		4	
Over Cert		141	WIP		0	

	Gross			Period		
	Allowance	Cost	Variance	Allowance	Cost	Variance
Labour	10	60	−50	5	3	2
Plant	30	40	−10	15	12	3
Materials	160	140	20	50	40	10
Subcontract	350	340	10	200	180	20
Staff	30	35	−5	5	4	1
Site Ohds	3	5	−2	1	1	0
Subtotal	583	620	−37	276	240	36
On-cost	6		6	3		3
NSV	589	620	−31	279	240	39

Cost to complete		Tender	Final account	Cash flow		
Valuation		2 567 800	2 952 970	Projected value		
Cost		2 362 376	2 787 604	Next month	month 2	month 3
Profit		205 424	165 366	754 500	85 500	94 000
%		8%	6%			

Figure 18.5   Exemplar of a completed CVR.

an organisation that was aware of the potential for misreporting and a corresponding additional reporting burden was placed on the site team as a result.

Obviously, senior commercial managers of organisations have had extensive experience of obscuring the picture when they were site based and are aware of the process and, in some cases, tacitly concur with the underlying reasons. This concurrence signals an acceptance that site teams require an element of flexibility and autonomy in decision making; however, most senior managers would require an indication as to the size of the contingency developed. The forms used for the reporting of costs and value are, consequently, designed to provide an accurate representation of financial performance, a means of auditing the site teams decision making and also the provision of records for company accounts.

18.11 Summary

This chapter has used a very simple example to illustrate the CVR process; the principles of application, overmeasure and development of net allowances would obviously be scaled up. The size and complexity of valuations can appear, on first sight, to be virtually unintelligible; however, the principle of calculating a net valuation remains at the heart of the CVR process.

A range of pro forma forms for reporting undermeasure, overmeaure and summarising the allowance performance on a contract have been used. An adjustment for rate loading has also been included to demonstrate that the contractor has to make further adjustments to net margin where monies have been allocated unevenly throughout the tender.

References

Barrett, F.R. (1992) *Cost Value Reconciliation*, 2nd edn. The Chartered Institute of Building, Ascot, UK.

ICAEW (1989) Statements of Standard Accounting Practice, No. 9. The Institute of Chartered Accountants in England and Wales (ICAEW), London.

Sizer, J. (1986) *An Insight into Management Accounting*. Penguin, Middlesex, UK.

Glossary

Accrual: An amount of money, representing the value of goods or services received but not invoiced or paid for, which is taken into account when completing a set of accounts.

Annual accounts: Documents that must be prepared and submitted to Her Majesty's Revenue and Customs (HMRC) and Companies House in the United Kingdom for each trading period, including a balance sheet, a profit and loss account and a statement of accounting policies adopted. The accounts may also include a movement of funds statement, a directors' report and an independent auditors' report.

Application: A formal written request for payment for work carried out to date accompanied by detailed calculations to justify the amount claimed. Payment applications may be made by a main contractor, where the contract conditions permit, or by a subcontractor.

Asset: Something owned by a business with a monetary value. It may be a fixed asset, such as land or buildings, or a current asset, such as debtors and cash at bank.

Attendance: Facilities provided to subcontractors by a main contractor such as the provision of power, water or off-loading facilities.

Auditor: an independent accountant who scrutinises the accounts and records of a business in order to verify that the accounts are free of any material mis-statements and state a true and fair view of the business.

BACS: Bankers Automated Clearing Service, which enables electronic Sterling money transfers to be made direct to a bank or building society usually within three working days.

Balance sheet: Part of the annual accounts which states the assets and liabilities of a business at a particular point in time.

Cash flow: The movement of money into and out of a business.

Financial Management in Construction Contracting, First Edition. Andrew Ross and Peter Williams.
© 2013 Andrew Ross and Peter Williams. Published 2013 by John Wiley & Sons, Ltd.

CDM: The Construction (Design and Management) Regulations 2007 are construction industry specific health and safety regulations which require the appointment of a CDM coordinator and a principal contractor for notifiable projects and which impose certain statutory duties on clients, designers, contractors and others aimed at the effective management of occupational health, safety and welfare on construction projects.

CECA: The Civil Engineering Contractors Association which has replaced the FCEC as the representative body for contractors working predominantly in the civil engineering and public works sector of the construction industry.

Certification: The process of checking a valuation for correctness and issuing a formal notice that identifies the amount to be paid according to the conditions of contract. Under the Housing Grants, Construction and Regeneration Act 2006, the issuing of certain statutory notices is part of the certification process.

CFA: The Continuous Flight Auger method of piling which enables *in situ* concrete frictional piles to be drilled and concreted in one continuous operation. The auger drills the hole and the soil-filled auger supports the sides of the excavation prior to introducing liquid concrete into the ground. This is done via a central Kelly bar in the auger which is then reverse-rotated to remove spoil and allow the concrete to fill the resulting hole. Reinforcement is placed immediately after the auger is removed.

CHAPS: The Clearing House Automated Payment System, which is a secure payment system that facilitates same-day electronic inter-bank Sterling transfers.

Cheque: A traditional form of payment for goods supplied or services rendered, with a 350-year history, whose definition and use is embodied in the Bills of Exchange Act 1882 and the Cheques Acts of 1957 and 1992. Cheques take up to six working days to clear the banking Clearing System and their use is in decline. Cheques are likely to be phased out in 2018.

Contra-charge: This is a non-contractual debt deducted from a payment application in consideration of the provision of goods or services that were not included in the contract, or the subcontract, or in consideration of general damages that one party to a contract believes are owed by the other party.

Contractor: An enterprise whose principal business activity is tendering for contracts with the intention of entering into a formal contract or subcontract agreement if the tender is accepted.

Cost: The amount of money, exclusive of overheads and profit, that is needed to supply the various factors of production required to complete a construction contract including all necessary labour, plant, materials, subcontractors and preliminaries items.

Credit: Deferred payment terms enjoyed by contractors in the process of procuring the necessary labour, materials, plant hire and subcontractors to complete a construction project or the deferred payment terms enjoyed by a client when acquiring a complete construction project.

Creditor: A current liability on the balance sheet which represents money owing to individuals or companies that have submitted an invoice for the provision of goods or services on credit.

Current asset: Non-tangible items with a value that are owned by the company, including debtors, stocks of materials, work in progress, cash.

Cut-off date: The reporting date for all contracts undertaken by a construction company. The date is set by the directors and the financial position of all contracts is reported at this date. This enables the profit/loss situation on all contracts to be evaluated at a common date on a like-for-like basis.

CVR: Cost value reconciliation is the process of matching the costs and values relating to a construction contract at a common date (the cut-off date). Value is determined by adjusting the external valuation, and adding work in progress. Costs are established from accounting records to which accruals are added.

Debtor: A current asset in the balance sheet representing monies owing to the company for goods and services rendered.

Domestic subcontractor: A contractor who carries out a standard building trade, such as groundworks, plastering and painting and decorating, and who will enter into a subcontract with a main contractor as part of a construction project.

Estimate: The output from the process of calculating the net cost of labour, materials, plant, subcontractors and on-costs with respect to a construction project or part thereof.

External valuation: The process of ascertaining the value of work in progress on a construction contract with a view to certification and payment for the work carried out to date.

FCEC: The Federation of Civil Engineering Contractors which is now replaced by the CECA.

Fixed asset: Property, plant, equipment or other tangible items with a value that are owned by a company or individual.

Fixed charge: A legal charge over a specified asset in order to secure a loan or an overdraft facility whilst retaining possession of the asset. The purpose is to ensure that a lender is protected against the default of a borrower.

Floating charge: A legal charge over a group of assets, whose value fluctuates, in order to secure a lender against default of a borrower whilst enabling the borrower to use, sell or trade the asset until such time as the charge crystallises (e.g. converts into a fixed charge).

FPS: Faster Payments Service which is an electronic, same-day, payment service offered by UK banking institutions.

Front-end loading: A euphemism for a variety of pricing strategies used by contractors aimed at improving cash flow and capitalising on opportunities to make money from mistakes in contract documentation. Techniques include front-end loading, back-end unloading, rate loading and preliminaries or 'on-cost' loading.

Gearing: The extent to which borrowings exceed share capital (capital gearing) or profit before interest and tax exceeds interest payments on borrowings (income gearing).

Gross rate: The net cost of carrying out a unit of work, with the addition of a margin for overheads and profit, entered by a contractor in a pricing document, such as bills of quantities or a schedule of rates, as part of a tender.

ICAEW: The Institute of Chartered Accountants in England and Wales which is the professional body that represents chartered accountants and monitors their standards of professional competence and ethics.

ICC: The Infrastructure Conditions of Contract, which is a suite of contracts intended for use in civil engineering works. Formerly known as the ICE Conditions of Contract, which it now replaces, the ICC Conditions suite includes a measurement version, a design and construct version, a small works version and so on.

ICE: The Institution of Civil Engineers.

Insolvency: There are two tests prescribed by the Insolvency Act 1986, Section 123 which are (1) the inability to pay debts as they fall due (cash flow insolvency) and (2) where the value of assets is less than the amount of liabilities (balance sheet insolvency).

Interim payment: The transfer of monies certified or authorised for payment at intervals defined in the conditions of contract to the relevant main contractor or subcontractor.

Internal valuation: The process of adjusting an external valuation for errors and inconsistencies and the evaluation of work in progress carried out after the external valuation date.

JCT: The Joint Contracts Tribunal, which publishes an extensive suite of contracts for the building industry. The JCT is made up of representatives from employers' bodies, professional institutions and contractors' organisations.

LADs: Liquidated and ascertained damages represent an agreed and genuine pre-estimate of the financial loss that may be incurred by the employer in the event that the contractor fails to complete the contract on or before the contract completion date or extended contract completion date. LADs may be recovered from the contractor by deduction from monies due, or to become due, to the contractor or as a debt.

Liability: Money owed or money that might be owed in the future or a claim against the assets of a business by shareholders.

Limited liability: The protection enjoyed by the shareholders of an incorporated company whose risk exposure to the debts of the company is limited to the amount of money that they have invested in the business.

Liquidity: A measure of the cash and liquid assets available to settle debts as and when they fall due.

Listed subcontractor: A specialist contractor, whose name is included on a limited list of approved names provided by the architect or engineer, to be chosen by the main contractor for the provision of a specialist trade required for the completion of a building or engineering project. The chosen listed subcontractor will become a domestic subcontractor of the main contractor.

Loan capital: Medium to long-term capital invested in a business, usually on terms which require the payment of interest, that may be subject to the observance of legally binding restrictions or rules (covenants) relating to the loan.

Margin: The amount that a contractor adds to his costs to allow for the cost of head office overheads and an element of profit.

"most contractors" ... "many subcontractors" and similar expressions: Judgements made by the authors based on first-hand experience and anecdotal evidence accumulated from a combined 70 years or so experience of the industry and exposure to the industry 'grapevine'.

Movement of funds statement: Part of the annual accounts of a business that states cash held at the beginning of the financial year, cash in-flows and out-flows during the financial year and cash held at the end of the financial year.

Named subcontractor: See Listed subcontractor.

NEC: The New Engineering Contract suite of contracts, published by ICE Publishing, which includes the Engineering and Construction Contract, Engineering and Construction Short Contract, the Engineering and Construction Subcontract, the Framework Contract and so on.

Net rate: A price representing the costs of a unit of work, including labour, materials, plant, subcontractors and on-costs prior to the addition of a margin for overheads and profit.

Net realisable value (NRV): Net realisable value is the contractor's valuation of the works, less foreseeable losses and future liabilities for defects and subcontracts, after stripping out overmeasure and adjusting for undermeasure and preliminaries taken in advance (e.g. front-end loading of the tender).

Nominated subcontractor: A specialist contractor, chosen by the architect or engineer, which enters into a subcontract with a main contractor for the purpose of carrying out a specialist trade required for the completion of a building or engineering project.

On-cost (1): The cost of employing labour, over and above the minimum cost determined by the National Working Rule Agreement, including employers' national insurance contributions, bonuses, travelling and skills allowances, employers' and public liability insurances and so on.

On-cost (2): See Preliminaries.

Overdraft: A facility offered by banks whereby the account holder may spend in excess of the amount on deposit up to a predetermined and agreed limit. It is a lending facility that can be withdrawn by summary notice.

Overheads: General expenses, including salaries, motor cars, stationary, postage, heating and lighting and so on, at head office.

Payment authorisation: A notification by the project quantity surveyor to the accounts department that a subcontractor's application for payment has been checked and adjusted where necessary and is now authorised for payment to the subcontractor.

Performance bond: A guarantee for the employer that the contractor will complete the project in accordance with the contract documents. If the contractor defaults, the bank or insurance company providing the bond will pay out to the employer an amount of money up to the value of the bond (usually 10% of the contract sum).

PERT: Acronym for Project Evaluation and Review Technique which is similar to critical path analysis but uses probabilistic estimates of time.

PQS: A Professional Quantity Surveyor, engaged by the employer to a construction contract, who is responsible for valuing work in progress along with other duties relating to the client's cost management of a project.

Preliminaries: The contractor's general cost items which do not form part of the permanent works. They include such items as site supervision, provision of temporary accommodation, the provision of temporary drainage, water, lighting and power, site transport and vans, small tools and tackle, scaffolding and temporary works, temporary access, hoardings and so on. Some preliminaries items are fixed costs (one-off charges), some are time-related costs (those that vary with time) and some are semi-variable costs (these may occur where extra work is ordered requiring additional supervision or plant not allowed for in the tender).

Price: See Tender.

Procurement: The process of tendering and selection undertaken in order to acquire a fixed asset such as a building, road, bridge or other building or engineering structure.

Profit: The amount of money left over from monies received when all costs and overheads have been paid. Profit is subject to taxation and limited companies pay over part of their profits to shareholders in the form of dividends.

Profit and loss account: Part of the annual accounts of a business that states turnover, cost of turnover, administrative costs, interest payments and profit for a trading year.

QS: An abbreviation for 'quantity surveyor' to be distinguished from the abbreviation PQS.

RICS: The Royal Institution of Chartered Surveyors, which is the primary professional body for quantity surveyors.

Risk capital: Long-term capital invested in a business by stockholders or shareholders who own stocks or shares in the business in return for lending their capital.

Set-off: The legal right to recover sums of money owing under a construction contract from sums owed under that contract.

Shareholder: An individual or institution, such as an insurance company, that has invested money by buying shares in a business and is usually entitled to vote at the annual general meeting of the business.

Sole trader: A business enterprise whose owner is responsible for the entire indebtedness of the business and whose private and personal assets are at risk.

SSAP9: Statement of Standard Accounting Practice No 9 which provides guidance on how the balance sheet current asset of 'stocks and work in progress' may be evaluated to ensure that contracts are reported prudently and that a true and fair view is given in the accounts.

Tender: The output from the process of converting an estimate for a construction project, or part thereof, into an offer capable of acceptance by a client taking into account head office overheads, profit margin, risk and market factors.

Valuation: A financial assessment of the value of construction work carried out to date based on the quantity of work done at the prices stated in the contract bills of quantities or other pricing document.

Value: Normally represents the payment that the contractor receives from the client during the construction phase of the project and is given by the equation: Value = Cost + Margin. 'Value' is also referred to as 'earned value'.

Work in progress: A current asset in the balance sheet that consists of work carried out but not yet valued or certified. It may be referred to in the accounts as 'stocks and long-term contracts' or 'inventories'. When work in progress has been valued and certified it becomes a debtor item. When work in progress has been paid for it becomes a progress payment on account.

Working capital: The cash required to operate a business pending receipt of payment for work invoiced. Working capital is used to pay wages, bills and overhead costs and to repay maturing short-term loans. It provides the liquidity necessary to settle debts as and when they fall due in order that a business may remain technically solvent.

Index

Financial Management in Construction Contracting, First Edition. Andrew Ross
and Peter Williams.
© 2013 Andrew Ross and Peter Williams. Published 2013 by John Wiley & Sons, Ltd.